Genetic Engineering and Biotechnology: Concepts, Methods and Biological Applications

Genetic Engineering and Biotechnology: Concepts, Methods and Biological Applications

Dr. Arvind Kumar

RANDOM
RANDOM PUBLICATIONS
NEW DELHI (INDIA)

Genetic Engineering and Biotechnology: Concepts, Methods and Biological Applications

---

ISBN 978-93-5111-605-9

Published in 2015 in India by

RANDOM PUBLICATIONS

4376-A/4B, Gali Murari Lal, Ansari Road
New Delhi-110 002
Phone : +9111-43580356, 011-23289044, 011-43142548
e-mail: sales@randompublications.com,
info@randompublications.com, randomexports@gmail.com

Reprinted 2026

Type Setting by : Friends Media, Delhi-110089

Preface

Genetic engineering alters the genetic make-up of an organism using techniques that remove heritable material or that introduce DNA prepared outside the organism either directly into the host or into a cell that is then fused or hybridized with the host. This involves using recombinant nucleic acid (DNA or RNA) techniques to form new combinations of heritable genetic material followed by the incorporation of that material either indirectly through a vector system or directly throughmicro-injection, macro-injection and micro-encapsulation techniques.

Genetic engineering does not normally include traditional animal and plant breeding, in vitro fertilisation, induction ofpolyploidy, mutagenesis and cell fusion techniques that do not use recombinant nucleic acids or a genetically modified organism in the process. However the European Commission has also defined genetic engineering broadly as including selective breeding and other means of artificial selection. Cloning and stem cell research, although not considered genetic engineering, are closely related and genetic engineering can be used within them. Synthetic biology is an emerging discipline that takes genetic engineering a step further by introducing artificially synthesized genetic material from raw materials into an organism.

The structure of DNA encodes all the information every cell needs to function and thrive. In addition, DNA carries hereditary information in a form that can be copied and passed intact from generation to generation. A gene is a segment of DNA. The biochemical instructions found within most genes, known as the genetic code, specify the chemical structure of a particular protein. Proteins are composed of long chains of amino acids, and the specific sequence of these amino acids dictates the function of each protein. The DNA structure of a gene determines the arrangement of amino acids in a protein, ultimately determining the type and function of the protein manufactured.

The book is intended primarily for undergraduates and postgraduates in biology, botany and biotechnology, whereas it can be used as a reference for researchers and post-doctoral students, university professors and instructors.

I would like to thank my team for standing beside me throughout my career and writing this book. My special thanks go to "Random Publications" who have published the book.

– Dr. Arvind Kumar

Contents

1

Introduction

The public discourse over biotechnology may be visualised as radiating from one seminal event–the recombinant DNA controversy of the mid-1970s. This singularly influential episode gave form to a constellation of issues that were subsequently cast as products and processes of the new genetics, including: the manufacture of biological weapons, genetic screening and discrimination, the creation of transgenic animals, genetic modification of human ova, microbial effluent from industrial applications, the deliberate release of genetically engineered organisms into the environment, and human genetic engineering. Concerning each of these areas of public concern one may ask:

- Are there important new public problems arising from innovations in biotechnology that call out for regulatory action or some other form of collective response?
- Can our existing laws, regulations, and institutions address the varied issues raised in the public arena?
- What institutions or authorities are responsible for the regulation or oversight of these issues?
- What role does the public play in contributing to the decisionmaking process?

Among the issues previously outlined, some have matured to the point where considerable government presence has already begun. Other issues are still in their embryonic stage of public awareness and debate. Some problems will never reach beyond a brief burst of media attention; others will activate litigious debate and aggressive legislation. For example, the international Biological Weapons Convention prohibits its signatories from developing organisms for offensive purposes. On May 22, 1990, President George Bush signed into law the Biological Weapons Anti-Terrorism Act of 1990, which makes it a civil crime to violate terms of the Biological Weapons Convention.

Environmental applications of GEOs have drawn considerable public debate and congressional study, but no new legislation. The policy issues associated with transgenic animals, cast mainly in legal and ethical terms, have been decided primarily by the courts with only modest congressional initiatives. One area of applied genetics where social oversight has probably not kept up with the intensity of public reaction is human genetic modification. The Department of Health and Human Services (HHS) regulates human experiments involving the modification of genes. However, HHS's role is constrained by two important factors. First, the agency's authority is limited to institutions that receive federal funding. And second, the values of HHS are very much in consonance with high-technology medical interventions. The chapters in this third section of the book focus on the needs and prospects for regulating different uses of genetic technologies.

INHERITANCE PATTERNS

Mendel was the first scientist to develop a method for predicting the outcome of inheritance patterns. He performed his work with pea plants, studying seven traits: plant height, pod shape, pod colour, seed shape, seed colour, flower colour, and flower location. Pea plants pollinate themselves. Therefore, over many generations, pea plants develop individuals that are homozygous for particular characteristics. These populations are known as *pure lines*.In his work, Mendel took pure-line pea plants and cross-pollinated them with other pure-line pea plants. He called these plants the *parent generation*. When Mendel crossed pure-line tall plants with pure-line short plants, he discovered that all the plants resulting from this cross were tall. He called this generation the *F1 generation* (first filial generation). Next, Mendel crossed the offspring of the F_1 generation tall plants among themselves to produce a new generation called the *F2 generation* (second filial generation). Among the plants in this generation, Mendel observed that three-fourths of the plants were tall and one-fourth of the plants were short.

MENDEL'S LAWS OF GENETICS

Mendel conducted similar experiments with the other pea plant traits. Over many years, he formulated several principles that are known today as Mendel's laws of genetics. His laws include the following:

1. Mendel's law of dominance: When an organism has two different alleles for a trait, one allele dominates.
2. Mendel's law of segregation: During gamete formation by a diploid organism, the pair of alleles for a particular trait separate, or segregate, during the formation of gametes (as in meiosis).
3. Mendel's law of independent assortment: The members of a gene pair separate from one another independent of the members of other gene pairs. (These separations occur in the formation of gametes during meiosis.)

MENDELIAN CROSSES

An advantage of genetics is that scientists can predict the probability of inherited traits in offspring by performing a genetic cross (also called a *Mendelian cross*). To predict the possibility of an individual trait, several steps are followed. First, a symbol is designated for each allele in the gene pair. The dominant allele is represented by a capital letter and the recessive allele by the corresponding lowercase letter, such as *E* for free earlobes and *e* for attached earlobes. For a homozygous dominant individual, the genotype would be *EE*; for a heterozygous individual, the genotype would be *Ee*; and for a homozygous recessive person, the genotype would be *ee*.

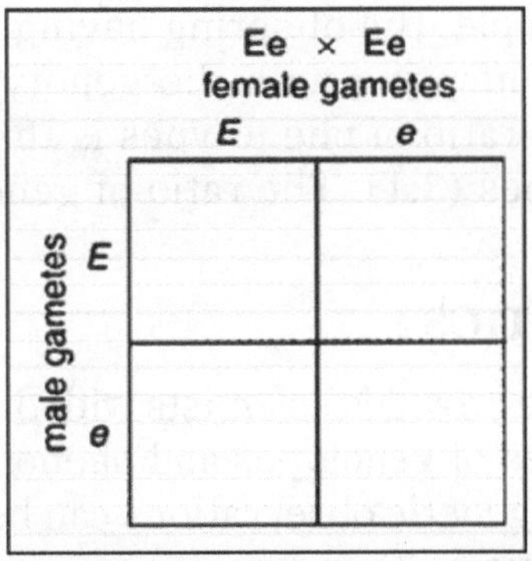

Fig. An Example of a Punnett Square

Continuing, all of the possible combinations of alleles are considered. This is done by filling in each square with the alleles above it and at its left. This is done as shown in Figure .

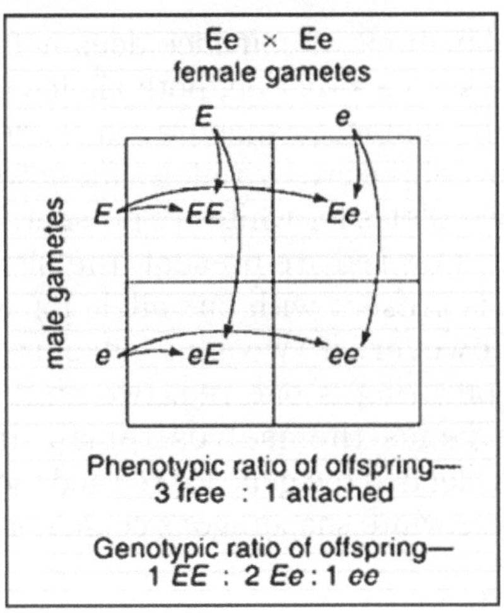

Fig. The Punnett Square is used to Determine the Probabilities of the Genotypes and Phenotypes in the Offspring of a Genetic Cross.

The next step in performing a genetic cross is determining the genotypes of the parents and the genotype of the gametes. A heterozygous male and a heterozygous female to be crossed have the genotypes of *Ee* and *Ee*. During meiosis, the allele pairs separate. A sperm cell contains either an *E* or an *e*, while the egg cell also contains either an *E* or an *e*. To continue the genetics problem, a Punnett square is used. A *Punnett square* is a boxed figure used to determine the probability of genotypes and phenotypes in the offspring of a genetic cross. The possible gametes produced by the female are indicated at the top of the square, while the possible gametes produced by the male are indicated at the left side of the square. Figure shows the Punnett square for the earlobe example.

From the Punnett square, the phenotype of each possible genotype can be determined. For example, the offspring having *EE, Ee*, and *Ee* will have free earlobes. Only the offspring with the genotype *ee* will have attached earlobes. Therefore, the ratio of phenotypes is three with free earlobes to one with attached earlobes (3:1). The ratio of genotypes is 1:2:1 (1 *EE* : 2 *Ee* : 1 *ee*).

PRINCIPLES OF GENETICS

Mendel's studies have provided scientists with the basis for mathematically predicting the probabilities of genotypes and phenotypes in the offspring of a genetic cross. But not all genetic observations can be explained and predicted based on Mendelian genetics. Other complex and distinct genetic phenomena may also occur. Several complex genetic concepts, described in this section, explain such distinct genetic phenomena as blood types and skin colour.

Incomplete Dominance

In some allele combinations, dominance does not exist. Instead, the two characteristics blend. In such a situation, both alleles have the opportunity to express themselves. For instance, snapdragon flowers display incomplete dominance in their colour. There are two alleles for flower colour: one for white and one for red. When two alleles for white are present, the plant displays white flowers. When two alleles for red are present, the plant has red flowers. But when one allele for red is present with one allele for white, the colour of the snapdragons is pink.However, if two pink snapdragons are crossed, the phenotype ratio of the offspring is one red, two pink, and one white. These results show that the genes themselves remain independent; only the expressions of the genes blend. If the gene for red and the gene for white actually blended, pure red and pure white snapdragons could not appear in the offspring.

Multiple Alleles

In certain cases, more than two alleles exist for a particular characteristic. Even though an individual has only two alleles, additional alleles may be present in the population. This condition is multiple alleles.

An example of multiple alleles occurs in blood type. In humans, blood groups are determined by a single gene with three possible alleles: A, B, or O. Red blood cells can contain two antigens, A and B. The presence or absence of these antigens results in four blood types: A, B, AB, and O. If a person's red blood cells have antigen A, the blood type is A. If a person's red blood cells have antigen B, the blood type is B. If the red blood cells have both antigen A and antigen B, the blood type is AB. If the red blood cells have neither antigen A nor antigen B, the blood type is O. The alleles for type A and type B blood are co-dominant; that is, both alleles are expressed. However, the allele for type O blood is recessive to both type A and type B. Because a person has only two of the three alleles, the blood type varies depending on which two alleles are present. For instance, if a person has the A allele and the B allele, the blood type is AB. If a person has two A alleles, or one A and one O allele, the blood type is A. If a person has two B alleles or one B and one O allele, the blood type is B. If a person has two O alleles, the blood type is O.

Polygenic Inheritance

Although many characteristics are determined by alleles at a single place on the chromosome, some characteristics are determined by an interaction of genes on several chromosomes or at several places on one chromosome. This condition is polygenic inheritance. An example of polygenic inheritance is human skin colour. Genes for skin colour are located in many places, and skin colour is determined by which genes are present at these multiple locations. A person with many genes for dark skin will have very dark skin colour, and a person with multiple genes for light skin will have very light skin colour. Many people have some genes for light skin and some for dark skin, which explains why so many variations of skin colour exist. Height is another characteristic probably reflecting polygenic characteristics.

Gene linkage

A chromosome has many thousands of genes; there are an estimated 100,000 genes in the human genome. Inheritance involves the transfer of chromosomes from parent to offspring through meiosis and sexual reproduction. It is common for a large number of genes to be inherited together if they are located on the same chromosome. Genes that are inherited together are said to form a linkage group. The concept of transfer of a linkage group is gene linkage. Gene linkage can show how close two or more genes are to one another on a chromosome. The closer the genes are to each other, the higher the probability that they will be inherited together. Crossing over occurs during meiosis, but genes that are close to each other tend to remain together during crossing over.

Sex Linkage

Among the 23 pairs of chromosomes in human cells, one pair is the sex chromosomes. (The remaining 22 pairs of chromosomes are referred to as autosomes.) The sex chromosomes determine the sex of humans. There are two types of sex chromosomes: the X chromosome and the Y chromosome. Females have two X chromosomes; males have one X and one Y chromosome. Typically, the female chromosome pattern is designated XX, while the male chromosome pattern is XY. Thus, the genotype of the human male would be 44 XY, while the genotype of the human female would be 44 XX (where 44 represents the autosomes).

In humans, the Y chromosome is much shorter than the X chromosome. Because of this shortened size, a number of sex-linked conditions occur. When a gene occurs on an X chromosome, the other gene of the pair probably occurs on the other X chromosome. Therefore, a female usually has two genes for a characteristic. In contrast, when a gene occurs on an X chromosome in a male, there is usually no other gene present on the short Y chromosome. Therefore, in the male, whatever gene is present on the X chromosome will be expressed.

DNA Defined

During the 1950s, a tremendous explosion of biological research occurred, and the methods of gene expression were elucidated. The knowledge generated during this period helped explain how genes function, and it gave rise to the science of molecular genetics. This science is based on the activity of deoxyribonucleic acid (DNA) and how this activity brings about the production of proteins in the cell. Genetic material is packaged into DNA molecules. DNA molecules relay the inherited information to messenger RNA (mRNA) which, in turn, codes for proteins. This chain of command is represented as:

DNA † ##mRNA † Protein

The flow of information from DNA to protein is known as the *Central Dogma* of molecular biology.In 1953, two biochemists, James D. Watson and Francis H.C. Crick, proposed a model for the structure of DNA. (In 1962, they shared a Nobel Prize for their work.) The publication of the structure of DNA opened a new realm of molecular genetics. Its structure provided valuable insight into how genes operate and how DNA can reproduce itself during mitosis, thereby passing on hereditary characteristics. Not only did the new research uncover many of the principles of protein synthesis, but it also gave rise to the science of biotechnology and genetic engineering.

DNA Structure

As proposed by Watson and Crick, deoxyribonucleic acid (DNA) consists of two long nucleotide chains. The two nucleotide chains twist around one another to form a double helix, a shape resembling a spiral staircase. Weak chemical bonds between the chains hold the two chains of nucleotides to one another. A nucleotide in the DNA chain consists of three parts: a nitrogenous base, a phosphate group, and a molecule of deoxyribose. The nitrogenous bases of each nucleotide chain are of two major types: purines and pyrimidines. Purine bases have two fused rings of carbon and nitrogen atoms, while pyrimidines have only one ring. The two purine bases in DNA are adenine (A) and guanine (G). The pyrimidine bases in DNA are cytosine (C) and thymine (T). Purines and pyrimidine bases are found in both strands of the double helix.

The phosphate group of DNA is derived from a molecule of phosphoric acid. The phosphate group connects the deoxyribose molecules to one another in the nucleotide chain. Deoxyribose is a five-carbon carbohydrate. The purine and pyrimidine bases are attached to the deoxyribose molecules, and the purine and pyrimidine bases are opposite one another on the two nucleotide chains. Adenine is always opposite thymine and binds to thymine. Guanine is always opposite cytosine and binds to cytosine. Adenine and thymine are said to be complementary, as are guanine and cytosine This is known as the principle of complementary base pairing.

DNA Replication

Before a cell enters the process of mitosis, its DNA replicates itself. Equal copies of the DNA pass into the daughter cells at the end of mitosis. In human cells, this means that 46 chromosomes (or molecules of DNA) replicate to form 92 chromosomes.The process of DNA replication begins when specialised enzymes pull apart, or "unzip," the DNA double helix. As the two strands separate, the purine and pyrimidine bases on each strand are exposed. The exposed bases then attract their complementary bases. Deoxyribose molecules and phosphate groups are present in the nucleus.

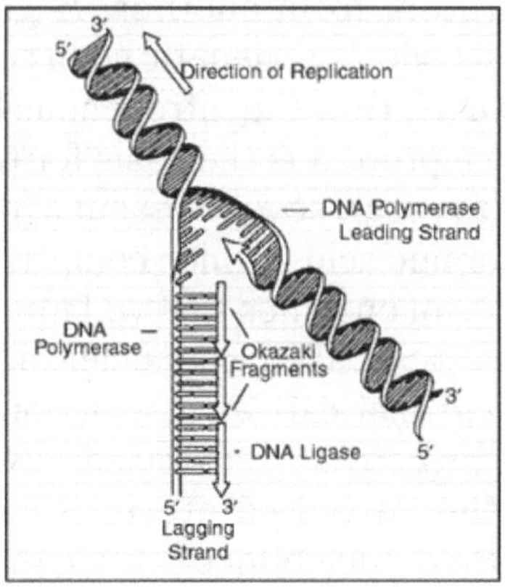

Fig. DNA replication. The double helix opens and a complementary strand of DNA is synthesised along each strand.

The enzyme DNA polymerase joins all the nucleotide components to one another, forming a long strand of nucleotides. Thus, the old strand of DNA directs the synthesis of a new strand of DNA through complementary base pairing. The old strand then unites with the new strand to reform a double helix. This process is called *semiconservative replication* because one of the old strands is conserved in the new DNA double helix. DNA polymerase joins nucleotides in a 52-32 direction on the leading strand, shown in Figure . However, DNA polymerase does not elongate a DNA strand in a 32-52 direction. Therefore, the 32-52 strand, called the lagging strand, is synthesised in short segments in a 52-32 direction.

These short segments placed on the lagging strand are Okazaki fragments and are ultimately joined together by the enzyme DNA ligase to form a new DNA strand. DNA replication occurs during the S phase of the cell cycle. After replication has taken place, the chromosomal material shortens and thickens. The chromatids appear in the prophase of the next mitosis. The process then continues, and eventually two daughter cells form, each with the identical amount and kind of DNA as the parent cell. The process of DNA replication thus ensures that the molecular material passes to the offspring cells in equal amounts and types

PROTEIN SYNTHESIS

During the 1950s and 1960s, it became apparent that DNA is essential in the synthesis of proteins. Proteins are used in enzymes and as structural materials in cells. Many specialised proteins function in cellular activities. For example, in humans, the hormone insulin and the muscle cell filaments are composed of protein. The hair, skin, and nails of humans are composed of proteins, as are all the hundreds of thousands of enzymes in the body

The key to a protein molecule is how the amino acids are linked. The sequence of amino acids in a protein is a type of code that specifies the protein and distinguishes one protein from another. A genetic code in the DNA determines this amino acid code. The genetic code consists of the sequence of nitrogenous bases in the DNA. How the nitrogenous base code is translated to an amino acid sequence in a protein is the basis for protein synthesis.

For protein synthesis to occur, several essential materials must be present, such as a supply of the 20 amino acids, which comprise most proteins. Another essential element is a series of enzymes that will function in the process. DNA and another form of nucleic acid called ribonucleic acid (RNA) are essential.

RNA is the nucleic acid that carries instructions from the nuclear DNA into the cytoplasm, where protein is synthesised. RNA is similar to DNA, with two exceptions. First, the carbohydrate in RNA is ribose rather than deoxyribose, and second, RNA nucleotides contain the pyrimidine uracil rather than thymine.

TYPES OF RNA

In the synthesis of protein, three types of RNA function. The first type is called *ribosomal RNA (rRNA)*. This form of RNA is used to manufacture ribosomes. Ribosomes are ultramicroscopic particles of rRNA and protein. They are the places (the chemical "workbenches") where amino acids are linked to one another to synthesise proteins. Ribosomes may exist along the membranes of the endoplasmic reticulum or in the cytoplasm of the cell.

A second important type of RNA is *transfer RNA (tRNA)*. Transfer RNA exists in the cell cytoplasm and carries amino acids to the ribosomes for protein synthesis. When protein synthesis is taking place, enzymes link tRNA molecules to amino acids in a highly specific manner. For example, tRNA molecule X will link only to amino acid X; tRNA Y will link only to amino acid Y.

The third form of RNA is *messenger RNA (mRNA)*. In the nucleus, messenger RNA receives the genetic code in the DNA and carries the code into the cytoplasm where protein synthesis takes place. Messenger RNA is synthesised in the nucleus at the DNA molecules. During the synthesis, the genetic information is transferred from the DNA molecule to the mRNA molecule. In this way, a genetic code can be used to synthesise a protein in a distant location. RNA polymerase, an enzyme, accomplishes mRNA, tRNA, and rRNA synthesis.

TRANSCRIPTION

Transcription is one of the first processes in the mechanism of protein synthesis. In transcription, a complementary strand of mRNA is synthesised according to the nitrogenous base code of DNA. To begin, the enzyme RNA polymerase binds to an area of one of the DNA molecules in the double helix. (During transcription, only one DNA strand serves as a template for RNA synthesis. The other DNA strand remains dormant.) The enzyme moves along the DNA strand and "reads" the nucleotides one by one. Similar to the process of DNA replication, the new nucleic acid strand elongates in a 52-32 direction, as shown in Figure . The enzyme selects complementary bases from available nucleotides and positions them in an mRNA molecule according to the principle of complementary base pairing. The chain of mRNA lengthens until a "stop" message is received.

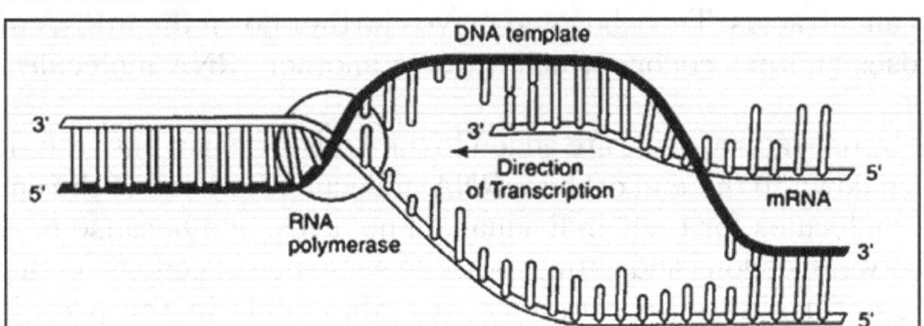

Fig. The process of transcription. The DNA double helix opens and the enzyme RNA polymerase synthesises a molecule of mRNA according to the base sequence of the DNA template.

The nucleotides of the DNA strands are read in groups of three. Each group is a *codon*. Thus, a codon may be CGA, or TTA, or GCT, or any other combination of the four bases, depending on their sequence in the DNA strand.

Each codon will later serve as a "code word" for an amino acid. First, however, the codons are transcribed to the mRNA molecule. Thus, the mRNA molecule consists of nothing more than a series of codons received from the genetic message in the DNA. After the "stop" codon is reached, the synthesis of the mRNA comes to an end. The mRNA molecule leaves the DNA molecule, and the DNA molecule rewinds to form a double helix. Meanwhile, the mRNA molecule passes through a pore in the nucleus and proceeds into the cellular cytoplasm where it moves towards the ribosomes.

TRANSLATION

The genetic code is transferred to an amino acid sequence in a protein through the translation process, which begins with the arrival of the mRNA molecule at the ribosome. While the mRNA was being synthesised, tRNA molecules were uniting with their specific amino acids according to the activity of specific enzymes. The tRNA molecules then began transporting their amino acids to the ribosomes to meet the mRNA molecule. After it arrives at the ribosomes, the mRNA molecule exposes its bases in sets of three, the codons. Each codon has a complementary codon called an anticodon on a tRNA molecule. When the codon of the mRNA molecule complements the anticodon on the tRNA molecule, the latter places the particular amino acid in that position. Then the next codon of the mRNA is exposed, and the complementary anticodon of a tRNA molecule matches with it. The amino acid carried by the second tRNA molecule is positioned next to the first amino acid, and the two are linked. At this point, the tRNA molecules release their amino acids and return to the cytoplasm to link up with new molecules of amino acid.

When it's time for the next amino acid to be positioned in the growing protein, a new codon on the mRNA molecule is exposed, and the complementary three-base anticodon of a tRNA molecule positions itself opposite the codon. This brings another amino acid into position, and that amino acid links to the previous amino acids. The ribosome moves further down the mRNA molecule and exposes another codon, which attracts another tRNA molecule with its anticodon.

One by one, amino acids are added to the growing chain until the ribosome has moved down to the end of the mRNA molecule. Because of the specificity of tRNA molecules for their individual amino acids, and because of the base pairing between codons and anticodons, the sequence of codons on the mRNA molecule determines the sequence of amino acids in the protein being constructed.

And because the codon sequence of the mRNA complements the codon sequence of the DNA, the DNA molecule ultimately directs the amino acid sequencing in proteins. The primary "start" codon on an mRNA molecule is AUG, which codes for the amino acid methionine. Therefore, each mRNA transcript begins with the AUG codon, and the resulting peptide begins with methionine.

Figure shows that the process of protein synthesis begins with the production of mRNA (upper right). The mRNA molecule proceeds to the ribosome where it meets tRNA molecules carrying amino acids (upper left). The tRNA molecule has a base code that complements the mRNA code and thereby brings a specific amino acid into position. The amino acids join together in peptide bonds (bottom), and the tRNA molecules are released to pick up additional amino acid molecules.

CONCEPTUAL CLARIFICATIONS

Few issues in the field of biotechnology produce as strong a public reaction as the prospect of human genetic engineering. The idea that some group of humans would set themselves up as the architects of our genetic makeup plays out as some kind of demonic plan whose adherents, should they exist, must be morally deranged. But the term "human genetic engineering" (HGE) has a broader range of meanings than what is signified above as a centralised programme of eugenics applied to fertilized human eggs. The most generalised meaning of HGE is any controlled intervention designed to affect the constitution or function of one or more human genes either directly in the mature organism or indirectly in human germ cells. This definition includes a number of activities that are summarised as follows.

- Use of chemicals to alter gene function in a human, human fetus, or fertilized egg.
- Transplantation of genes in the somatic cells of humans (all cells other than germline or reproductive cells).
- Cloning of a human being.
- Transplantation of genes to the germ cells of humans (sperm or egg cells; *in vivo* or *in vitro*).
- Deletion or addition of genes to the fertilized human egg prior to implantation or gestation.
- Deletion or addition of genes to the human fetus.

The ethical questions surrounding HGE are contingent on which of its several meanings is being advanced. Human gene therapy applies to those uses of HGE that are aimed at curing disease. Proposals introduced by members of the scientific community generally substitute the term "gene therapy" for "genetic engineering" in order to cleanse the term of any negative connotations. Notwithstanding such efforts at conceptual "sterilisation" however, the two terms are still used interchangeably by the media and in popular discourse. Somatic cell therapy is distinguished from germline therapy in that the latter change, unlike the former, can be passed on to future generations. Eugenics is a concept that is much broader than HGE. It means "cleaning up," "purifying," or improving the human gene pool. Genetic engineering potentially can be used to carry out a eugenics programme. But there are many other paths to eugenics, including a national policy of selective breeding as well as the widespread use of genetic screening in conjunction with selective abortion. Sometimes eugenics is defined as "the science of breeding," while elsewhere it is understood as a national policy directed at "purifying" the race. Individual decisions unrelated to a pattern of state-controlled procreation for the purpose of "improving the species" generally do not qualify as eugenic activities. However, if the state supports or mandates genetic screening (amniocentesis) for identification of certain traits of the fetus (mongolism, Tay Sachs), and encourages or provides incentives for abortion when one or more traits are discovered, then a eugenics policy is being advanced.

Certain forms of human genetic engineering have nothing to do with eugenics whether or not these activities are supported by the state. Gene therapy of human somatic cells that does not affect the germline and therefore will not be passed on to subsequent generations, does not qualify under the term eugenics.

HUMAN GENE THERAPY

The prospect of altering human genes to treat disease was widely discussed by biologists around the mid-to late 1960s. At that time geneticists discovered the mechanisms through which viruses carry DNA to the cells they infect. These discoveries prompted a series of articles on the application of this knowledge to human genetics.

In 1966, following a series of symposia and books on the control of human evolution, geneticist Joshua Lederberg, then at Stanford University, postulated for further discussion several ways that genetic engineering might be used to alter our genes. In the area of controlled human evolution, Lederberg distinguished between eugenics, which he defined as selective breeding, and algeny, a term he coined and derived from the fusion of "genetic" and "alchemy," under which he included clonal reproduction (cloning) and genetic engineering. Robert Sinsheimer is another distinguished biologist who wrote positively about the "designed genetic change" of humankind. He postulated a scenario in which a virus would carry an expressible gene that could cure a serious form of diabetes. At the time Sinsheimer was optimistic about science taking control of the human gene pool. He saw it as an opportunity to improve upon our flawed species nature.Ironically, Sinsheimer emerges a few years later as a moral conscience for the biological community on the issue of genetic engineering. His optimism for science expressed so poignantly in his 1969 essay turns to a critical skepticism after the discovery of recombinant DNA technology.

Other scientists cast the promise of altering human genes into political terms. They wrote for a much wider audience. Several biologists felt a sense of responsibility to forewarn society about aspects of their work that presented ethical problems. It was reinforced by the moral condemnation of elite science that took place during the Vietnam War period and the anti-positivistic movements that flourished in the 1960s.

Salvador Luria published his essay "Modern Biology: A Terrifying Power" in the *Nation*–an important venue for New Left analysis and critique–and warned us about the technological imperative: "Once the scientific principles are established, technological application is almost certain to come. Thus faced with the prospect of a new technology, we must ask ourselves as soon as possible whether and how it will be used and what we can do about it." Despite his Huxlian scenario of how biology might be misused ("We may even be tempted to manufacture many identical copies of a supposedly 'superior' individual"), Luria opposed any form of moratorium to prevent potential misuses of biology. He called for "rational machinery" to deal with research and its uses and a heightened moral responsibility among scientists to "tell society, in a forceful and persistent manner, what science is discovering and what the technological consequences are likely to be."

The period of the late 1960s and early 1970s was accompanied by an explosion of moral self-examination on the part of biologists. In 1969 Jonathan Beckwith of the Harvard Medical School led a team of biologists in isolating the first gene. At a press conference announcing the feat, Beckwith and his colleague James Shapiro spoke of the frightening possibilities that their discovery raised. Subsequently, Shapiro left science because of his concern that genetic research might be used for immoral purposes. But his departure was not permanent. After a stay in Cuba, Shapiro took a post at the University of Chicago where he continued working in bacterial genetics.In 1972, reports began appearing in the scientific literature describing the steps required for the modification of a mammalian cell. The prospect of human gene therapy was advanced by successes in the use of viruses as vectors for gene transfer. The more formidable problems included putting the transported gene in the correct site and initiating its expression. Along with the excitement about progress in this field, there were also expressions of ambivalence or caution about the eventual application of human genetic modification. Theodore Friedmann and Richard Roblin were early pioneers in examining the difficult problem of proposing norms for the use of human gene therapy. Their 1972*Science* essay with the interrogative title "Gene Therapy for Human Genetic Disease?" is emblematic of the early sensitivity that some scientists felt towards this issue. The authors described the first documented human gene therapy experiments that were undertaken in 1958 without publicity or public controversy. Two children suffering from a hereditary enzyme deficiency were treated with injections of a virus. At the time it was believed that the virus carried an essential piece of DNA that, when transplanted into human cells, could put into proper balance the body's enzyme chemistry.

Performed without public oversight and prior to the establishment of institutional human subjects committees, the experiment failed to achieve its desired outcome. The authors reported that the first effort at human gene therapy was premature and utilised a technique that was not sufficiently demonstrated on animal models. In 1971, under stricter controls and more accountability for human experiments, scientists working at NIH infected defective human cells with a virus they hoped would carry a missing gene, the absence of which is responsible for the hereditary disease galactosernia. Without the gene a person is unable to utilise galactose, a simple sugar common to dairy products. A baby with this defect that is not put on a dairy-free diet will become mentally retarded and eventually die.

The experiment transferred into human cells a normal bacterial gene that was thought to provide the same sugar-utilizing enzyme as the missing human gene. Now this experiment did *not* supply a normal *human* gene to an abnormal human cell.

John Fletcher discussed one of the earliest collaborative efforts at modifying the genes of an adult individual utilizing viral vectors. Between 1970 and 1973 an American scientist assisted a German physician in treating three German sisters who suffered from hyperargininemia–a hereditary disorder involving a deficiency in the enzyme arginase.

Without treatment, individuals build up excessive amounts of the amino acid arginine in the blood, resulting in severe mental retardation. The sisters were infected with the Shope papilloma virus, which the physicians believed would stimulate the production of the enzyme and reduce the concentration of the amino acid. The treatment was unsuccessful.

In their 1972 paper, Friedmann and Roblin frame an ethical discussion around the issue of misusing a promising technology. They defined misuse as the premature application of gene therapy or the application of it for anything that does not provide primary benefit to the patient. The ethical issues are cast in a form that renders them best handled by the medical community, those institutions with closely shared goals, or patient advocates. Another group of scientists who responded to the prospect of human genetic engineering in this early period focused on broad social and philosophical issues. The essays avoided tackling the specific implications of any new technological development on the grounds that the sensitive experiments were still too far off to address with any degree of specificity. In 1970, after a frog had been successfully cloned, Joshua Lederberg's comments captured this perspective. Such experiments, with laboratory animals, will surely be very fruitful of basic scientific knowledge if the techniques can be developed. It would also have enormous value in livestock breeding, just as cloning (propagation by cuttings) is a mainstay of horticulture. Until such experiments have been pursued in some depth, with other animals, it is merely a speculative game to discuss applying such reproductive novelties to man.

A rather different approach was taken by MIT biologist Ethan Signer, who viewed scientists as the instrument of the power elite. Without a fundamental restructuring of society's values and institutions, science, including genetic engineering, Signer argued, will be exploited by the powerful and used against the powerless.

He wrote about scientific elitism, class interests, and technological hegemony. He contended that scientific objectivity was a myth perpetrated by one class interest on another. "Science can be used in conflicting ways in the interest of different sectors of society, and that scientists, rather than participating only in a disinterested search for truth, are by the research they choose to do implicitly advocating the interests of one sector or another." Signer's analysis hit a sensitive nerve among critical mainstreamers–those whose framework of analysis did not include a full dressing-down of the capitalist state. But there were some areas of common ground. Signer argued that scientists bear a responsibility to analyse and communicate to the public how their work will be used.

This was one proposal embraced by the bioethics community. He fulfilled his own recommendation that university science courses should consider the social effects of the subject matter. Like other scientists reaching out to society during this period, Signer was more adept at exploring the possible applications of genetic engineering than he was at postulating norms for its controlled use.

Without concrete cases of human genetic manipulation for consideration, the realisation of public policy regarding its misapplication was reduced to a speculative game. The discovery of recombinant DNA in 1973 brought the possibilities closer at hand and gave rise to a new chapter in the public debate over the ethics of human genetic engineering.

GENE-SPLICING AND HUMAN GENE TRANSFER

As previously noted, the first wave of scientific concern about human genetic engineering came in the aftermath of discoveries that viruses could be used to transfer foreign DNA into mammalian cells. The development of recombinant DNA molecule technology greatly simplified the process of gene exchange by using extra-chromosomal rings of DNA called plasmids as the transporting entities. Once plasmid-mediated DNA exchange was refined, the simplicity of its method made it available to a broad sector of scientists, and not just to virologists. The process promised to be more controllable and predictive than viral-mediated DNA transfer.

The international Asilomar conference was held January 24-27, 1975, to address some of the potential risks of rDNA molecular technology: Organisers of the meeting restricted the agenda to laboratory biohazards since it was those concerns that prompted the call for a voluntary moratorium for certain classes of experiments. From Asilomar came an explosion of media attention. Among the first generation of popular genetic engineering accounts targeted for a mass market, the one with the broadest public distribution was Who Should Play God? by Ted Howard and Jeremy Rifkin.

One of the central themes in this work is that lurking behind genetics research is the ideology of eugenics which, like an opportunistic pathogen, waits for the proper moment to infect society: "Bioengineering a better, or even 'perfect' human specimen has long been the goal of geneticists. In this regard, genetics, unlike the other branches of science, has always had an ideological wing, which has been as indispensable to its development as the scientific discoveries themselves."According to the authors, whatever the original intentions and applications of genetic engineering, whatever its scale of use or its intended humanitarian directions, it will be led inexorably by this ideology to creating new forms of human life. Howard and Rifkin make the bold assumption that to take the first step is to permit the final consequence. This argument presupposes that science and the society within which it functions are part of a deterministic process. Any theory that views social behaviour through a deterministic lens must be approached with skepticism. But we should not dismiss these ideas entirely without examining several forms of the deterministic thesis.

We may distinguish historical determinism from an absolute fatalism. In the former, the social decision matrix consists of critical nodal points where "free" agents choose from among alternative pathways to the future. As an example, under the historical determinist thesis, once we unleashed the first atomic bomb, we were driven by the iron laws of necessity to a global arms race. Others who are more pessimistic see the final outcome as nuclear war. Had another choice been made at the critical juncture, we might have been spared the massive buildup of nuclear weapons.This view of historical determinism assumes that societies can select the initial state or starting point (node) along a specific historical path. The node represents the initiating event of some more or less deterministic path leading to a final state. Let us assume for a moment that such deterministic pathways existed in technological developments.

In the case of genetic engineering, where would the critical node be? Is it the point at which the first microbe was genetically modified? Might it be the first manipulation of a human gene? Alternatively, is the starting point the discovery of the genetic code in 1953, restriction enzymes in the late 1960s, or recombinant DNA in 1973? Furthermore, if Pandora's box of human eugenics has been opened and the action is irreversible, what is the point of contravention? If it must happen because the critical node has been trespassed, we are left only with the option of informing without the possibility of altering the inevitable outcome.Historical determinism as described above is not a useful framework for anticipating the direction of science and technology.

It has no predictive efficacy. A technology has many potential uses but not all are realised. All too often we are made aware of the abuses, but there are clearly many undesirable things that can be done in the name of technology that are proscribed. On the other hand, it is far more difficult to control atomic energy once the atom has been split or to ban biological weapons once they have been created. The Howard-Rifkin position could also be interpreted as a variant of the slippery-slope argument where the acceptance of some forms of genetic engineering will allow more problematic forms to gain acceptance. Slowly, by small incremental steps, human eugenics may be adopted as public policy. The slippery-slope argument may be framed in a deterministic or probabilistic form. However, it is more convincing as a probabilistic thesis.

Suppose there are n types of human genetic engineering (HGE) scaled such that the first level (HGE-1) is least objectionable and the nth level (HGE-n) is most objectionable (according to current norms), with increasing states of "objectionableness" between 1 and n. The fatalist position is that we will reach the nth state regardless of what interventions we make once the initiating discovery has taken place. The fatal flaw is in the discovery itself. Once this takes place, human weakness draws civilization to disaster. A deterministic but non-fatalistic position holds that once HGE-1 is permitted, we will inevitably draw closer to HGE-n. In this case, HGE-1 is a nodal point that links a precipitating event to a series of escalating applications of HGE. We can avoid HGE-n only if we avoid HGE-1.

The probabilistic thesis maintains that any decision, say HGE-k (where kj"n), makes HGE-n more likely, but does not causally determine it. It allows for the possibility of human intervention at any point in the chain. Thus, while there is no law of iron necessity that links the treatment of a thalassernic patient by transplanting genetically engineered cells and the implantation of genes in a human egg, these two events are united by a similar technological process and impelled by similar economic forces and professional motivation. The former event gives shape to the latter event without "determining" it. The probabilistic thesis implies that we are treading on ethically sensitive ground. For this reason there must be clarity about the justification for HGE-k and the ethical boundaries between each class in the series of possible human genetic engineering events.

Strong advocates of the first HGE experiments dismiss the deterministic thesis and give little if any credence to the probabilistic thesis. Each event is treated on its own terms. Some opponents of these experiments reject this form of historical reductionism but put in its place a naive technological determinism.

No one has successfully argued that the decision to execute the first group of HGE experiments will have no effect on the next generation of experimental interventions into the human genome. It is not unreasonable to hold the probabilistic view, particularly for those who have no confidence in the moral gatekeepers. The next section describes a highly publicised human gene therapy experiment that shows how fragile the system of oversight is capable of becoming.

THE MARTIN CLINE EXPERIMENT

In July 1990, Martin Cline, a scientist at the University of California at Los Angeles (UCLA), transplanted genetically engineered cells into the bone marrow of two women. One was a 16 year-old Italian and the other a 21 year-old Israeli. Both were stricken with an inherited blood disease called beta thalassernia major. The experiments were performed in their respective countries at the Haddasah Hospital in Jerusalem and the University Poly Clinic in Naples.

Each woman had some bone marrow cells extracted and treated with a corrected gene for hemoglobin and a gene designed to augment the cells' production when they were transplanted back into the patient.These experiments represent the first use of recombinant DNA techniques on human beings. The story was followed by the media for a period of months during that year. The ethical issues surrounding the case were complicated by the following factors. UCLA refused Cline permission to do similar experiments in California on the grounds that more animal research was needed. Permission had been granted by hospital officials in the two countries.

Patients gave their informed consent. The two women had a life-threatening form of the disease that brings death to most patients by their early twenties. Cline's work was supported by the NIH, which had guidelines on the use of rDNA techniques. The guidelines were applicable to projects done outside the United States if they were funded at least in part by the NIH.

When the story broke there was a rash of questions. Was Cline morally justified in carrying out this experiment? Did he violate federal regulations? Were patients and hospital officials adequately informed about the nature of the experiment? Should U.S. guidelines for experimental therapies apply to other countries? The moral issues were obfuscated in a cloud of negative publicity. The general criticism of the two experiments fell into four categories.

- The experiments were premature. There was not enough evidence to conclude that the experiment had any chance of working. The animal data were not deemed satisfactory. The experiments were not designed to provide useful information if they failed. And too little was known about the side effects. Some speculated that placing genetic material in a cell could cause cancer or create new disorders, making patients more ill than they were before the treatment.
- The experimenter violated the NIH rules, albeit on a narrow technicality. Cline had approval from federal authorities to insert two genes separately into human cells. However, he went a step further and introduced them in combined form (a recombinant molecule). Cline argued that the separated genes tend to recombine within the cell naturally so that there is no substantive difference between the approved experiment and the one he actually conducted. Cline decided on his own to alter the protocols of the Israeli experiment and inject rDNA-altered cells into humans in violation of the January 1980 NIH guidelines. The collaborating physician at Haddasah Hospital stated that Cline used recombinant DNA "without letting anyone here know this before, or even whilst the experiment was carried out."

- The experiments violated the protocols on human subjects. Strictly speaking, the guidelines for the protection of human subjects, initially adopted in 1974, and modified thereafter, apply to investigators engaged in federally funded research at U.S. institutions. But investigators at a U.S. university engaging in human experiments at a foreign institution must obtain approval from the Institutional Review Board or its equivalent at both their host and collaborating institution.
- The experiment was a step in the direction of onerous forms of human genetic engineering. This final argument was not part of the dialogue that took place in the scientific community, but formed part of the misplaced antieugenic rhetoric.

When the negative publicity over the Cline experiment erupted, there was great concern among members of the biomedical community that a few premature and unworthy experiments might bring a rash of new laws restricting this line of enquiry for the treatment of disease. They felt it was time for scientists to reach a consensus about human genetic engineering and then to educate the public on the results of the consensus. Both a presidential commission report and a National Academy of Sciences study provided the framework for further discussion and public outreach.

SCIENTIFIC CONSTRUCTION OF THE PROBLEM

A meeting was convened at the Cold Spring Harbour Laboratory in 1982 at which 45 biomedical scientists discussed the scientific and ethical issues associated with gene therapy. The proceedings of that meeting, titled *Gene Therapy: Fact and Fiction*, were published as a public information report. The report advanced two principal ideas. First, many human diseases are genetic in origin and cannot be treated by conventional approaches (*i.e.*, drugs, surgery, diet, or environmental factors). Second, if human gene therapy becomes possible it will be an important part of the physician's treatment regime.

The majority of those gathered at Cold Spring Harbour accepted these premises and were optimistic about the new form of medical intervention. The ethical issues were reduced to the queries: How close are we to clinical trials and how do you decide when you know enough to go ahead and act?

When is it justifiable to use conceptually new, untried, and possibly dangerous procedures on human subjects, procedures that are aimed at developing therapy for them when no previous therapy has existed?

Is it ill-advised to use an experimental procedure on a patient when it is unlikely to benefit that patient, but may provide breakthroughs for future cases? Similar questions were asked during the period when experimental organ transplants were initiated.

Conventional ethics does not offer an unambiguous response to this query. One tradition in ethics holds that people should be treated as ends and not as means. Using people as experimental subjects solely for the purpose of helping others is an example of the latter and therefore warrants ethical condemnation. But what if these individuals are fully informed and wish to serve as experimental subjects for the benefit of others–a form of philanthropy?

While these issues were raised, the scientists' primary concerns were that research in human genetic engineering not be derailed and that information about human gene therapy must be appropriately managed to avoid public overreaction. The frame of reference, after all, was the public's response to the issue of recombinant DNA research in university laboratories. That convinced many scientists to exercise much greater caution in communicating risks and in bringing to a public forum the dissonances that are expressed within their professional societies.

David Baltimore's response to a query about what is permissible is a striking example of the "we and they" discourse that pervaded the meeting, and the translation of ethics to political expediency. We have to be extremely careful about what is tried in humans. This is crucial because any attempts at human therapy that are less than totally defensible on scientific grounds are going to muddy the political atmosphere in which we do our science.

Biomedical scientists sought ways to depoliticise human genetic engineering despite its powerful political and social symbolism. Several strategies were advanced to achieve this outcome. First, the concept of human genetic engineering was subdivided into two parts, one bearing a medical association and the other a political one.

This was accomplished through the distinction between somatic cell and germline genetic engineering. The former was identified as medical therapy, while the latter was seen by some as a form of eugenics. Advancing this distinction was important for minimizing the impact of the slippery-slope argument and for undercutting the deterministic thesis, both of which were gaining adherents. Among some scientists, there was skepticism about whether people could be taught to accept the distinction.

People will never understand the difference between gene therapy of somatic cells and modification of the germ line. That is a fundamental distinction. We all understand it and we have no difficulty with it. But although we will keep trying–we have to keep trying–believe we will never actually get it through to the general population because these concepts are too close to each other, and they require too much scientific understanding to make those distinctions.

It is not complex science that fixes this distinction, but rather a concept quite accessible to a lay audience. In somatic cell HGE, the subject alone is modified by the procedure. In germline HGE, the changes are carried through multiple generations. Pollsters used the distinction quite handily in attitudinal surveys on human genetic engineering. A second approach for depoliticizing science is to reconstruct the terminology. Words carry political messages, some positive and some negative.

The earlier genetics debates provide a few examples. The term "breaching species barriers" sheds a negative light on recombinant DNA research. It projected a scientific hubris that played poorly on people's sensibility. Eventually, this terminology lost favour among scientists.

To cite another example, the term "mutant" has a common usage in biology, as in "mutant organism" or "mutant genes. During the community debates over the field test of ice minus, scientists railed against citizens and some newspapers for using the provocative term "mutant bug" instead of the more neutral expression "genetically modified organism."

Similarly, when the use of HGE was being considered as a treatment for human disease, the term "therapy" and not "engineering" was deemed appropriate. This usage would reinforce the public's positive associations with biomedical science. A third effort at depoliticizing HGE was aimed at dramatizing the idea that its therapeutic use was the *only* available cure for a dread disease. Powerful emotional overtones accompanied the message that a patient, holding on to life by a thread, might be offered a reprieve with gene therapy. This imagery over-shadowed any political baggage or ethical liability that might be carried by the procedure.

While the biomedical scientists and medical practitioners joined forces to build a solid case for human genetic engineering, the issues were eventually addressed by broader constituencies of bioethicists, theologians, and lawyers. One of the most significant meeting grounds for this confluence of interests came in the form of a presidential commission.

FRACTIONALISATION OF HGE

Modern science follows rather judiciously Descartes' principle that large problems are solved by dividing them up into manageable parts. The same method has been applied to ethical problems. Rather than viewing human genetic engineering as a cluster definition, scientists and bioethicists fractionated the concept, explicating its subtle components. At first, HGE was bifurcated into somatic cell and germline manipulation. When that proved to be inadequate, finer divisions were created. By treating each division of the concept as a special category, the concerns about a slippery slope can be more readily managed.

The National Academy of Sciences and its affiliate the Institute of Medicine sponsored a workshop on human gene therapy in October 1986. Two years later a book based upon these meetings and written by medical/science writer Eve Nichols was published under the academy's imprimatur. Following conventions discussed at the meeting, the book disaggregates HGE into four categories of experiments.

1. Somatic Cell Gene Therapy
2. Germline Gene Therapy
3. Enhancement Genetic Engineering

4. Eugenic Genetic Engineering

The principal division is between gene therapy and genetic engineering, where the former connotes the curing of disease and the latter the "improvement" of human beings. Somatic cell gene therapy is defined rather narrowly as "the insertion of a single gene into the somatic cells of an individual with a lifethreatening genetic disease."

This restricted form of HGE is cited by Nichols as "the only technique now considered appropriate for use in human beings." Germline therapy is defined as the insertion of a healthy gene into a human egg with a specific genetic defect. This definition builds on a controversial distinction between healthy and unhealthy genes. Are extremely short-statured people endowed with an "unhealthy" gene? Are people who are hypersensitive to certain toxic chemicals endowed with an "unhealthy" gene? Efforts to rid the species of "unwanted traits" that are not associated with disease are referred to by ethicists as positive eugenics. The third category employs the term genetic engineering rather than gene therapy. The implication is that the intervention has moved beyond medical treatment. By "enhancement genetic engineering" Nichols means the alteration of a specific characteristic of a human being "in a preferred direction." These characteristics would be altered in the fully formed person through modification of somatic cells. The techniques for the attainment of (1) and (3), namely somatic cell HGE, may be equivalent. The categories are not distinguished by the science involved but rather by the social/medical objectives

Table: Nichols-NAS Typology on Human Genetic Engineering

	Experimental		*Single or*	*Cure or*
	Subject	Tareet Cell	Multiple Genes	Enhancement
Somatic Cell Gene Therapy	persons	somatic	single	cure
Germ Line Gene Therapy	eggs/fetus	germline	single	cure
Enhancement Gene Engineering	persons esp. children	somatic	single/multiple	enhancement
Eugenic Gene Engineering	eggs	germline	multiple	enhancement

In (1) the objective is the elimination of the clinical manifestations of a disease, whereas in (3) it is the improvement of a clinically normal individual. This distinction is easily bluffed. Medical abnormalities come in all forms from mild to severe.

Many abnormalities are not medical although they might be interpreted as such. Some phenotypes in conjunction with environmental factors are associated with a higher likelihood of clinical pathologies. And there are cases where the medical community is not in agreement on whether a physical condition should be characterised as a disease (*e.g.*, fibrocystic breasts).

The final category, eugenic genetic engineering, is defined as the alteration of complex traits involving multiple genes, such as intelligence and organ formation.

This definition of eugenics is far narrower than the more common usage where it signifies "cleaning up" the gene pool of defective or inferior genes (negative eugenics) or, conversely, improving the gene pool by increasing the frequency of superior genes (positive eugenics). It is left unclear whether this form of intervention is carried out on fertilized eggs (including zygotes and fetuses) or persons, although it is most likely the former. The four-part classification developed by Nichols in cooperation with NAS is unmistakably opportunistic. It stipulates new definitions that treat one class of experiments (type 1) as unambiguously permissible. But the only distinction that is defensible on scientific grounds is that between somatic and germline HGE. The boundary between enhancement and cure is socially constructed. Positing well-defined categories of illness and normalcy trivialises the gray areas and neglects the important distinction between potentiality and actuality in the onset of disease.

EUGENICS

Much of the ethical burden on HGE is carried by the prospect of altering human germline cells. Government reports and science-based studies acknowledge the special moral context of such interventions. In many of the sources have cited, the possibility is left open for future experiments that would alter the human germline. Some advocates of germline engineering defend it on the grounds that a programme of negative eugenics is morally acceptable. Simply put: By engineering the fertilized egg, why not rid the human gene pool of those genes responsible for undesirable traits?

The argument against any form of germline intervention rests on two premises. First, there is no natural demarcation between "unhealthy" and "healthy" genes. There may be broad consensus for cases at the extremes (*e.g.*, the gene responsible for Severe Combined Immune Deficiency or SCID, is deleterious), but there are other cases where a gene may serve positive and negative roles (the sickle-cell trait reduces the likelihood of malarial infection).

Second, once a programme of negative eugenic engineering gets initiated, research into positive eugenic engineering will be forthcoming. This argument is based, to some extent, on the first premise. The likelihood of moving from negative to positive eugenics is increased when the dividing line is blurred. These and other considerations were taken up during the next critical period, when oversight for human genetic engineering was assumed by the National Institutes of Health.

After the publication of "Splicing Life", the NIH's Recombinant DNA Advisory Committee set up a study panel in April 1983 to determine whether this committee should play a role in the review of experiments involving the modification of genes. A year later the NIH accepted such a role and established a 15-member Working Group on Human Gene Therapy (hereinafter Working Group). in January 1985 the NIH published a set of draft guidelines for gene therapy experiments developed by the Working Group. After public review and revision, the final document, titled "Points to Consider in the Design and Submission of Human Somatic Cell Gene Therapy Protocols," was adopted by the RAC in September 1985.

The protocols were subject to broad public and scientific review. While focusing on somatic cell modifications, the protocols were mindful of germline consequences. Investigators seeking permission to do such experiments were responsible for providing information on the likelihood that the proposed somatic cell gene therapy would inadvertently affect reproductive cells. "Points to Consider" made explicit reference to the possibilities of eugenic uses of genetics. "While research in molecular biology could lead to the development of techniques for germ line intervention or for the use of genetic means to enhance human capabilities rather than to correct defects on patients, the working group does not believe that these effects will follow immediately or inevitably from experiments with somatic-cell gene therapy." According to the document the RAC would not "at present" entertain proposals for germline alterations, which left the future somewhat uncertain.

Two citizen initiatives advocated an outright ban on germline experiments. Jeremy Rifkin's role in publicizing signatures of scientists and religious leaders supporting such a ban was discussed earlier. The Committee (now Council) for Responsible Genetics (CRG)–a national public interest group located in Boston–issued a policy position in response to the draft protocols. The CRG's proposal asked the RAC not to review or approve two types of experiments; first, human genetic therapy that:

- Is not aimed solely at the relief of a lifethreatening disease or severely disabling condition,
- Could alter germline cells; second, *in vitro* recombinant DNA experiments that alter human germline cells or early human embryos.

CRG's principal justification for the first prohibition was: The establishment of restricted zones of application of HSCT [Human Somatic Cell Therapy] at the outset will make it possible to support the most urgent and humane applications of HSCT with less fear that such support gives tacit approval to other uses that, propelled by economic and professional pressures, may slip through the review process.

The CRG also requested the NIH to take a position against the use of HSCT for enhancement of human characteristics, arguing that such uses will result in the medicalisation of social problems. This was not a fixed and unyielding opposition but one based on the need for a broad public consensus and wider democratic participation.

The CRG also asked the NIH to make an explicit statement that it would not approve any protocols involving the genetic manipulation of human germline cells through the addition or deletion of genes in the sperm, egg, or zygote. The CRG argued that this would be tantamount to experimentation on future generations with no possibility of informed consent and that it would establish the foundation for a programme of eugenics.

The RAC rejected the CRG proposal and decided against making any a priori judgements about a class of experiments that should be proscribed. In a memorandum to the RAC, the Working Group on Human Gene Therapy stated that specific proposals are a useful way to discuss public policy. Therefore, the prohibition of any types of proposals on genetic therapy was deemed detrimental to the general policy discussion. While the Working Group accepted the boundary between somatic and germline HGE, it did not wish to close off a case-by-case review for the latter. "With respect to germ-line cells and embryos this subcommittee has not specifically addressed the complex issues involved but expresses its concern that the CRG recommendation would impede research and clinical advance in reproductive biology." LeRoy Walters, bioethicist, RAC member, and chairman of the Working Group, argued in print that successful somatic cell therapy will give justification to germline therapy.

Assume that HSCT is used to cure single-gene defects like cystic fibrosis or sickle-cell anemia. The treated patients will grow to adulthood and will be capable of reproducing. As a result, they are homozygous carriers of genetic disease. Each subsequent generation could be treated or, as Walters proposes, some patients might find it more acceptable to prevent the transmission of the disease by germline therapy. Another reason Walters gives for pursuing germline therapy is that HSCT might not work. "Early intervention that affects all the cells of the future organism, including the germ cells, may be the only means available for treating cells or tissues which are not amenable to genetic repair after birth."

For the most part, scientists have been cautious about germline therapy, but there is no consensus about establishing a Maginot line between somatic and germ cells. Bob Williamson at the University of London sees a role for embryo therapy in cases like cystic fibrosis where he believes it may be desirable to "alter the inheritance ('genotype') of the individual as well as the way his body works ('phenotype')."

In response to the prospect of human genetic engineering, the NIH preferred to establish procedural recommendations exclusively, leaving open all future possibilities. Any attempt to fix boundaries was opposed by the RAC Working Group on the grounds that it would "impede research and clinical advance in human reproductive biology." Although it was an explicit practice of the NIH not to fund germline manipulation, the agency refused to formalise that practice in response to petitions by public groups.

The debate over HGE has engendered three types of responses: those who advocate distinct moral boundaries (such as between somatic and germline, or between therapy and enhancement); those who support pragmatic boundaries (such as what is feasible and acceptable versus what is still debated); and those who are opposed to any articulated boundaries (each experiment is taken on its merits regardless of its domain).

A particular classification scheme sets the framework for policy discussions. With finer distinctions set within the main classification of somatic and germline interventions, the morally suspect applications of HGE have been assigned to narrower and narrower categories. Harvard bacteriologist and former RAC member Bernard Davis has argued that the ethical concerns of HGE should be directed at any future efforts to improve or "blueprint" people according to somebody's plan. For Davis, the only justified ethical line to draw is that between medical and eugenic uses of genetic engineering, but not between somatic and gerrnline cells.

Eugenics, in Davis' terminology, is synonymous with embryo enhancement. The removal or suppression of genes in an embryo associated with some clinical abnormality is not, in this view, a eugenic intervention. W. French Anderson is a physician at the Laboratory of Molecular Hematology at the National Heart, Lung, and Blood Institute, NIH. On January 19, 1989, the director of NIH approved Anderson's clinical protocol to insert a foreign gene into the immune cells of cancer patients. The technique is identical to those planned for use in human gene therapy.

Anderson, a pioneer in human gene therapy, entered the public sphere for several years engaging ethicists, religious leaders, and the general public in a dialogue that guided him towards a position on the ethics of HGE and cleared a path for the first series of human experiments. He reached a conclusion that a sharp line can and should be drawn to distinguish acceptable from unacceptable forms of human gene therapy. Genetic engineering, he contends, should be used on humans only for the treatment of serious diseases and not for improvements or enhancements of the human condition. Anderson also accepts the use of gene therapy on the germ cells of adult individuals who want their children to receive a normal gene, but only under the conditions that:

- The effectiveness and safety of the techniques are established first with somatic cells;
- The techniques are reproducible and reliable;
- Public awareness and approval is assured.

Anderson has approached the issue of *in vitro* germline gene therapy cautiously in his published essays. While he offers no moral criteria for proscribing human germline manipulations, he does suggest safety and pragmatic grounds.

Genetic enhancement therapy, according to Anderson, should be avoided because it would threaten human values by both creating a medical hazard and requiring moral decisions that would lead to inequality and discrimination. Furthermore, he believes that it would be difficult, if not impossible, to establish a boundary once enhancement HGE had begun. In contrast, David Suzuki and

Peter Knudtson define eugenics as "a strategy of trying to orchestrate human evolution through programmes aimed at encouraging the transmission of 'desirable' traits and discouraging the transmission of 'undesirable ones.' They include medical-type traits such as "mutant hemoglobin molecule" in their criteria.

The authors advocate a prohibition of germline therapy "without the consent of all members of society." How can we ever obtain the consent of all members of society? Could they possibly mean every last member or simply some majority? There has never been an ethical debate resolved by appealing to the consent of "all members of society."

At the very least, their call for the consent of the governed on the moral issues of HGE implores the policyrnakers to expand the arena of participation. Public attitudes towards HGE will be discussed in the following section. The positions on HGE of Davis, Anderson, and others are based on a distinction between medical and non-medical genetic traits. Figure provides a two-dimensional matrix of prospective medical and non-medical applications of HGE. At the extremes are conditions that are sharply defined as medical (bottom) and non-medical (top).

Genetic traits in the middle signify conditions with both social and medical components. For example, phenylketonuria (PKU) is a hereditary disease that is caused by the lack of an enzyme that converts one amino acid (phenylalanine) into another. Without the enzyme and with no treatment, a child becomes brain damaged.

Once screened at birth, the disease is successfully treated by a diet low in phenylalanine. However, even in its successful treatment, IQ deficits are observed in older children. Since an IQ deficit in this case is correlated with a medical abnormality (although IQ in itself is not considered a medical problem), PKU falls into a gray zone. From this example, legal scholar Julie Gage concludes:

The suggestion that gene therapy for enhancement purposes will not follow inevitably from gene therapy to correct defects shallowly skims the issue. There is no easy line between enhancement and curing defects. Genetic markers can reveal an individual's predisposition to disease.

As an illustration, the AT trait (ataxia telangiectasia is a recessive and often fatal genetic disease) predisposes its carriers to a higher risk of cancer. Recognizing that predisposition is a statistical concept, is predisposition on par with having a disease for the purpose of distinguishing between therapy and enhancement?

Is preventing a predisposition "therapy" or "enhancement?" In July 1990 the New England Journal of Medicine reported a successful experiment in which elderly men were treated with human growth hormone (HGH). The study indicated that compared to controls, those treated appeared more vigourous and showed physiological signs that belied their age. The body's production of human growth hormone declines precipitously after age 60. One theory maintains that continued production of HGH may slow the aging process. If this can be achieved by turning the dormant growth hormone genes on, would this be considered therapy or enhancement? Is aging a disease? Without a clear distinction between the medical and social interpretation of a trait or physical state, a consensus position on germline genetic modification does not seem close at hand. This is borne out by public opinion polls.

2

Emergence of the Science of Genetic Engineering

INTRODUCTION

Advances in cytology, the science of the structure and function of cells, enabled scientists to more deeply appreciate Mendel's work. In 1902 American biologist Walter S. Sutton and German cell biologist Theodor Boveri separately noted the parallels between Mendel's units and chromosomes. The demonstration of the chromosomal basis of inheritance gave rise to the modern science of genetics. The term *genetics* itself was coined in 1905 by British biologist William Bateson. The terms *gene* and *genotype* were contributed in 1909 by German scientist Wilhelm Johannsen.

In 1905 American biologists Edmund B. Wilson and Nettie Stevens independently discovered and identified the sex chromosomes. Wilson discovered the X chromosome in a butterfly, and Stevens discovered the Y chromosome in a beetle. The discoveries of the X and Y chromosomes helped scientists begin to unravel new patterns of inheritance. Foremost among this research was the work of American biologist

Thomas Hunt Morgan on fruit flies. In 1910 Morgan identified the first proof of a sex-linked trait, an eye-colour characteristic that resides on the X chromosome of fruit flies. With this finding, Morgan became the first scientist to pin down the location of a gene to a specific chromosome. Morgan was also the first to explain the implications of linkage, unusual patterns of inheritance that occur when multiple genes found on the same chromosome are inherited together.

A student of Morgan's, American biologist Alfred Sturtevant, found early evidence of the mechanisms of crossing over, the phenomenon in which chromosomes interchange genes. More definitive proof emerged in the 1930s with work by American geneticists Harriet Creighton and Barbara McClintock. The pair demonstrated gene recombination with experiments on seed colour in corn. McClintock later gained notice for her work on transposable elements, large genetic segments that move within a chromosome or even between chromosomes. Her research into these elements, commonly known as jumping genes, earned McClintock the 1983 Nobel Prize in physiology or medicine.

BREAKTHROUGHS IN DNA STUDIES

While cytologists and geneticists were studying the properties and location of genes on chromosomes, other scientists focused their studies on the composition of genes. In 1928 British microbiologist Frederick Griffith ran a series of experiments on two strains of bacteria, one that kills mice and another that is harmless to them. When Griffith injected mice with killed cells of the virulent bacteria, all of the mice survived. But in a second trial, when Griffith injected a combined cocktail of dead virulent bacteria and live "harmless" bacteria, the mice all died. He concluded that something in the dead virulent cells "transformed" the hereditary material of normally harmless bacteria so that they became killers. Most scientists at the time theorized that the transforming factor was composed of a protein. The real identity of the transforming factor in this experiment was not identified until 1944, when American geneticists Oswald Avery, Colin MacLeod, and Maclyn McCarty revisited Griffith's research. After isolating different molecular components from dead bacterial cells, Avery and his colleagues determined that DNA was the agent that transformed the live harmless bacteria into killers.

Despite a growing body of evidence about the function of DNA, many scientists were not ready to reject proteins as the hereditary material. The debate was largely quieted in 1952 by American geneticists Alfred Hershey and Martha Chase. Hershey and Chase showed that when a type of virus called a bacteriophage infects a bacterium, it is the virus's DNA—not protein—that enters the bacterium to cause infection. Their studies confirmed that DNA contained the virus's genetic information, which triggered viral replication within the bacteria.

The experiments of Hershey and Chase convinced most scientists that DNA was the molecule of heredity, but many questions about the structure and mechanisms of DNA remained. In the early 1950s researchers began to apply techniques of X-ray diffraction to learn about the basic structure of DNA. X-ray diffraction can determine molecular structures by measuring patterns of scattered X rays after they pass through a crystalline substance. British physical chemist Rosalind Franklin and British biophysicist Maurice Wilkins used X-ray diffraction to obtain DNA images of unprecedented clarity.

Yet the exact three-dimensional structure of DNA remained unclear. The groundbreaking work of American biochemist James Watson and British biophysicist Francis Crick solved that mystery. In 1953 the two proposed a model of DNA that is still accepted today: A double helix molecule formed by two chains, each composed of alternating sugar and phosphate groups, connected by nitrogenous bases. Watson and Crick (along with Wilkins) were awarded the 1962 Nobel Prize in physiology or medicine for their discoveries.

Watson and Crick speculated that the structure of DNA provided some obvious clues about how the molecule could replicate itself. They proposed a replication model in which each strand of DNA serves as a template for making exact copies. This model of replication, called semi-conservative replication, was demonstrated in 1958 by American molecular biologists Matthew Meselson and Franklin Stahl. Their experiments demonstrated the mechanisms of replication by tracking DNA containing a heavy nitrogen isotope through a series of replications.

With DNA's structure and replication mechanisms largely solved, scientists turned their attention to identifying the genetic code—learning how a gene's nucleotide sequence determines what type of protein is made. In the late 1950s, South African geneticist Sydney Brenner and other scientists confirmed that RNA acted as an intermediary between DNA and protein production. Researchers still were uncertain how the sequence of nucleotides in DNA corresponded to the production of specific amino acids. In 1961 Crick and Brenner determined that groups of three nucleotides, now known as codons, code for the 20 amino acids that form the foundation of proteins. The exact relationship between codons and amino acids was clarified after several important discoveries.

American biochemists Marshall Nirenberg and J. Heinrich Matthaei synthesized repeated nucleotide sequences that led to the production of repeated single amino acids. They identified how certain codon combinations code for a specific amino acid. A process developed by American geneticist Har Gobind Khorana helped scientists create a "dictionary" of codons that defined specific amino acids, thus resolving the remaining ambiguities in the genetic code. Only 12 years after the structure of DNA was deduced, the genetic code was solved.

Learning to Manipulate DNA

After scientists had unraveled the structure and replication mechanisms of DNA, many felt that the major discoveries of genetic research were resolved. They predicted that the only task left in genetics was to sort out the molecular details of how genes work. But in the process of studying gene function, researchers developed powerful new molecular techniques, enabling them to analyze and manipulate genes with a speed and precision never before possible.

A number of discoveries made during the 1960s and 1970s shed light on how distinct fragments of DNA could be isolated. The work of Swiss molecular biologist Werner Arber focused on specialized enzymes that digest, or "restrict," the DNA of viruses infecting bacteria. These enzymes were subsequently dubbed restriction enzymes. In the following decade, scientists learned that restriction enzymes could also act like molecular scissors to cut DNA. In 1970 American molecular biologist Hamilton Smith and colleagues determined that restriction enzymes could cleave DNA molecules at precise and predictable locations. Hamilton concluded that the enzymes were able to recognize specific nucleotide sequences. Scientists quickly realized that restriction enzymes could be used in the laboratory to manipulate DNA. In 1973 American biochemist Herb Boyer used restriction enzymes to produce a DNA molecule with genetic material from two different sources. This splicing technique is now known as recombinant DNA. Boyer inserted foreign genes into plasmids and observed that the plasmids could replicate to make many copies of the inserted genes. In subsequent experiments, Boyer, American biochemist Stanley Cohen, and other researchers demonstrated that inserting a recombinant DNA molecule into a host bacteria cell would lead to extremely rapid replication and the production of many identical copies of the recombinant DNA.

This process, known as cloning, gave scientists the power to make many copies of desired DNA for molecular study. The speed and efficiency of DNA cloning were vastly improved in the 1980s with the invention of polymerase chain reaction (PCR). Developed by American biochemist Kary Mullis, PCR enables scientists to produce large amounts of DNA sequences in a test tube. In a matter of hours, the process can produce millions of cloned DNA molecules. Yet all of the advances in isolating and replicating DNA would not be possible or be of much use if researchers could not determine the nucleotide sequence of genetic material. In the late 1970s and early 1980s, British biochemist Frederick Sanger and his associates developed DNA sequencing techniques. Sanger's methods, which used special compounds called dideoxy nucleotides, rapidly yielded the exact nucleotide sequence of a desired sample. With the use of automated equipment, the new techniques transformed genetic sequencing into a speedy, routine laboratory procedure.

Many of the new techniques for isolating, sequencing, and replicating DNA have been put to practical use through the field of genetic engineering. The Human Genome Project and the new field of proteomics have both benefited from continuing technical advances and have accelerated the development of new genetic technologies. Modern genetics is poised to radically change the practice of medicine and the biotechnology industry.

DEVELOPMENT OF GENETICS

The significance of Mendel's laws was not recognised by his contemporaries but became evident 35 years later, just after the turn of the twentieth century. At this time other important events marked the progress of genetics: Sir Archibald Garrod, a physician, made important contributions to the chemistry of diseases especially porphyria, cystinuria and alkaptonuria, noting their recurrence among sibs in the same family and the effect of consanguineous marriage. He introduced the term "inborn errors of metabolism" and the concept of "one gene, one enzyme" and thus opened the way to biochemical genetics, which became one of the most important aspects in the development of genetics.

Theodor Boveri and Walter's. Sutton in 1903 independently proposed the chromosome theory of inheritance, stating that the chromosomes carried the "hereditary factors" proposed by Mendel and that their behaviour during cell division provided an explanation for their segregation during meiosis.

In the twentieth century the fruit fly, Drosophila was the organism most extensively used in the study of genetics. It was extremely useful because it was easily bred in the labouratory, reproduced very rapidly (about 20 times in one year) and produced thousands of offspring that could be analysed. Besides, it had only four chromosome pairs that were exceptionally large in the salivary glands and showed a pattern of transverse bands on which individual genes could be localised.

THOMAS HUNT MORGAN is one of the most famous of the hundreds of scientists who worked on Drosophila. He described sex-linked characteristics that were manifest in the male flies. He is best known for proposing methods for estimating distances between genes located on the same chromosome through the phenomenon of recombination. The unit of gene distance, the centimorgan, is named after him. Morgans work opened up the important field of gene mapping.

Advances in human genetics did not parallel the great advances in the genetics of Drosophila since the beginning of the twentieth century.

This was mainly due to:

- The inability to perform breeding experiments in man;
- The very long generation time
- The small size and relatively large number of human chromosomes, which made their microscopic examination very difficult.

However, several hundreds of human genes were known to exist through the visible characteristics they produced. In the 1950s two great advances completed the phase of growth of classical genetics and paved the way for a new era of genetics.

The molecular structure of DNA, which explained how the molecule could replicate itself and how the same molecule could represent an almost infinite variety of different genes. From then the gene could be visualised as a structural entity and was no longer an invisible hereditary particle that could be identified only by its effects on the characteristics of an organism. The discovery of the molecular structure of DNA thus paved the way for the study of the gene itself and the direct analysis of DNA in the new era of molecular genetics that was to follow.TIJO and LEVAN devised a technique for the study of human chromosomes and established that the human chromosome number was 46. Earlier techniques were suitable for organisms that had few chromosomes but not for human chromosomes, which were small and numerous. This was immediately followed by the discovery of chromosomal abnormalities and marked the beginning of human cytogenetics. It was one of the most important advances that brought human and medical genetics into prominence.

THE NEW GENETICS

A new era of genetics began in the 1970s with the development of three new techniques, which together enabled the direct analysis and manipulation of genes. At this stage studies on DNA were performed mainly on bacteria as they had a very short generation time and very simple DNA. Most of the new discoveries arose from the study of bacterial genetics. Hamilton Smith discovered that bacteria produced "restriction endonucleases", enzymes that had the ability to cut open the DNA molecule at specific sites. ED Southern at Edinburgh University developed a technique by which specific genes could be isolated from the whole complement or genome of DNA. The process was called Southern blotting.

Stanley Cohen at Stanford University developed the use of plasmids, which are naturally occurring circular loops of DNA capable of entering bacterial cells, as vectors for carrying foreign DNA into bacterial cells. These three techniques of cutting, isolating and inserting genes form the basis of recombinant DNA technology, popularly called genetic engineering. They also formed the basis of modern genetic technology. We are now on the threshold of a new era that promises the treatment of genetic diseases.

SOME BASIC DEFINITIONS

The genetic material is contained within the nucleus of all cells. It consists of extremely fine thread-like structures composed of DNA and associated proteins. Within it is stored the genetic information, a whole library of genes which constitute the genome. These thread-like structures become visible as chromosomes during cell division as they gradually coil up and become thick and prominent.

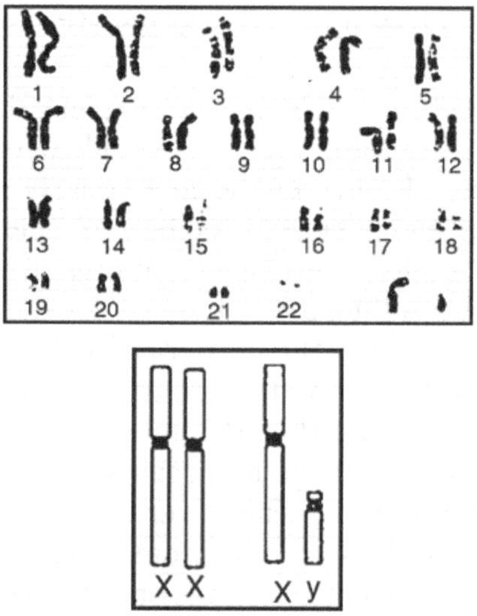

Fig. The sex Chromosomes are XX in Females and XY in Males.

The Y chromosome is much smaller than the X chromosome. It carries the genes responsible for the male sexual characteristics. The X chromosome carries genes that are not related to sex and are equally important in both males and females. The genetic material is made of DNA.

The DNA molecule consists of two spirally arranged chains in which each unit consists of a sugar (deoxyribose), a phosphate group and a nitrogenous base. The base can be Adenine, Guanine, Thymine or Cytosine (A,G,T or C).A gene is a segment of DNA that carries a genetic message. One gene carries instructions for the synthesis of one protein or enzyme.Enzymes are the functioning units of the cell.

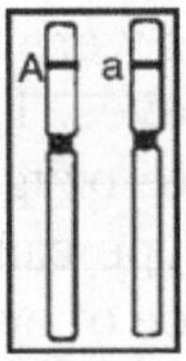

Fig. The Genes are Situated on the Chromosomes.

A locus is the point on a chromosome at which a gene is situated.

Genes occur in pairs that occupy corresponding loci on a pair of chromosomes.

The genes at a locus are called alleles.An individual is homozygous if the two alleles are identical *e.g.* AA or aa.An individual is heterozygous if the two alleles are different *e.g.* Aa.

A A	A a	a a
Homozygous	Heterozygous	Homozygous

The genotype of an individual indicates the type of alleles present (heterozygous or homozygous).The phenotype of an individual indicates the identifiable characteristics present in an individual. These may be visible or biochemical characteristics. Examples of phenotypic characteristics are eye colour, height, blood group. A dominant characteristic is one that manifests itself in heterozygotes (*i.e.* only one gene is required for its manifestation in the phenotype). A recessive characteristic is one that appears only in homozygotes (*i.e.* two identical alleles are required for its manifestation in the phenotype).An X-linked gene is one that is situated on the X chromosome.

Alleles	B - gene for brown eyes b - gene for blue eyes		
Genotype	B B Homozygote	B b Heterozygote	b b Homozygote
Phenotype	Brown eyes	Brown eyes	Blue eyes

An X-linked recessive characteristic is usually manifest in males while females are carriers of the gene. An X-linked dominant characteristic is manifest in both sexes. The following is an example of an inherited trait for brown or blue eyes"Brown eyes" is the dominant characteristic because it appears in the heterozygote (Bb). "Blue eyes" is the recessive characteristic because it appears only in the homozygote (bb).

Genotypes of Parents

	B	b
B	BB	Bb
b	Bb	bb

It is possible to work out how genes are inherited by constructing a Punnett square. In this example both parents were heterozygous. The phenotypes of the offspring were 3 brown: 1 blue eyes. The genotypes of the offspring were 1 dominant homozygote (BB): 2 heterozygote: 1 recessive homozygote (bb)

The inheritance of the phenotype can be illustrated as a pedigree. The following pedigree shows inheritance of eye colour.

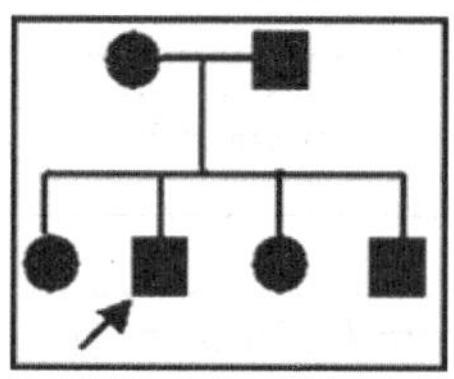

Fig. Males are Represented by Squares, Females by Circles.

The arrow indicates the propositus or proband *i.e.* the individual with a particular characteristic through whom the family came to be investigated. In this case the propositus was a boy with blue eyes.

PRESENCE OF A GENETIC EFFECT

The presence of a genetic effect was originally established for activity through selective breeding of animals showing high activity scores, low activity scores, or both on a standardized measuring technique. One of the earliest attempts to establish the hereditary basis of activity using this method was a selection study by Rundquist, in which he selected rats for high and low activity following testing in an activity wheel. His results were disappointing initially, because of an error in the method of selection he employed: the mating of high active and low active offspring from nonselected parents.

That is, rats showing high activity scores were selected not only from parents that also had high activity scores, but from those yielding low activity scores as well, with the reverse holding true for rats with low activity scores. As a result, there was no differentiation between groups from generations F_1 to F_4. When this error was corrected, differentiation into high and low active groups did occur and laid the foundation for further research in this area.

Hall performed selection experiments in the early 1930s to establish the genetic basis of emotionality in rats using as phenotypes the Behaviour of the rats in an open field. Although he was not concerned with activity per se, his selection procedures and his use of the open field test have served as the prototype for many of the modem studies on this problem. In the 1950s, it became possible to use another method for determining hereditary basis for activity and for any other phenotype. This method involves finding differences in activity between strains of genetically controlled animals, usually highly inbred strains of mice and rats, but also established breeds of dogs and inbred strains of various species of *Drosophila.* Thompson tested five strains of inbred mice on two activity measures, the open field and the Y maze. His results showed marked differences in the activity levels of the five strains, and the two C_{57} strains showed much higher activity levels in both open field test and the Y maze than did the BALB/c mice.

In mouse studies, McClearn has shown that there is a clear relationship between locomotor activity and the per centage of C_{57}BL/Crgl genes present (from backcrosses with A/Jax mice). Thompson illustrated that there is considerable variability between strains in exploratory activity, with C_{57}BR/a, C_{57}BL/6, and C_{57}BL/10 having the highest mean score (for activity) and A, AK/e, and BALB/c strains having the lowest.

These orderings were fairly invariant with another testing apparatus (enclosed arena versus Y maze). There are differences in the ordering of mice on exploratory activity with different environmental conditions. For example, he found that in bright white light, A/Crgl mice showed less exploratory activity than in dim red light. In the C_{57}BL/Crgl mice, activity increased in white light and decreased in dim red light. Although he found no change in the ordering of the strains, McClearn did find a significant interaction between strain and illumination.

Of particular historic and current interest are the MR and MNR rats originated by Broadhurst. Using a procedure essentially similar to that used by Hall more than two decades earlier for selection for emotionality, Broadhurst made several important improvements; the open field was more uniformly and brightly lighted and initial responses of the subjects were weighted more heavily, giving less emphasis to the possible adaptation of the rats to a familiar environment. MR and MNR rats clearly differ on open field defecation scores, an index of emotionality, and are often regarded as "benchmark" strains for this category of Behaviour. However, on several *activity* measures, these strains do not differ greatly. Selection studies illustrating genetic differences in activity have also been carried out in species other than rats and mice.

The comparison of the normal Behaviour of different established breeds of dogs provides an excellent example; for "everyday" differences in activity levels, one need only compare a bassett hound and a fox terrier. Dogs have also been used as subjects in more systematic genetic studies, with different breeds treated analogously to inbred strains of rodents. A prime example of this sort of research is the 19 year study conducted by Scott and Fuller that dealt with, among other things, the emotional Behaviour of several breeds of dogs, together with their hybrids and backcrosses. Clear genetic differences were found in the five breeds on a variety of reactivity tests, including open field Behaviour.

Some work on activity and its implications has also been done with the fruit fly. In one study flies were selected for mating speed. Activity differences for these two selected strains (fast mating speed and slow mating speed) were measured by counting the number of squares a fly entered in an open field arena over a given period of time. It was found that slow mating lines exhibited more open field activity than did fast mating lines. In contrast, Ewing, in selecting for spontaneous activity, found that the less active lines displayed *more* sexual Behaviour than did the more active lines. Although these results appear to contradict Manning's findings, the discrepancy appears to have been due to an "apparatus effect," as he used an apparatus other than the open field used by Manning. When Ewing replicated his study using the type of open field apparatus used by Manning, no significant differences between the selected lines were found.

Besides being alterable by apparatus effects and other environmental changes, activity is readily changed, often differentially for different genotypes, by internal events, especially pharmacological agents. A thorough review of drug effects on activity, and the genetic factors involved, has recently been prepared by Broadhurst. The studies we have cited are typical of many others to be found in the literature, in that many of the findings are contradictory with one another, with varying differences among genotypes (whether inbred strains or selected lines) depending on procedural details, types of apparatus, and so on. Therefore, generalizations on activity levels of specific genetically controlled animals should be treated cautiously.

An excellent example illustrating this point can be found in an article by Lassalle and Le Pape. They demonstrated that the environmental conditions associated with testing of two inbred mouse strains (C_{57}BL/6 and BALB/c) can reverse the general finding we referred to earlier that C_{57}s are more active than BALB/cs. Under seminatural and breeding cage conditions, the BALB/cs showed a higher level of activity than did the C_{57}s, and in the seminatural condition, the BALB/cs were more active in the dark).

In an issue related to the genetic determination of activity, several studies have attempted to determine heritabilities and degrees of genetic determination. However, as these findings are limited to the specific genetic population measured within a specific type of apparatus and a specific set of environmental conditions, these data are of rather limited use in the generalizable information they provide for an understanding of the genetic basis of activity.

POLYGENIC MODELS

There are various avenues of research that have suggested that a polygenic model can best account for activity. One form of evidence for this type of genetic determination can be found in score distributions; typically, a continuous, unimodal distribution of scores has been found for most types of activity measurement when genetically heterogeneous animals were tested.

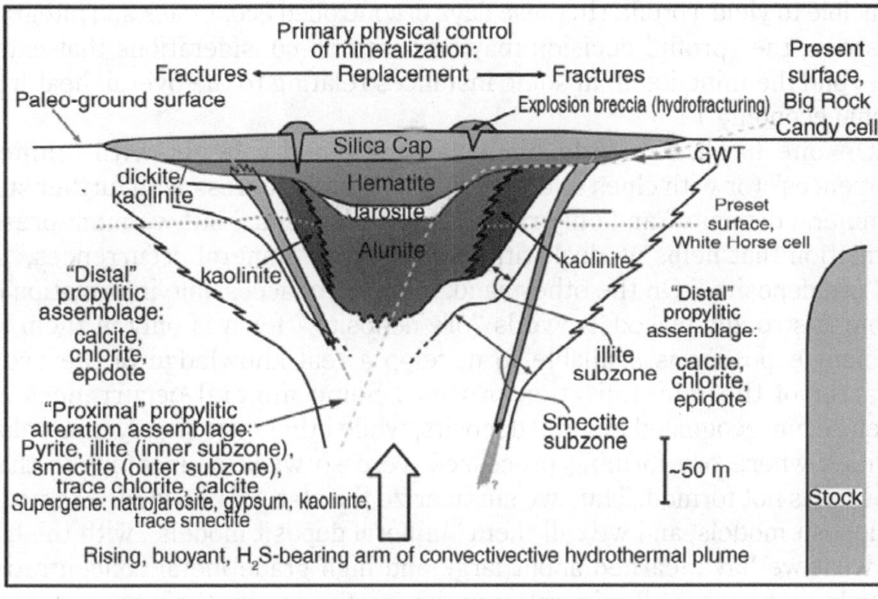

Fig. Genetic Models

Tryon suggests a polygenic model on the basis of this form of evidence alone. An excellent example of a more thorough analysis supporting a polygenic model can be found in the series of studies reported by DeFries and Hegmann. They tested large numbers of two inbred strains of mice (C_{57}BL/6J and BALB/cJ), as well as their reciprocal F_1 and F_2 hybrids and backcrosses.

The mice were tested for 3 minutes on each of 2 successive days in an automated, white Plexiglas open field, 90 cm^2. The means and variances of the transformed scores. The degree of increase in variance from the F_1 to the F_2 generation is inversely related to the number of loci affecting a trait. The variance between the F_1 and F_2 generations is very slight, especially when compared with the difference in variances between the parental strains and the F_1 generation. (One can discount the variance differences between the reciprocal F_2 s as a maternal effect, especially as it diminishes in later generations.) DeFries and Hegmann conclude that activity is controlled by genes at several loci, suggesting *at least* 3.2 loci from an earlier analysis by Hegmann.

A "mineral occurrence" is a concentration of a mineral (usually, but not necessarily, considered in terms of some commodity, such as copper, barite or gold) that is considered valuable by someone somewhere, or that is of scientific or technical interest. In rare instances (such as titanium in a rutile bearing black sand), the commodity might not even be concentrated above its average crustal abundance.

A "mineral deposit" is a mineral occurrence of sufficient size and grade that it might, under the most favourable of circumstances, be considered to have economic potential.An "ore deposit" is a mineral deposit that has been tested and is known to be of sufficient size, grade, and accessibility to be producible to yield a profit. (In these days of controlled economies and integrated industries, the "profit" decision may be based on considerations that extend far beyond the mine itself, in some instances relating to the overall health of a national economy.)

On one hand, the field observations usually begin with "mineral occurrences" (or with clues to their existence) and progress with further study to "mineral deposits" and only rarely to "ore deposits," but we must present information that helps us deal with all classes of "mineral occurrences," not just "ore deposits." On the other hand, in terms of accessible information our sample is strongly biased Towards "ore deposits," for it is only in them that sufficient exposure is available to develop a real knowledge of the overall character of the mineralization process. Some mineral occurrences are, therefore, unrecognized mineral deposits, while others are simply mineralized localities where ore forming processes were so weak or incomplete that a deposit was not formed. Thus we summarize the state of knowledge regarding ore deposit models, and we call them "mineral deposit models" with the hope that what we have learned about large and high grade metal concentrations will help us sort out all mineral occurrences to identify their true character and, we hope, to recognize which have potential to constitute ore deposits.

The attributes or properties of a mineral occurrence are, of course, those features exhibited by the occurrence. When applied to a model, these terms refer to those features possessed by the class of deposits represented by the model. It is useful to consider attributes on at least two scales: the first deals with local features that may be observed directly in the field (mineralogy, zonal patterns, local chemical haloes, and so on); the second is those features concerning the regional geologic setting and which must be interpreted from the local studies or may be inferred from global tectonic considerations (for instance, that the rock sequence under study represents a deep water, back arc rift environment, or that the area is underlain by anomalously radioactive high silica rhyolite and granite). Two of the most prominent attributes, the commodities/geochemical patterns and the mineralogy, are cross indexed to model types.

To the greatest extent possible, models were constructed so as to be independent of site specific attributes and therefore contain only those features which are transferable from one deposit to another. This goal is difficult to attain, because we do not always know which features are site specific.The term "model" in an earth science context elicits a wide variety of mental images, ranging from the physical duplication of the form of a subject, as in a scale model of the workings of a mine, to a unifying concept that explains or describes a complex phenomenon. In this context we shall apply only the latter usage. Therefore, let us propose a working definition of "model" in the context of mineral deposits, the overriding purpose being to communicate information that helps mankind find and evaluate mineral deposits. A mineral deposit model is the systematically arranged information describing the essential attributes (properties) of a class of mineral deposits. The model may be empirical (descriptive), in which instance the various attributes are recognized as essential even though their relationships are unknown; or it may be theoretical (genetic), in which instance the attributes are interrelated through some fundamental concept.

One factor favouring the genetic model over the simply descriptive is the sheer volume of descriptive information needed to represent the many features of complex deposits. If all such information were to be included, the number of models would escalate until it approached the total number of individual deposits considered.Thus we should no longer have models, but simply descriptions of individual deposits. Therefore, the compilers must use whatever sophisticated or rudimentary genetic concepts are at their disposal to distinguish the critical from the incidental attributes. It is commonly necessary to carry some possibly superficial attributes in order not to preclude some permissible but not necessarily favoured, multiple working concepts.

One of the commonly accepted attributes of the model for the carbonate hosted lead zinc deposits of the Mississippi Valley type is the presence of secondary dolomite. But do we know that this is essential? Suppose a deposit were found in limestone; would we reject its assignment to the Mississippi Valley class? Or could it be correct that the critical property is permeability and that the formation of dolomite either:

- Enhances permeability (and thereby makes the ground more favourable),
- Reflects pre existing permeability that is exploited by both the dolomite and the ore?

MODEL NAMES

Each model has been assigned a name that is derived either from the special characteristics of the classes or from a type locality. The latter strategy was employed to avoid excessively long descriptive names.

The use of type names derived from specific deposits does produce confusion in some readers, however, who may feel, for example, that a deposit that does not look "exactly" like Comstock cannot be represented by a "Comstock epithermal vein" model.

This confusion may be minimized by realizing that most models are blends of attributes from a large number of deposits and that the names are only conveniences, not constrictions. The contributors to this report and the literature in general are not without disagreements regarding nomenclature (as well as genetic aspects and some facets of the groupings made here), but provision for alternative names is made in the model format under the heading of approximate synonyms.

DESCRIPTIVE MODELS

Because every mineral deposit, like every fingerprint, is different from every other in some finite way, models have to progress beyond the purely descriptive in order to represent more than single deposits.

Deposits sharing a relatively wide variety and large number of attributes come to be characterized as a "type," and a model representing that type can evolve. Generally accepted genetic interpretations play a significant role in establishing model classes. Here we shall emphasize the more descriptive aspects of the deposits because our goal is to provide a basis for interpreting geologic observations rather than to provide interpretations in search of examples. The attributes listed are intended to be guides for resource assessment and for exploration, both in the planning stage and in the interpretation of findings.

The descriptive models have two parts. The first, the "Geological Environment," describes the environments in which the deposits are found; the second gives the identifying characteristics of the deposits. The headings "Rock Types" and "Textures" cover the favourable host rocks of deposits as well as source rocks believed to be responsible for hydrothermal fluids which may have introduced epigenetic deposits. "Age" refers to the age of the event responsible for the formation of the deposit. "Tectonic Setting" is concerned with major features or provinces (perhaps those that might be portrayed only at 1:1,000,000 or smaller scale), not ore control by structures that are local and often site specific. "Associated Deposits" are listed as deposits whose presence might indicate suitable conditions for additional deposits of the type portrayed by the model.

The second part of the model, the "Deposit Description," provides the identifying characteristics of the deposits themselves, particularly emphasizing aspects by which the deposits might be recognized through their geochemical and geophysical anomalies.

In most cases the descriptions also contain data useful in project planning for mineral assessment or exploration; this aspect is especially important where limited financial and manpower resources must be allocated to the more significant tasks.

PRINCIPLES OF GENETICS

The site where genes work is the cell. Some organisms, such as paramecia or amoebas, are made up of a single cell. Other organisms are made of many kinds of cells, each having a different function. For instance, a tree contains some cells that form the root system and other cells that form leaves. Each cell's function within an organism is determined by the genetic information encoded in DNA. In animals, plants, and other eukaryotes (organisms whose cells contain a nucleus), DNA resides within membrane-bound structures in the cell. These structures include the nucleus, the energy-producing mitochondria, and, in plants, the chloroplasts (structures where photosynthesis takes place). In prokaryotes, one-celled organisms and bacteria that lack internal membrane-bound structures, DNA floats freely within the cell body.

Cell Division and Reproduction

Organisms could not grow or function properly if the genetic information encoded in DNA was not passed from cell to cell. DNA is packaged into structures called chromosomes within a cell. Every chromosome in a cell contains many genes, and each gene is located at a particular site, or locus, on the chromosome. Chromosomes vary in size and shape and usually occur in matched pairs called homologues. The number of homologous chromosomes in a cell depends upon the organism—for example, most cells in the human body contain 23 pairs of chromosomes, while most cells of the fruit fly *Drosophila* contain 4 pairs.

Within all organisms, cells divide to produce new cells, each of which requires the genetic information found in DNA. Yet simply splitting the DNA of a dividing cell between two new cells would lead to disaster—the two new cells would have different instructions and each subsequent generation of cells would have less and less genetic information to work with. Imagine how chaotic it would be to rip an architectural blueprint in two, give each half to different contractors, and tell them to construct identical buildings. Just as each contractor would require a full copy of the blueprint to construct a complete building, each new cell needs a complete copy of an organism's genetic information to function properly.

Organisms use two types of cell division to ensure that DNA is passed down from cell to cell during reproduction. Simple one-celled organisms and other organisms that reproduce asexually—that is, without the joining of cells from two different organisms—reproduce by a process called mitosis. During mitosis a cell doubles its DNA before dividing into two cells and distributing the DNA evenly to each resulting cell. Organisms that reproduce sexually use a different type of cell division. These organisms produce special cells called gametes, or egg and sperm. In the cell division known as meiosis, the chromosomes in a gamete cell are reduced by half. During sexual reproduction, an egg and sperm unite to form a zygote, in which the full number of chromosomes is restored.

Mitosis

Mitosis occurs in five stages: prophase, prometaphase, metaphase, anaphase, and telophase. During prophase, the start of mitosis, the DNA of each chromosome replicates. Each chromosome then reorganizes into paired structures called sister chromatids, with each member of the pair containing a full copy of the DNA sequence.

The sister chromatids condense, thickening until they appear joined at a single site, known as the centromere. Prometaphase is marked by the disintegration of the nuclear membrane. The sister chromatids line up in the middle of the cell halfway between the poles. In metaphase, exactly half of the chromatids face one pole, and half face the other pole. This equilibrium position is called the metaphase plate. In anaphase, the chromatid pairs split apart at the centromere, and each half of the pair then moves towards opposite poles of the cells. In telophase, the final stage of mitosis, a nuclear membrane forms around the chromosomes at each pole of the cell. Mitosis ends with the formation of two new cells, each with a matching full set of chromosomes as well as an identical complement of cellular structures.

Meiosis

During meiosis, two cell divisions occur to produce four daughter cells from the original parent cell. Each resulting cell has half the chromosomal DNA of the parent cell. A half set of chromosomes in an organism is known as the haploid number. In the first cell division of meiosis the chromosomes of a gamete cell duplicate and join in pairs. The paired chromosomes align at the equator of the cell, and then separate and move to opposite poles in the cell. The cell then splits to form two daughter cells. As meiosis proceeds, the two daughter cells undergo another cell division to form four cells, each of which bears half of the number of chromosomes found in the other cells of the organism.

Meiosis ensures that reproduction will produce a zygote that has received one set of chromosomes from the male parent and one set of chromosomes from the female parent to form a full set of chromosomes. The entire set of chromosomes in an organism is known as the diploid number. Once formed, the zygote continues to divide and grow through the process of mitosis.

Patterns of Inheritance

In life forms that reproduce asexually, such as bacteria and amoebas, all offspring share the exact same genes and are identical to their parents. The genetic transmission that occurs in organisms that reproduce sexually is far more complex. An individual that forms by the union of two gametes inherits its chromosomes from two distinct parents. Consequently, sexual reproduction guarantees that offspring with new combinations of genes will continually arise.

Certain patterns of inheritance were evident long before scientists discovered the molecular structure of DNA and chromosomes. Throughout history, people have recognized that certain traits, whether in humans, animals, or agricultural crops, could be passed from generation to generation. Yet for centuries, people were unable to reconcile many confusing observations about the mechanisms of inheritance. The first person to make sense of this complex subject was Austrian monk Gregor Mendel, who conducted a series of experiments on pea plants beginning in the 1850s. Mendel observed the results of crossbreeding plants with different characteristics, such as height, flower colour, and seed shape. His conclusions from these experiments led him to develop explanations for how traits are transmitted from generation to generation. Mendel's theories form the foundation of modern genetics.

Mendel's Rules

In his research, Mendel observed that characteristics were inherited as separate units, each of which was inherited independently of the others. Mendel suggested that each parent has pairs of these units but contributes only one of each pair to offspring. The units that Mendel described were later given the name *genes*. Mendel recognized that a gene can exist in different forms.

Today these alternate forms are known as alleles. For example, pea seeds, the edible part of the plant we call peas, have a texture trait controlled by a single gene. This gene occurs in two alleles: one corresponding to round (smooth) peas, the other to wrinkled peas. Although an individual can carry only two alleles for a particular gene, each gene may have dozens of different alleles. Mendel's experiments focused on interbreeding different strains of pea plants and then observing the traits that appeared in subsequent generations. When he crossbred plants with round peas and those with wrinkled peas, he discovered that all of the resulting offspring had round peas.

Today we know that peas are round and smooth when they contain the right amount of sugar. If peas are missing the gene that produces a protein called starch branching enzyme 1 (SBE1), the peas make too much sugar, causing the peas to swell and then wrinkle and shrivel as they dry. Mendel concluded that when an organism has two different alleles corresponding to the same genetic trait, one of the two may be dominant.

The other allele is said to be recessive, meaning that its presence will be detectable only if an organism has inherited the recessive gene from both parents. For convenience, geneticists designate alleles by a single letter—the dominant allele is represented by a capital letter and the recessive allele by a small letter. In the pea texture example, a plant inherits one allele for pea texture from each parent. The dominant allele that produces SBE1, resulting in round, smooth peas, is designated as *R*, while the recessive allele that does not produce SBE1 and produces wrinkled peas is designated as *r*.

To determine the set of alleles an organism has for a given trait just by visual observation can often be difficult. In the pea plant example, for instance, plants with smooth peas might be carrying two dominant alleles for that characteristic (*RR*) or one dominant and one recessive allele (*Rr*). Geneticists use the term *genotype* to refer to the combination of genes that code for a trait, while the term *phenotype* describes the physical manifestation of that trait.

Therefore, the presence of two dominant alleles for pea texture (*RR*) would reflect the genotype while a smooth pea indicates the phenotype. Mendel did not limit his experiments to testing the rules of inheritance of single traits. He also studied plant traits involving multiple pairs of genes, breeding plants that have round, yellow seeds with plants that produce wrinkled, green seeds. Such experiments demonstrated that the patterns of inheritance he observed in his experiments with single traits also apply to cases involving more complex gene combinations.

Exceptions to Mendel's Rules

Mendel published his studies in a science journal in 1865, at which time no other scientist commented on his work. Since that time, geneticists have learned that sometimes genes do not easily conform to so-called Mendelian patterns of inheritance.

Incomplete Dominance

In cases of incomplete dominance, the inheritance of a dominant and a recessive allele results in a blending of traits to produce intermediate characteristics. For example, four-o'clock paint plants may have red, white, or pink flowers. Plants with red flowers have two copies of the dominant allele *R* for red flower colour (*RR*). Plants with white flowers have two copies of the recessive allele *r* for white flower colour (*rr*). Pink flowers result in plants with one copy of each allele (*Rr*), with each allele contributing to a blending of colours.

Quantitative Inheritance

Mendel focused his studies on traits determined by a single pair of genes, and the resulting phenotype was easy to distinguish. A tall plant can be markedly different from a short one, and a green pea can easily be distinguished from a yellow one. There are some traits, however, that are not easy to distinguish. Human skin colour, for example, may be any of a wide variety of shades. Traits such as skin colour differ from the ones Mendel studied because they are determined by more than one pair of genes. In this form of inheritance, known as quantitative inheritance, each pair of genes has only a slight effect on the trait, while the cumulative effect of all the genes determines the physical characteristics of the trait. At least four pairs of genes control human skin colour. Multiple genes also control many traits important in agriculture, such as milk production in cows and ear length in corn.

Multiple Alleles

Another exception to Mendelian genetics involves genes with multiple alleles. Certain traits are controlled by multiple alleles that have complex rules of dominance. In humans, for example, the gene for blood type has three alleles: I_A, I_B, and i. With three alternatives for each member of a gene pair, there are six possible combinations of these genes (I_AI_A, I_BI_B, ii, I_Ai, I_Bi, I_AI_B). Although there are six possible combinations, humans have only four major blood types: A, B, AB, and O. This results because both I_A and I_B dominate over i, but not over each other, so a person with a gene combination of I_AI_A or I_Ai has blood type A. The gene combinations I_BI_B and I_Bi both produce blood type B. I_AI_B results in a blood type AB, and ii results in blood type O.

Gene Linkage

In his experiments, Mendel was careful to study traits in pea plants where one trait did not appear to influence another, such as the plant's height or the pea's texture. These two phenotypes (height and texture) occur randomly with respect to one another in a manner known as independent assortment. Today scientists understand that independent assortment occurs when the genes affecting the phenotypes are found on different chromosomes. An exception to independent assortment develops when genes appear near one another on the same chromosome. When genes occur on the same chromosome, they are inherited as a single unit. Genes inherited in this way are said to be linked. For example, in fruit flies the genes affecting eye colour and wing length are inherited together because they appear on the same chromosome.

But in many cases, genes on the same chromosome that are inherited together produce offspring with unexpected allele combinations. This results from a process called crossing over. Sometimes at the beginning of meiosis, a chromosome pair (made up of a chromosome from the mother and a chromosome from the father) may intertwine and exchange sections of chromosome. The pair then breaks apart to form two chromosomes with a new combination of genes that differs from the combination supplied by the parents. Through this process of recombining genes, organisms can produce offspring with new combinations of maternal and paternal traits that may contribute to or enhance survival.

Sex-Linked Traits

Most chromosome pairs consist of identical, or homologous, partners. In many species, including humans, there is one pair of chromosomes in which the partners noticeably differ from each other. These are called the sex chromosomes because they determine the differences between males and females. Genes located on the sex chromosomes display different patterns of inheritance than genes located on other chromosomes.

In human females, the sex chromosomes consist of two X chromosomes, while males have an X chromosome and a shorter Y chromosome with many fewer genes. In males the X chromosome contains many genes that have no corresponding gene on the Y chromosome. A male's X chromosome may contain a recessive allele associated with a genetic disorder, such as hemophilia or Duchenne muscular dystrophy. In this case, males do not have a normal second copy of the gene on the Y chromosome to mask the effects of the recessive gene, and disease typically results. Additional examples of sex-linked traits include red-green colour blindness in humans and eye colour in fruit flies.

THE GENETIC CODE

The structure of DNA encodes all the information every cell needs to function and thrive. In addition, DNA carries hereditary information in a form that can be copied and passed intact from generation to generation. A gene is a segment of DNA. The biochemical instructions found within most genes, known as the genetic code, specify the chemical structure of a particular protein. Proteins are composed of long chains of amino acids, and the specific sequence of these amino acids dictates the function of each protein. The DNA structure of a gene determines the arrangement of amino acids in a protein, ultimately determining the type and function of the protein manufactured.

DNA Structure

DNA molecules form from chains of building blocks called nucleotides. Each nucleotide consists of a sugar molecule called deoxyribose that bonds to a phosphate molecule and to a nitrogen-containing compound, known as a base. DNA uses four bases in its structure: adenine (A), cytosine (C), guanine (G), and thymine (T). The order of the bases in a DNA molecule—the genetic code—determines the amino acid sequence of a protein.In the cells of most organisms, two long strands of DNA join in a single molecule that resembles a spiraling ladder, commonly called a double helix. Alternating phosphate and sugar molecules form each side of this ladder. Bases from one DNA strand join with bases from another strand to form the rungs of the ladder, holding the double helix together.

The pairing of bases in the DNA double helix is highly specific—adenine always joins with thymine, and guanine always links to cytosine. These base combinations, known as complementary base pairing, play a fundamental role in DNA's function by aiding in the replication and storage of genetic information. Complementary base pairing also enables scientists to predict the sequence of bases on one strand of a DNA molecule if they know the order on the corresponding, or complementary, DNA strand. Scientists use complementary base pairing to help identify the genes on a particular chromosome and to develop methods used in genetic engineering.

Genes line up in a row along the length of a DNA molecule. In humans a single gene can vary in length from 100 to over 1,000,000 bases. Genes make up less than 2 per cent of the length of a DNA molecule. The rest of the DNA molecule is made up of long, highly repetitive nucleotide sequences. Once dismissed as "junk" DNA, scientists now believe these nucleotide sequences may play a role in the survival of cells. Identifying the function of these sequences is a thriving field of genetics research.

DNA Replication

In order for inherited traits to be transmitted from parent to child, the genetic information encoded in DNA must be copied with great precision during cell division. The accuracy of DNA replication depends upon the complementary pairing of bases. During replication, the DNA double helix unwinds and bonds joining the base pairs break, separating the DNA molecule into two separate strands. Each strand of DNA directs the synthesis of another complementary strand. The unpaired bases of each DNA strand attach to bases floating within the cell. But the DNA strand's unpaired bases bond only with specific, complementary bases—for example, an adenine base will bond only with a thymine base and a cytosine bases will pair only with a guanine base.

Once all of the bases of a DNA strand bond to complementary bases, the complementary bases then link to each other, forming a new DNA double-helix molecule. Thus the original DNA molecule replicates into two DNA molecules that are exact duplicates.

PROTEIN SYNTHESIS

DNA replication ensures that the genetic instructions encoded in DNA can be used continuously through generations to produce the proteins that build and operate the cells of an organism. The process of tapping the genetic code to create proteins, known as protein synthesis, has two crucial steps: transcription and translation.

Transcription

Transcription transfers the genetic code from a molecule of DNA to an intermediary molecule called ribonucleic acid (RNA). The basic nucleotide structure of RNA resembles that of DNA, but the two compounds have three critical differences. First, the structure of RNA incorporates the sugar ribose rather than deoxyribose, the sugar in DNA. Second, RNA uses the base uracil (U) instead of thymine (T). In RNA uracil binds with adenine just as thymine does in DNA. Third, RNA usually exists as a single strand, unlike the double-helix structure that normally characterizes DNA.

Transcription involves the production of a special kind of RNA known as messenger RNA (mRNA). The process begins when the two strands of a DNA molecule separate, a task directed by the enzyme RNA polymerase. After the double helix splits apart, one of the strands serves as a template, or pattern, for the formation of a complementary mRNA molecule. Free-floating individual bases within the cell bind to the bases on the DNA template using complementary base pairing. The individual bases then link together to form a strand of mRNA.

In eukaryotes (organisms whose cells have a nucleus), the mRNA strand undergoes an additional step before the next stage of protein synthesis can occur. The mRNA strand consists of coding regions called exons separated by regions called introns. The introns do not contribute to protein synthesis. Special enzymes in the nucleus remove the introns from the mRNA strand. The remaining exons then link together to form an mRNA strand that contains the entire code for making a protein.

Translation

Once transcription is complete and the genetic code has been copied onto mRNA, the genetic code must be converted into the language of proteins. That is, the information coded in the four bases found in mRNA must be translated into the instructions encoded by the 20 amino acids used in the formation of proteins. This process, called translation, takes place in cellular organelles called ribosomes. In eukaryotes, mRNA travels out of the nucleus into the cell body to attach to a ribosome. In prokaryotes (organisms without a nucleus), the ribosome clasps mRNA and starts translation before these strands have finished transcription and separated from the DNA. In both eukaryotes and prokaryotes, the ribosome acts like a workbench and clamp that holds the mRNA strand and coordinates the activity of enzymes and other molecules essential to translation.

Another form of RNA called transfer RNA (tRNA) is found in the cytoplasm of the cell. There are many different types of tRNA, and each type binds with one of the 20 amino acids used in protein formation. One end of a tRNA binds with a specific amino acid. The other end carries three bases, known as an anticodon. The tRNA with an amino acid attached travels to the ribosome where the mRNA is stationed. The anticodon of the tRNA undergoes complementary base pairing with a series of three bases on the mRNA, known as the codon. The mRNA codon codes for the type of amino acid carried by the tRNA.

A second tRNA bonds with the next codon on the mRNA. The resident tRNA transfers its amino acid to the amino acid of the incoming tRNA and then leaves the ribosome. This process continues repeatedly, with new tRNA receiving the growing chain of amino acids, known as a polypeptide chain, from a resident tRNA. The ribosome moves the mRNA strand one codon at a time, making new codons available to bind with tRNAs. The process ends when the entire sequence of mRNA has been translated. The polypeptide chain falls away from the ribosome as a newly formed protein, ready to go to work in the cell.

MUTATIONS

Occasionally mistakes occur during DNA replication and protein synthesis. Any alteration in the structure of a gene results in a mutation. Mutations occur during DNA replication when the chemical structure of genes undergoes random modifications. Once a change has occurred, the altered genes continue to replicate in their changed form unless another mutation occurs. Sometimes mutations occur during transcription or translation, causing protein synthesis to go awry. Although mutations may occur in any living cell, they are most important when they occur in gametes because then the change affects the traits of following generations. Most mutations harm an organism. If a mutation occurs in a gene sequence that codes for a particular protein, the mutation may result in a change in the amino acid sequence directed by the gene. This change, in turn, may affect the function of the protein. The implications can be significant: The amino acid sequence distinguishing normal hemoglobin from the altered form of hemoglobin responsible for sickle-cell anemia differs by only a single amino acid.

Some mutations may be neutral or silent and do not affect the function of a protein. Occasionally a mutation benefits an organism. Over the course of evolutionary time, however, mutations serve the crucial role of providing organisms with previously non-existent proteins. In this way, mutations are a driving force behind genetic diversity and the rise of new or more competitive species better able to adapt to changes, such as climate variations, depletion of food sources, or the emergence of new types of disease.

Mutations can produce a change in any region of a DNA molecule. In a *point mutation*, for example, a single nucleotide replaces another nucleotide. Although a point mutation produces a small change to the DNA sequence, it may cause a change in the amino acid sequence, and thus the function, of a protein.

Far more serious are mutations that involve the addition or deletion of one or more bases from a DNA molecule. Adding or subtracting even a single base from a normal sequence during transcription can disrupt translation by shifting the "reading frame" of every subsequent codon. For example, an mRNA strand may include two codons in the following sequence: AUG UGA. The addition of a cytosine base at the beginning of this sequence shifts the "spelling" of these codons so that they read: CAU GUG. This may result in an incorrect amino acid sequence during translation, or the protein may be truncated. Known as *frameshift mutation*, this type of alteration could result in the production of a protein with no real function or one with a harmful effect.

Sometimes mutations are caused by *transposition*, in which long stretches of DNA (containing one or more genes) move from one chromosome to another. These jumping genes, called transposons, can disrupt transcription and change the type of amino acids inserted into a protein. Transposons rearrange and interrupt genes in a way that generally improves the genetic variation of a species.

While mutations can occur spontaneously, some can be caused by exposure to physical or chemical agents in the environment called mutagens. Common environmental mutagens include ultraviolet rays from the sun and various chemicals, such as asbestos, cigarette smoke, and nitrous acid. High-energy radiation, such as medical X rays, can cause DNA strands to break, leading to the deletion of potentially important genetic information. Radiation damage can also affect an entire chromosome, disrupting the function of many genes. In chromosomal translocation, a piece of one chromosome breaks off and merges with another chromosome. In some cases, large sections of chromosomes may break off and be lost. The cell has highly effective self-repair mechanisms that can correct the harmful changes made by mutations and prevent some mutations from being passed on. Some 50 specialized enzymes locate different types of faulty sequences in the DNA and clip out those flaws. Another repair mechanism scans DNA after replication and marks mismatched base pairs for repair.

GENE REGULATION

The processes that enable information to be copied from genes and then used to synthesize proteins must be regulated if an organism is to survive. Different cells within an organism share the same set of chromosomes. In each cell some genes are active while others are not.

For example, in humans only red blood cells manufacture the protein hemoglobin and only pancreas cells make the digestive enzyme known as trypsin, even though both types of cells contain the genes to produce both hemoglobin and trypsin. Each cell produces different proteins according to its needs so that it does not waste energy by producing proteins that will not be used.

A variety of mechanisms regulate gene activity in cells. One method involves turning on or off gene transcription, sometimes by blocking the action of RNA polymerase, an enzyme that initiates transcription. Gene regulation may also involve mechanisms that slow or speed the rate of transcription, using specialized regulatory proteins that bind to DNA. Depending on an organism's particular needs, one regulatory protein may spur transcription for a particular protein, and later, another regulatory protein may slow or halt transcription.

Prokaryotes

The bacterium *Escherichia coli* (commonly referred to as *E. coli*), found in the intestines of humans and other mammals, provides a good example of gene regulation. *E. coli* uses three enzymes to digest lactose, the primary sugar found in milk. The bacterium produces great quantities of lactose-digesting enzymes when lactose is present and saves energy by not synthesizing the enzymes when the sugar is not available. *E. coli* prefers consuming glucose to lactose, so the bacterium produces these enzymes in the presence of lactose only when no glucose is available.

A region of DNA known as an operon controls this gene regulation process. In *E. coli*, the operon includes at least five genes: Three genes, called lac genes, code for the enzymes that digest lactose; one gene encodes for a regulatory protein, called a repressor, that can sense the presence or absence of lactose; and one gene, called an operator, activates transcription, the first step in the synthesis of the lactose-digesting enzymes. In the absence of lactose, the repressor protein binds with the operator gene to block transcription. This prevents the lac genes from being transcribed and halts production of the lactose-digesting enzymes. If lactose is present, the repressor protein binds to the sugar, leaving the operator gene free to trigger transcription of the lac genes. The transcription of the lac genes produces the mRNA that will direct the production of the three lactose-digesting enzymes.

Eukaryotes

Gene regulation in eukaryotes is more complex than in bacteria and other prokaryotes. Even the simplest eukaryotes have far more genes than prokaryotes, and these genes must be turned on or off as conditions dictate. Most multicellular organisms contain different types of cells that serve specialized functions. The cells of an animal's heart, blood, skin, liver, and muscles all contain the same genes. But in order to carry out their specific functions within the body, each cell must produce different proteins and respond to changing environmental stimuli, such as glucose levels in the blood or body temperature. Such specialization is possible only with sophisticated gene regulation.

Eukaryotes use a variety of mechanisms to ensure that each cell uses the exact proteins it needs at any given moment. In one method, eukaryotic cells use DNA sequences called enhancers to stimulate the transcription of genes located far away from the point on the chromosome where transcription occurs. If a specific protein binds to an enhancer site on the DNA, it causes the DNA to fold so that the enhancer site is brought closer to the site where transcription occurs. This action can activate or speed up transcription in the genes surrounding the enhancer site, thereby affecting the type and quantity of proteins the cell will produce. Enhancers often exert their effects on large groups of related genes, such as the genes that produce the set of proteins that form a muscle cell.

Gene regulation can also take place after transcription has occurred by interfering with the steps that modify mRNA before it leaves the nucleus to take part in translation. This process typically involves removing exons (segments that code for specific proteins) and introns. These sections of the mRNA can be modified in more than one way, enabling a cell to synthesize different proteins depending on its needs.

GENES IN DEVELOPMENT

Gene regulation helps individual cells within an organism function in a specialized way. Other regulatory mechanisms coordinate the genes that determine how cells develop. All of the specialized cells in an organism, including those of the skin, muscle, bone, liver, and brain, derive from identical copies of a single fertilized egg cell. Each of these cells has the exact same DNA as the original cell, even though they have vastly different appearances and functions. Genes dictate how these cells specialize.

Early in an organism's embryonic development the overall body plan forms. Individual cells commit to a particular layer and region of the embryo, often migrating from one location to another to do so. As the organism grows, cells become part of a particular body organ or tissue, such as skin or muscle. Ultimately, most cells become highly specialized—not only to develop into a neuron rather than a muscle cell, for example, but to become a sensory neuron instead of a motor neuron. This process of specialization is called differentiation. At each stage of the differentiation process, specific genes known as developmental control genes actively turn on and switch off the genes that differentiate cells.

One class of developmental control genes, known as homeotic genes, directs the formation of particular body parts. Activating one set of homeotic genes instructs part of an embryo to develop into a leg, for example, while another set initiates the formation of the head. If a homeotic gene becomes altered or damaged, an organism's body development can be dramatically disrupted. A change in a single gene in some insects, for instance, can cause a leg to grow where an antenna belongs.Homeotic genes work by regulating the activity of other genes. Homeotic genes code for the production of a regulatory protein that can bind to DNA and thus affect the transcription of one or more genes. This enables homeotic genes to initiate or halt the development and specialization of characteristics in an organism.

Nearly identical homeotic genes have been identified in varied organisms, such as insects, worms, mice, birds, and humans, where they serve similar embryonic development functions. Scientists theorize that homeotic genes first appeared in a single ancestor common to all these organisms. Sometime in evolutionary history, these organisms diverged from their common ancestor, but the homeotic genes continued to be passed down through generations virtually unchanged during the evolution of these new organisms.

TRANSMISSION GENETICS ANALYSIS

In 1900, three European scientists independently discovered an obscure research paper that had been published nearly 35 years before.

When he cross-bred purple-flowered pea plants with white-flowered ones, the next generation had only purple flowers. But directions for making white Written by Gregor Mendel, an Austrian monk who was also a scientist, the report described a series of breeding experiments performed with pea plants growing in his abbey garden. Mendel had studied how pea plants inherited the two variant forms of easy-to-see traits. These included flower Colour (white or purple) and the texture of the peas (smooth or wrinkled). Mendel counted many generations of pea plant offspring and learned that these characteristics were passed on to the next generation in orderly, predictable ratios. flowers were hidden somewhere in the peas of that generation, because when those purple-flowered plants were bred to each other, some of their offspring had white flowers. What's more, the second-generation plants displayed the colors in a predictable pattern. On average, 75 percent of the second-generation plants had purple flowers and 25 percent of the plants had white flowers. Those same ratios persisted, and were reproduced when the experiment was repeated many times over.

Trying to solve the mystery of the missing Colour blooms, Mendel imagined that the reproductive cells of his pea plants might contain discrete "factors," each of which specified a particular trait, such as white flowers. Mendel reasoned that the factors, whatever they were, must be physical material because they passed from parent to offspring in a mathematically orderly way. It wasn't until many years later, when the other scientists unearthed Mendel's report, that the factors were named genes.

Early geneticists quickly discovered that Mendel's mathematical rules of inheritance applied not just to peas, but also to all plants, animals and people. The discovery of a quantitative rule for inheritance was momentous. It revealed that a common, general principle governed the growth and development of all life on Earth.

DNA

Up until the 1950s, scientists knew a good deal about heredity, but they didn't have a clue what

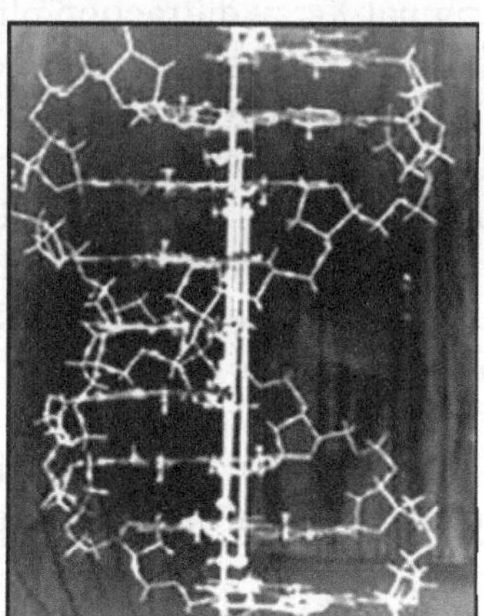

DNA looked like. In order to learn more about DNA and its structure, some scientists experimented with using X rays as a form of molecular photography. *In 1953, Watson and Crick created their historic model of the shape of DNA*: The double helix.

Rosalind Franklin, a physical chemist working with Maurice Wilkins at King's College in London, was among the first to use this method to analyze genetic material. Her experiments produced what were referred to at the time as "the most beautiful X-ray photographs of any substance ever taken."

Other scientists, including zoologist James Watson and physicist Francis Crick, both working at Cambridge University in the United Kingdom, were trying to determine the shape of DNA too.

Ultimately, this line of research revealed one of the most profound scientific discoveries of the 20th century: that DNA exists as a double helix.

The 1962 Nobel Prize in physiology or medicine was awarded to Watson, Crick and Wilkins for this work. Although Franklin did not earn a share of the prize due to her untimely death at age 38, she is widely recognized as having played a significant role in the discovery.

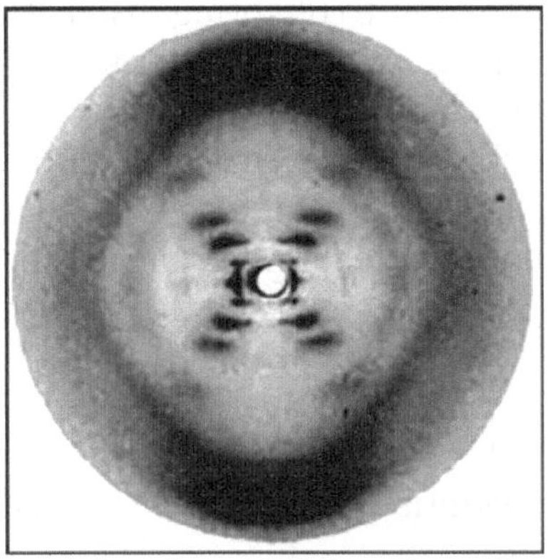

Rosalind Franklin's original X-ray diffraction photo revealed the physical structure of DNA. The spiral staircase-shaped double helix has attained global status as the symbol for DNA. But what is so beautiful about the discovery of the twisting ladder structure isn't just its good looks. Rather, the structure of DNA taught researchers a fundamental lesson about genetics.

It taught them that the two connected strands—winding together like parallel handrails—were complementary to each other, and this unlocked the secret of how genetic information is stored, transferred and copied.

In genetics, complementary means that if you know the sequence of nucleotide building blocks on one strand, you know the sequence of nucleotide building blocks on the other strand: A always matches up with T and C always links to G.

Long strings of nucleotides form genes, and groups of genes are packaged tightly into structures called chromosomes. Every cell in your body except for eggs, sperm and red blood cells contains a full set of chromosomes in its nucleus.If the chromosomes in one of your cells were uncoiled and placed end to end, the DNA would be about 6 feet long. If all the DNA in your body were connected in this way, it would stretch approximately 67 billion miles! That's nearly 150,000 round trips to the Moon.

DNA STRUCTURE

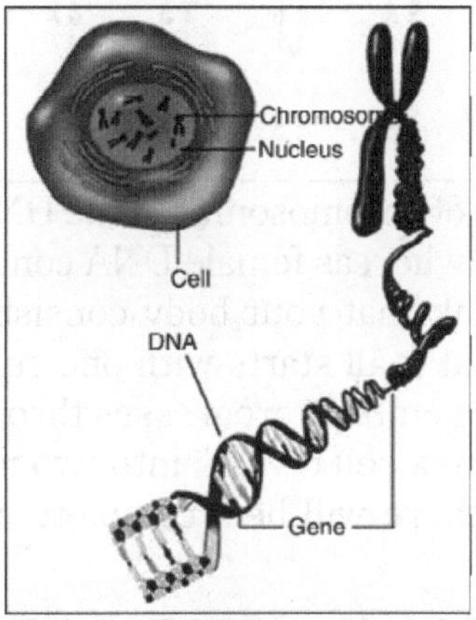

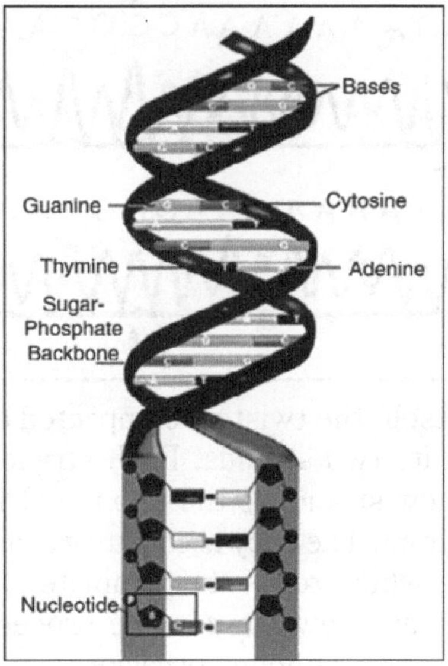

The long, stringy DNA that makes up genes is spooled within chromosomes inside the nucleus of a cell. (Note that a gene would actually be a much longer stretch of DNA than what is shown here.) DNA consists of two long, twisted chains made up of nucleotides.

Each nucleotide contains one base, one phosphate molecule and the sugar molecule deoxyribose. The bases in DNA nucleotides are adenine, thymine, cytosine and guanine.

COPYCAT

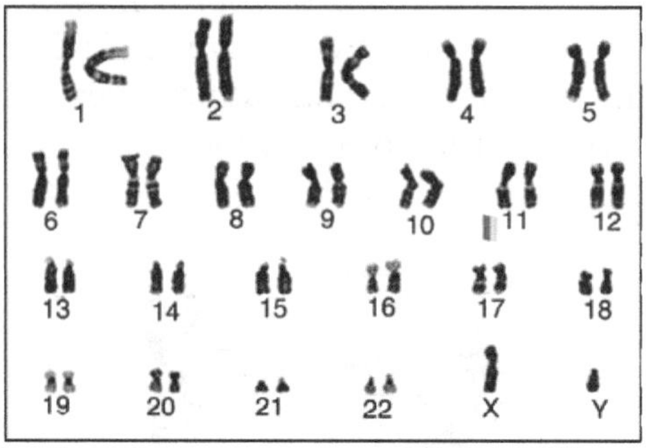

Humans have 23 pairs of chromosomes. Male DNA (pictured here) contains an X and a Y chromosome, whereas female DNA contains two X chromosomes.

It's astounding to think that your body consists of trillions of cells. But what's most amazing is that it all starts with one cell. How does this massive expansion take place?As an embryo progresses through development, its cells must reproduce. But before a cell divides into two new, nearly identical cells, it must copy its DNA so there will be a complete set of genes to pass on to each of the new cells.

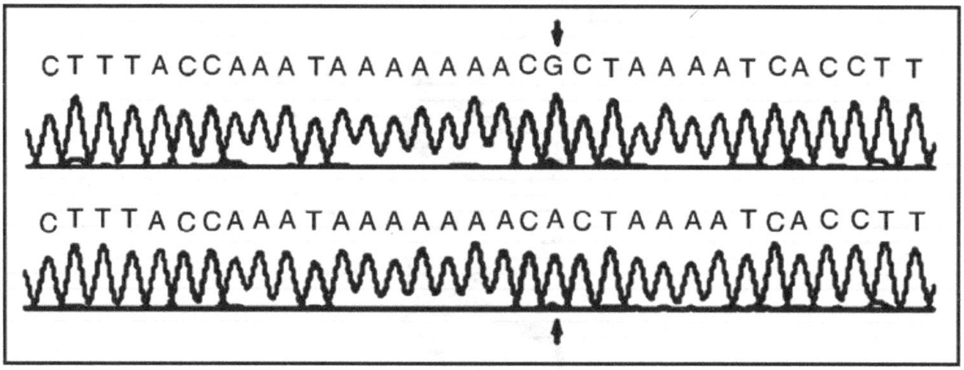

To make a copy of itself, the twisted, compacted double helix of DNA has to unwind and separate its two strands. Each strand becomes a pattern, or template, for making a new strand, so the two new DNA molecules have one new strand and one old strand.The copy is courtesy of a cellular protein machine called DNA polymerase, which reads the template DNA strand and stitches together the complementary new strand. The process, called replication, is astonishingly fast and accurate, although occasional mistakes, such as deletions or duplications, occur. Fortunately, a cellular spell-checker catches and corrects nearly all of these errors.

When DNA polymerase makes an error while copying a gene's DNA sequence, the mistake is called a mutation. In this example, the nucleotide G has been changed to an A.

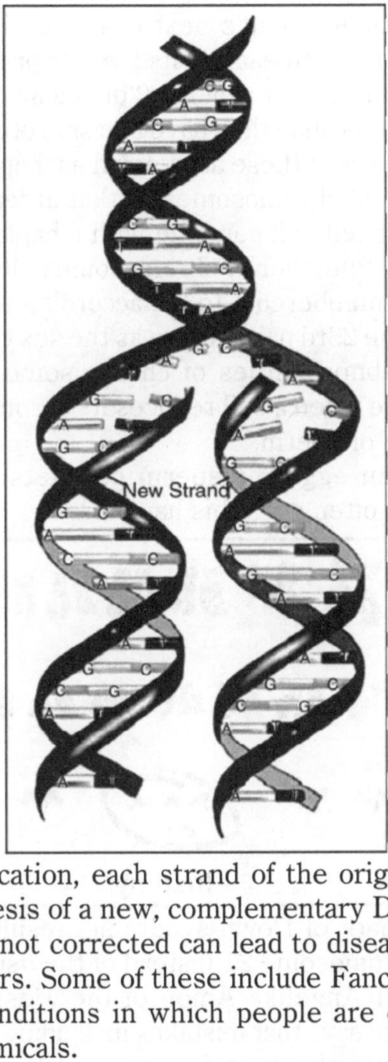

During DNA replication, each strand of the original molecule acts as a template for the synthesis of a new, complementary DNA strand.

Mistakes that are not corrected can lead to diseases such as cancer and certain genetic disorders. Some of these include Fanconi anemia, early aging diseases and other conditions in which people are extremely sensitive to sunlight and some chemicals.

DNA copying is not the only time when DNA damage can happen. Prolonged, unprotected sun exposure can cause DNA changes that lead to skin cancer, and toxins in cigarette smoke can cause lung cancer.

It may seem ironic, then, that many drugs used to treat cancer work by attacking DNA. That's because these chemotherapy drugs disrupt the DNA copying process, which goes on much faster in rapidly dividing cancer cells than in other cells of the body.

The trouble is that most of these drugs do affect normal cells that grow and divide frequently, such as cells of the immune system and hair cells. Understanding DNA replication better could be a key to limiting a drug's action to cancer cells only.

Even

After copying its DNA, a cell's next challenge is getting just the right amount of genetic material into each of its two offspring.

Most of your cells are called diploid ("di" means two, and "ploid" refers to sets of chromosomes) because they have two sets of chromosomes (23 pairs). Eggs and sperm are different; these are known as haploid cells. Each haploid cell has only one set of 23 chromosomes so that at fertilization the math will work out: A haploid egg cell will combine with a haploid sperm cell to form a diploid cell with the right number of chromosomes: 46.

Chromosomes are numbered 1 to 22, according to size, with 1 being the largest chromosome. The 23rd pair, known as the sex chromosomes, are called X and Y. In humans, abnormalities of chromosome number usually occur during meiosis, the time when a cell reduces its chromosomes from diploid to haploid in creating eggs or sperm.

What happens if an egg or a sperm cell gets the wrong number of chromosomes, and how often does this happen?

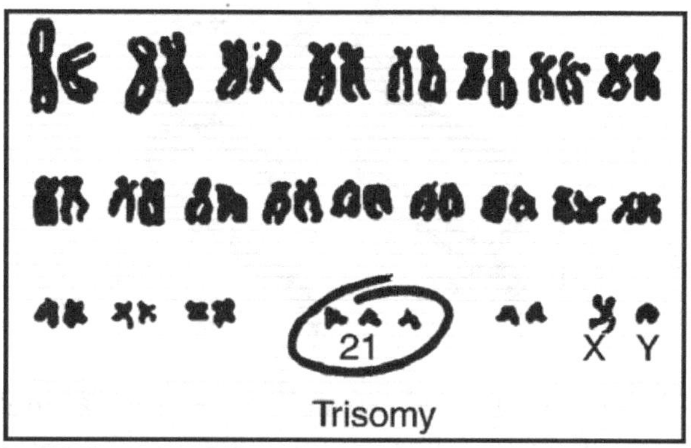

Trisomy, the hallmark of Down syndrome, results when a baby is born with three copies of chromosome 21 instead of the usual two.

Molecular biologist Angelika Amon of the Massachusetts Institute of Technology in Cambridge says that mistakes in dividing DNA between daughter cells during meiosis are the leading cause of human birth defects and miscarriages. Current estimates are that 10 percent of all embryos have an incorrect chromosome number. Most of these don't go to full term and are miscarried.

In women, the likelihood that chromosomes won't be apportioned properly increases with age. One of every 18 babies born to women over 45 has three copies of chromosome 13, 18 or 21 instead of the normal two, and this improper balancing can cause trouble. For example, three copies of chromosome 21 lead to Down syndrome.

To make her work easier, Amon—like many other basic scientists—studies yeast cells, which separate their chromosomes almost exactly the same way human cells do, except that yeast do it much faster. A yeast cell copies its DNA and produces daughter cells in about 1 1/2 hours, compared to a whole day for human cells.The yeast cells she uses are the same kind bakeries use to make bread and breweries use to make beer!

Amon has made major progress in understanding the details of meiosis. Her research shows how, in healthy cells, gluelike protein complexes called cohesins release pairs of chromosomes at exactly the right time. This allows the chromosomes to separate properly.

MEIOSIS

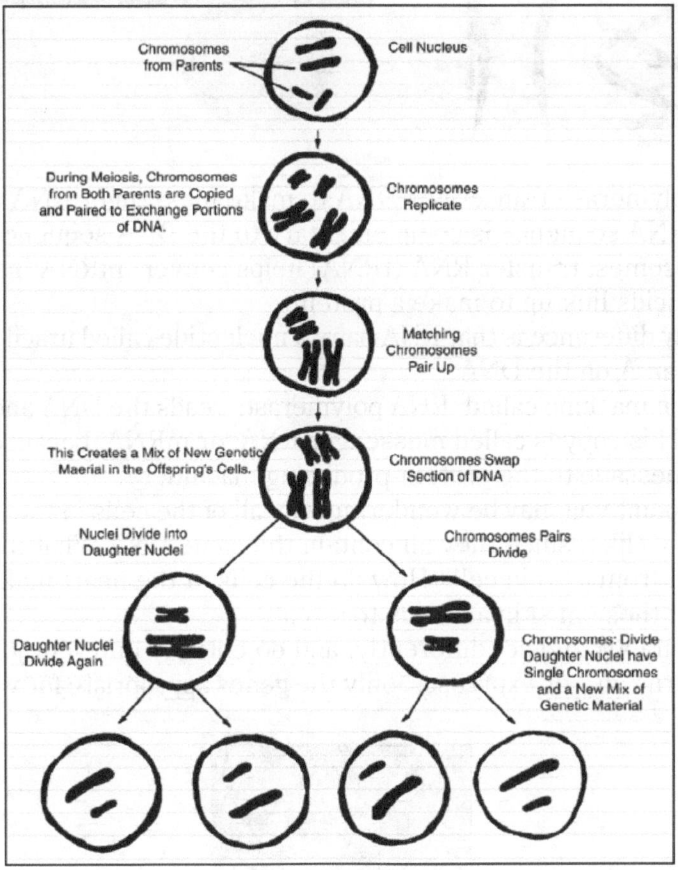

GETTING THE MESSAGE

So, we've described DNA—its basic properties and how our bodies make more of it. But how does DNA serve as the language of life? How do you get a protein from a gene?

There are two major steps in making a protein. The first is transcription, where the information coded in DNA is copied into RNA. The RNA nucleotides are complementary to those on the DNA: a C on the RNA strand matches a G on the DNA strand.

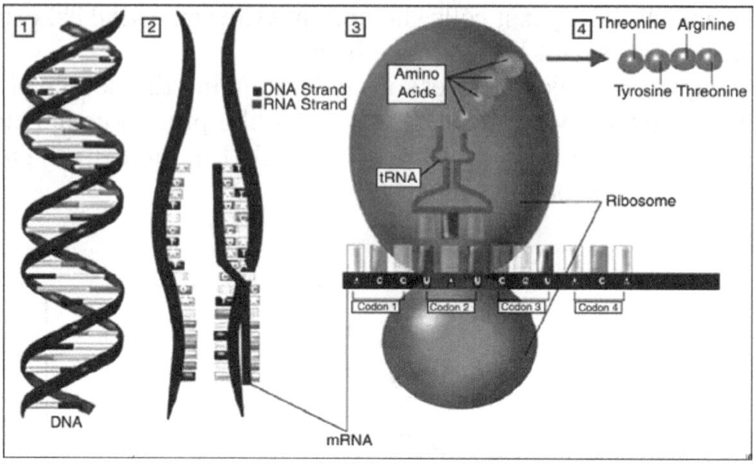

RNA polymerase transcribes DNA to make messenger RNA (mRNA).
The mRNA sequence is complementary to the DNA sequence.
On ribosomes, transfer RNA (tRNA) helps convert mRNA into protein.
Amino acids link up to make a protein.

The only difference is that RNA pairs a nucleotide called uracil (U), instead of a T, with an A on the DNA.

A protein machine called RNA polymerase reads the DNA and makes the RNA copy. This copy is called messenger RNA, or mRNA, because it delivers the gene's message to the protein-producing machinery.

At this point you may be wondering why all of the cells in the human body aren't exactly alike, since they all contain the same DNA.What makes a liver cell different from a brain cell? How do the cells in the heart make the organ contract, but those in skin allow us to sweat?

Cells can look and act differently, and do entirely different jobs, because each cell "turns on," or expresses, only the genes appropriate for what it needs to do.

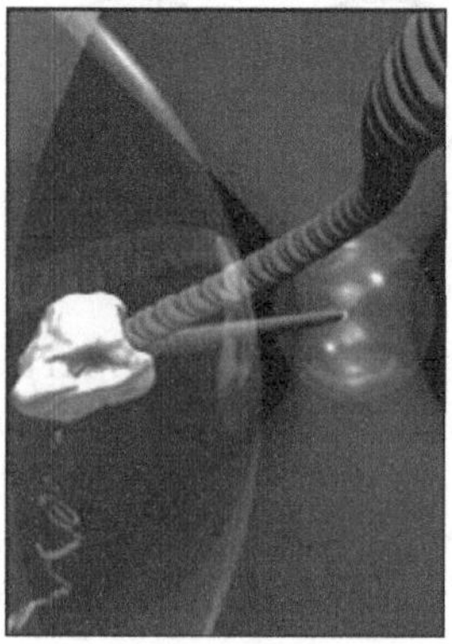

RNA polymerase and one end of a DNA strand are attached to clear beads pinned down in two optical traps. As RNA polymerase moves along the DNA, it creates an RNA copy of a gene, shown here as a pink strand.

Steven Block

That's because RNA polymerase does not work alone, but rather functions with the aid of many helper proteins. While the core part of RNA polymerase is the same in all cells, the helpers vary in different cell types throughout the body.You'd think that for a process so essential to life, researchers would know a lot about how transcription works. While it's true that the basics are clear— biologists have been studying gene transcribing by RNA polymerases since these proteins were first discovered in 1960— some of the details are actually still murky.The biggest obstacle to learning more has been a lack of tools. Until recently, researchers were unable to get a picture at the atomic level of the giant RNA polymerase protein assemblies inside cells to understand how the many pieces of this amazing, living machine do what they do, and do it so well.

But our understanding is improving fast, thanks to spectacular technological advances. We have new X-ray pictures that are far more sophisticated than those that revealed the structure of DNA.

Roger Kornberg of Stanford University in California used such methods to determine the structure of RNA polymerase. This work earned him the 2006 Nobel Prize in chemistry. In addition, very powerful microscopes and other tools that allow us to watch one molecule at a time provide a new look at RNA polymerase while it's at work reading DNA and producing RNA.

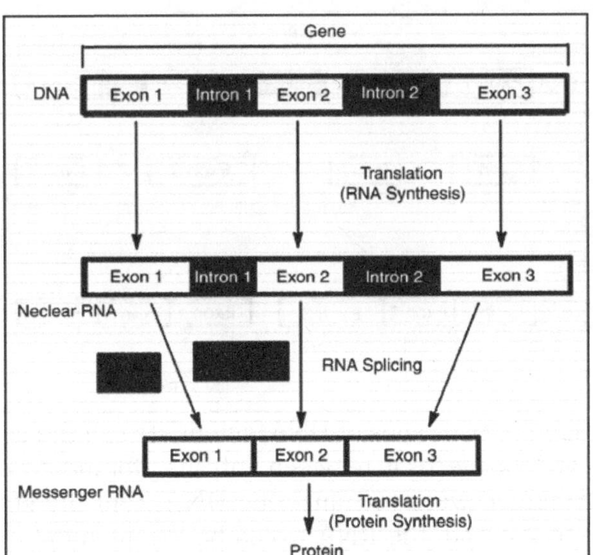

For example, Steven Block, also of Stanford, has used a physics technique called optical trapping to track RNA polymerase as it inches along DNA. Block and his team performed this work by designing a specialized microscope sensitive enough to watch the real-time motion of a single polymerase traveling down a gene on one chromosome.

The researchers discovered that molecules of RNA polymerase behave like battery-powered spiders as they crawl along the DNA ladder, adding nucleotides one at a time to the growing RNA strand. The enzyme works much like a motor, Block believes, powered by energy released during the chemical synthesis of DNA.

White are often interrupted by stretches of DNA (introns, Black) that do not contain instructions for making a protein. The DNA segments that do contain protein-making instructions are known as exons.

Several types of RNA play key roles in making a protein. The gene transcript (the mRNA) transfers information from DNA in the nucleus to the ribosome's that make protein. Ribosomal RNA forms about 60 percent of the ribosomes. Lastly, transfer RNA carries amino acids to the ribosomes. As you can see, all three types of cellular RNAs come together to produce new proteins.

But the journey from gene to protein isn't quite as simple as we've just made it out to be. After transcription, several things need to happen to mRNA before a protein can be made. For example, the genetic material of humans and other eukaryotes (organisms that have a nucleus) includes a lot of DNA that doesn't encode proteins. Some of this DNA is stuck right in the middle of genes.

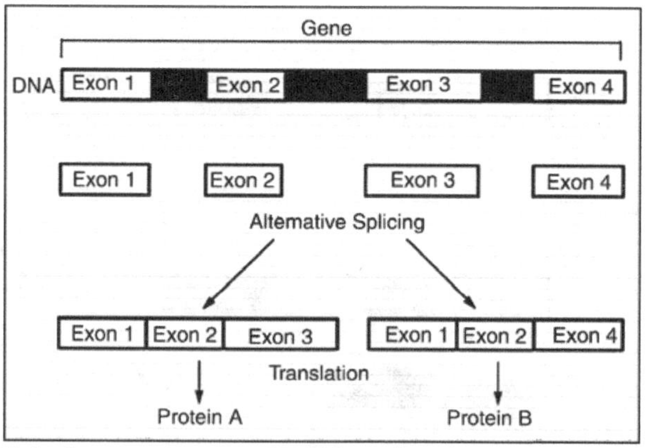

To distinguish the two types of DNA, scientists call the coding sequences of genes exons and the pieces in between introns (for intervening sequences).

If RNA polymerase were to transcribe DNA from the start of an intron-containing gene to the end, the RNA would be complementary to the introns as well as the exons.To get an mRNA molecule that yields a working protein, the cell needs to trim out the intron sections and then stitch only the exon pieces together. This process is called RNA splicing. Arranging exons in different patterns, called alternative splicing, enables cells to make different proteins from a single gene.

Splicing has to be extremely accurate. An error in the splicing process, even one that results in the deletion of just one nucleotide in an exon or the addition of just one nucleotide in an intron, will throw the whole sequence out of alignment. The result is usually an abnormal protein—or no protein at all. One form of Alzheimer's disease, for example, is caused by this kind of splicing error.

Molecular biologist Christine Guthrie of the University of California, San Francisco, wants to understand more fully the mechanism for removing intron RNA and find out how it stays so accurate.

She uses yeast cells for these experiments. Just like human DNA, yeast DNA has introns, but they are fewer and simpler in structure and are therefore easier to study. Guthrie can identify which genes are required for splicing by finding abnormal yeast cells that mangle splicing.

So why do introns exist, if they're just going to be chopped out? Without introns, cells wouldn't need to go through the splicing process and keep monitoring it to be sure it's working right.

As it turns out, splicing also makes it possible for cells to create more proteins.

Think about all the exons in a gene. If a cell stitches together exons 1, 2 and 4, leaving out exon 3, the mRNA will specify the production of a particular protein. But instead, if the cell stitches together exons 1, 2 and 3, this time leaving out exon 4, then the mRNA will be translated into a different protein

By cutting and pasting the exons in different patterns, which scientists call alternative splicing, a cell can create different proteins from a single gene. Alternative splicing is one of the reasons why human cells, which have about 20,000 genes, can make hundreds of thousands of different proteins.

Until recently, researchers looked at genes, and the proteins they encode, one at a time. Now, they can look at how large numbers of genes and proteins act, as well as how they interact. This gives them a much better picture of what goes on in a living organism.

Already, scientists can identify all of the genes that are transcribed in a cell—or in an organ, like the heart. And although researchers can't tell you, right now, what's going on in every cell of your body while you read a book or walk down the street, they can do this sort of "whole-body" scan for simpler, single-celled organisms like yeast.

Using a technique called genome-wide location analysis, Richard Young of the Massachusetts Institute of Technology unraveled a "regulatory code" of living yeast cells, which have more than 6,000 genes in their genome. Young's technique enabled him to determine the exact places where RNA polymerase's helper proteins sit on DNA and tell RNA polymerase to begin transcribing a gene.

Since he did the experiment with the yeast exposed to a variety of different conditions, Young was able to figure out how transcription patterns differ when the yeast cell is under stress (say, in a dry environment) or thriving in a sugary-rich nutrient solution. Done one gene at a time, using methods considered state-of-the-art just a few years ago, this kind of analysis would have taken hundreds of years.

After demonstrating that his technique worked in yeast, Young then took his research a step forward. He used a variation of the yeast method to scan the entire human genome in small samples of cells taken from the pancreases and livers of people with type 2 diabetes. He used the results to identify genes that aren't transcribed correctly in people with the disease.

This information provides researchers with an important tool for understanding how diabetes and other diseases are influenced by defective genes. By building models to predict how genes respond in diverse situations, researchers may be able to learn how to stop or jump-start genes on demand, change the course of a disease or prevent it from ever happening.

TRANSLATION

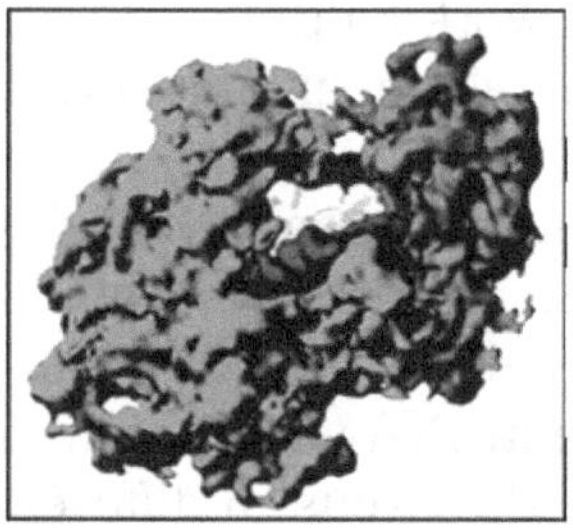

A ribosome consists of large and small protein subunits with transfer RNAs nestled in the middle.

After a gene has been read by RNA polymerase and the RNA is spliced, what happens next in the journey from gene to protein? The next step is reading the RNA information and fitting the building blocks of a protein together. This is called translation, and its principal actors are the ribosome and amino acids.

Ribosomes are among the biggest and most intricate structures in the cell. The ribosomes of bacteria contain not only huge amounts of RNA, but also more than 50 different proteins. Human ribosomes have even more RNA and between 70 and 80 different proteins!

Harry Noller of the University of California, Santa Cruz, has found that a ribosome performs several key jobs when it translates the genetic code of mRNA. As the messenger RNA threads through the ribosome protein machine, the ribosome reads the mRNA sequence and helps recognize and recruit the correct amino acid-carrying transfer RNA to match the mRNA code

. The ribosome also links each additional amino acid into a growing protein chain. For many years, researchers believed that even though RNAs formed a part of the ribosome, the protein portion of the ribosome did all of the work. Noller thought, instead, that maybe RNA, not proteins, performed the ribosome's job. His idea was not popular at first, because at that time it was thought that RNA could not perform such complex functions.

Some time later, however, the consensus changed. Sidney Altman of Yale University in New Haven, Connecticut, and Thomas Cech, who was then at the University of Colorado in Boulder, each discovered that RNA can perform work as complex as that done by protein enzymes. Their "RNA-as-an-enzyme" discovery turned the research world on its head and earned Cech and Altman the 1989 Nobel Prize in chemistry.

Noller and other researchers have continued the painstaking work of understanding ribosomes. In 1999, he showed how different parts of a bacterial ribosome interact with one another and how the ribosome interacts with molecules involved in protein synthesis. These studies provided near proof that the fundamental mechanism of translation is performed by RNA, not by the proteins of the ribosome.

RNA

But which ribosomal RNAs are doing the work? Most scientists assumed that RNA nucleotides buried deep within the ribosome complex—the ones that have the same sequence in every species from bacteria to people—were the important ones for piecing the growing protein together.

However, recent research by Rachel Green, who worked with Noller before moving to Johns Hopkins University in Baltimore, Maryland, showed that this is not the case. Green discovered that those RNA nucleotides are not needed for assembling a protein. Instead, she found, the nucleotides do something else entirely: They help the growing protein slip off the ribosome once it's finished.

Noller, Green and hundreds of other scientists work with the ribosomes of bacteria. Why should you care about how bacteria create proteins from their genes?

One reason is that this knowledge is important for learning how to disrupt the actions of disease-causing microorganisms. For example, antibiotics like erythromycin and neomycin work by attacking the ribosomes of bacteria, which are different enough from human ribosomes that our cells are not affected by these drugs.

As researchers gain new information about bacterial translation, the knowledge may lead to more antibiotics for people.

New antibiotics are urgently needed because many bacteria have developed resistance to the current arsenal. This resistance is sometimes the result of changes in the bacteria's ribosomal RNA. It can be difficult to find those small, but critical, changes that may lead to resistance, so it is important to find completely new ways to block bacterial translation. Green is working on that problem too. Her strategy is to make random mutations to the genes in a bacterium that affect its ribosomes. But what if the mutation disables the ribosome so much that it can't make proteins? Then the bacterium won't grow, and Green wouldn't find it.

Using clever molecular tricks, Green figured out a way to rescue some of the bacteria with defective ribosomes so they could grow. While some of the rescued bacteria have changes in their ribosomal RNA that make them resistant to certain antibiotics (and thus would not make good antibiotic targets) other RNA changes that don't affect resistance may point to promising ideas for new antibiotics.

DEVELOPMENT

In the human body, one of the most important jobs for proteins is to control how embryos develop. Scientists discovered a hugely important set of proteins involved in development by studying mutations that cause bizarre malformations in fruit flies.

The most famous such abnormality is a fruit fly with a leg, rather than the usual antenna, growing out of its head. According to Thomas C. Kaufman of Indiana University in Bloomington, the leg is perfectly normal—it's just growing in the wrong place.

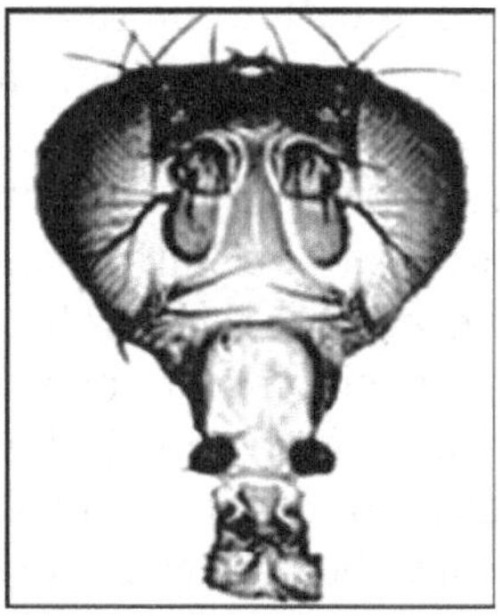

Fig. Normal Fruit Fly Head.

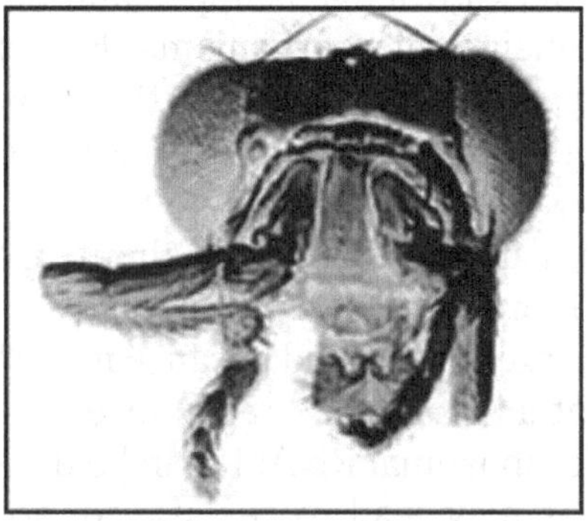

Fig. Fruit Fly Head Showing the Effects of the Antennapedia gene. This Fly has Legs where its Antennae should be.

In this type of mutation and many others, something goes wrong with the genetic Programmes that directs some of the cells in an embryo to follow developmental pathways, which are a series of chemical reactions that occur in a specific order. In the antenna-into-leg problem, it is as if the cells growing from the fly's head, which normally would become an antenna, mistakenly believe that they are in the fly's thorax, and therefore ought to grow into a leg. And so they do.

Thinking about this odd situation taught scientists an important lesson—that the proteins made by some genes can act as switches. Switch genes are master controllers that provide each body part with a kind of identification card. If a protein that normally instructs cells to become an antenna is disrupted, cells can receive new instructions to become a leg instead.

Scientists determined that several different genes, each with a common sequence, provide these anatomical identification card instructions. Kaufman isolated and described one of these genes, which became known as *Antennapedia*, a word that means "antenna feet."

Kaufman then began looking a lot more closely at the molecular structure of the *Antennapedia* gene. In the early 1980s, he and other researchers made a discovery that has been fundamental to understanding evolution as well as developmental biology. The scientists found a short sequence of DNA, now called the home box, that is present not only in *Antennapedia* but in the several genes next to it and in genes in many other organisms. When geneticists find very similar DNA sequences in the genes of different organisms, it's a good clue that these genes do something so important and useful that evolution uses the same sequence over and over and permits very few changes in its structure as new species evolve.

Researchers quickly discovered nearly identical versions of homeobox DNA in almost every non-bacterial cell they examined—from yeast to plants, frogs, worms, beetles, chickens, mice and people.

Hundreds of homeobox-containing genes have been identified, and the proteins they make turn out to be involved in the early stages of development of many species. For example, researchers have found that abnormalities in the homeobox genes can lead to extra fingers or toes in humans.

MUTATION, SEGREGATION, AND RECOMBINATION

The process of natural selection is often characterized as "blind" and this is largely because the source of all genetic variation is largely a stochastic process. The ultimate source of genetic variation is mutations; however, segregation, and recombination provide an stochastic mechanisms for randomizing genetic variation in a population. While mutations alter DNA by changing base pairs, segregation and recombination do not alter the material content of DNA. Instead they provide powerful mechanisms for mixing up the DNA during sexual reproduction.

MUTATION

The process of mutation is probabilistic. We describe the process in terms of the probability of a mutation occurring in an individual during its lifetime. Typical rates of mutation are between 1 in 10,000 (10^{-4}) and 1 in 1,000,000 (10^{-6}) for many organisms. Either this means that a long time must pass before a mutation occurs in a population, or the population must be very large.

The population must number in excess of 10^6 or one million members in order to see an average of one mutation per generation for a gene with a mutation rate of 10^{-6}. Most mutations are detrimental, and perhaps only 1 in 1,000 is beneficial. Thus, in this population of a million we might have to wait for one thousand years for a specific genetic locus to throw us a beneficial mutation (of course there are thousands of possible loci in an organism so the waiting time for any beneficial mutation is less). Even if a mutation arises, there is no guarantee that natural selection will act on it.

We can calculate the probability that a beneficial mutation makes it through meiosis and segregation. A beneficial mutation occurs within a diploid parent which has two copies of each gene.

The beneficial mutation has a 50per cent chance of being transmitted to any one offspring and a 50per cent chance of not being passed on. With an organism that has four progeny, the probability of a single offspring not getting the beneficial mutation is independent of the other offspring not getting the mutant allele. By using the laws of probability we can multiply successive independent events to compute the probability that none of the four progeny gets the beneficial mutation:

$$1/2 \times 1/2 \times 1/2 \times 1/2 = 1/16.$$

If the unique beneficial mutation is not passed on to the progeny, then more time will be required for a new beneficial mutation to arise in the population. At this point, natural selection has a chance of promoting its spread through the population.

SEGREGATION

Genes are found on chromosomes. The first random event in sexual reproduction occurs during meiosis when homologous pairs of chromosomes line up on the equatorial plate in preparation for the cell division that reduces a diploid (2N) cell of the germ line to an haploid (N) gamete.

During the process of segregation different homologues will randomly be distributed between the two daughters chromosomes, gametes end up with very different chromosome complements.

Given that the total number of chromosomes is C, then there are 2 raised to the power of C different gametes that could be produced from segregation of chromosomes (in humans that would be 2^{46}). Segregation can produce a vast number of different gametes and is a potent mixer of variation in sexual organisms.

RECOMBINATION

Recombination between homologous pairs of chromosomes during meiosis can generate even more gamete types. The chromosome is loaded with recombinational hotspots. Any given chromosome will typically recombine with its homologue at one or two points during a given meiotic event (recombination rate on a given chromosome is a function of chromosome length). However, the recombination points for two different meiotic events can be different. Given that gametes such as sperm are generated by millions of different meiotic events, the number of potential gamete types from a single parent is vast. If one considers all possible parents in a modest-sized breeding population such as humans (5 billion), there are easily more potential recombination products than molecules in the known universe.

ADDITIVE GENETIC VARIATION VERSUS DOMINANCE VARIATION

SOURCE OF HERITABLE VARIATION

The resemblance between parents and offspring is entirely due to the additive effect that genes have on phenotype. Not all genetic variation is additive. For example, dominance variation arises from the degree to which one allele at a locus alters the effect of the allele on the complementary chromosome. A useful way to think about additive genetic variation is to consider the proportion of the phenotype that can be predicted from a knowledge of the number of *A* versus *a* alleles that are present in the genotype — or the genotypic value. Consider an *aa* individual to have zero *A* alleles, *Aa* has one*A* allele, and *AA* has two *A* alleles. Only in the idealized case when alleles at a genetic locus are co-dominant, are the genes said to act in a purely additive fashion. Dominance reduces the additive genetic variation, increasing the dominance variation at the same time.

Co-dominant Alleles

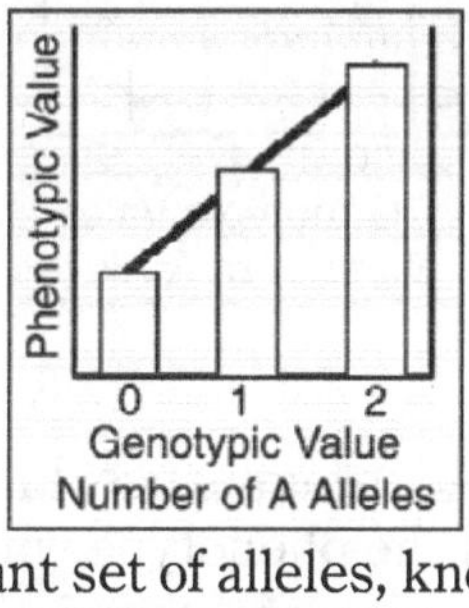

In the case of a co-dominant set of alleles, knowledge of genotype allows us to perfectly predict the value of the phenotype. A regression of phenotypic value on the number of *A*alleles or the genotypic value yields a perfect predictive line.

Dominant and Recessive Alleles

In the case of a dominant *A* allele, knowledge of genotypic value allows us to only imperfectly predict the value of the phenotype. Because heterozygotes and dominant homozygotes have the same phenotypic value, the fit between the regression of phenotypic value on genotypic value has error (black arrows). There is still a significant regression slope showing that some of the variation is due to additive genetic causes and some is due to dominance variation (black arrows).

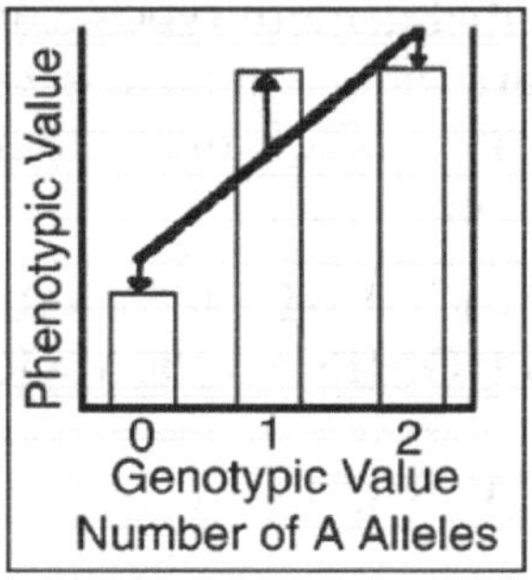

Overdominance

In the case of overdominance, knowledge of genotype does not allow us to confidently predict the value of the phenotype. There is no slope to the regression line so none of the variation is due to "additive genetic effects" of alleles. All variation in our model is due to error in our model which arises from dominance variation (black arrows). Consider two parents that are homozygous for alternative alleles (*aa* and *AA*). If Behaviour were strongly determined by an overdominant allele then the heterozygous progeny would not at all resemble the homozygous parents.

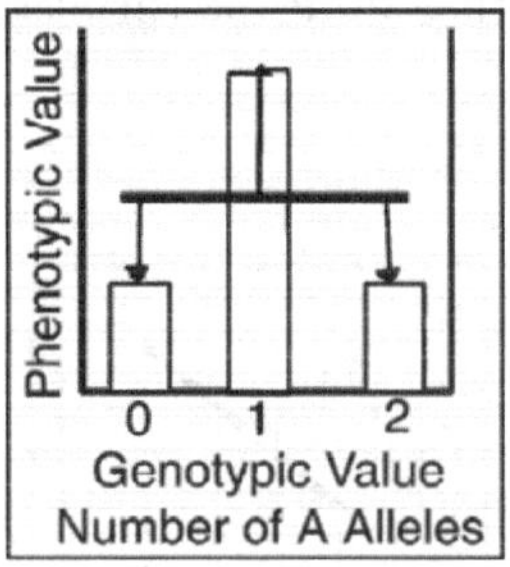

Underdominance

Likewise, in the case of underdominance, knowledge of genotype does not allow us to predict the value of the phenotype with any confidence. There is no slope to the regression line so none of the variation is due to "additive genetic effects" of alleles. All variation in our model is due to error in our model which arises from dominance variation (black arrows).

TRAITS

We will explore many of these issues of genetics through examples. I have chosen a diverse set of behavioral traits to introduce many concepts in animal Behaviour which will be covered in greater detail in later chapters. Each example, illustrates one of the genetic terms discussed above. In many cases model organisms studied in the laboratory facilitate our study of behavioral causes because of the large sample sizes possible in the laboratory allow for necessary statistical power required to assess genetic bases of behavioral traits. Other genetic examples are drawn from natural populations. While it is more difficult to collect genetic data on animals in the wild, these larger animals provide answers to natural selection because we can follow them through their lives. Moreover, it is essential to study animals in the wild so that we can infer how their actions are reflected in the process of natural or sexual selection. By long-term observation of animals in the wild, we are able to reconstruct present day patterns of selection that may have been historically significant during the origin of behavioral traits. A first step in selection analyses is determining the genetic basis of behavioral traits. Classic genetic crosses and pedigree analyses are the standard tools of a behavioral geneticist.

BEHAVIORAL MUTANTS IN THE LABORATORY: ROVING VERSUS SITTING IN DROSOPHILA LARVAE

Many of the basic issues of mendelian inheritance of animal Behaviour are found in the example of roving versus sitting larvae in *Drosophila*. Marie Sokolowski and her colleagues generated two strains of larvae and then carried out some illuminating crosses. If you cross parental rover males and sitter females, each of the ensuing strains is pure and homozygous for either the rover or sitter allele.

Crossing two pure parental lines (P0) produces progeny which are referred to as the F1. In theory, if the phenotype rests entirely on a single gene with two alternative alleles (a dominant rover allele and a recessive sitter allele). The F1 cross produces all rovers with no sitters.

Table A. Theoretical Results from F1 × F1 cross

<table>
<tr><td colspan="2"></td><td colspan="2">Genotype of first parent</td></tr>
<tr><td colspan="2"></td><td>R</td><td>S</td></tr>
<tr><td>Genotye of</td><td>R</td><td>R|R</td><td>R|S</td></tr>
<tr><td>Second parent</td><td>S</td><td>R|S</td><td>S|S</td></tr>
</table>

In practice, this cross produces nearly all rovers, suggesting that sitter is recessive to rover (or rover is dominant to sitter). The telling bit of evidence arises from a cross between F1 progeny, which should produce 3 genotypes, but only two phenotypes, owing to dominance of the R allele over the S allele. These offspring produce rovers and sitters in nearly a 3:1 ratio — exactly the pattern one expects if the F1 are all heterozygous for rover and sitter alleles.

Table B. Observed Phenotype in F2.

	Hypothesized genotype	Predicted Ratio
Rover	R|R	1
Rover	R|S	2
Sitter	S|S	1

BEAK SIZE AND SEED PREFERENCE IN AFRICAN FINCHES

Like the behavioral morphs of fruitflies, species of African Finch that inhabits Cameroon, *Pyrenestes ostrinus*, displays a discrete set of bill morphs with a simple mendelian pattern of inheritance. In particular, a small-beaked (s) and large-beaked (L) form co-occur throughout much of their geographic range. The small- and large-beaked forms are often found building nests together. These two forms differ dramatically in the feeding behaviors and preferences on large versus small seeds. In the wild, the small-billed morph has a very strong preference for the seeds from a species of sedge that produces small seeds while the large-billed morph prefers the seeds from a large-seeded sedge species. Given their propensity to feed on seeds, the birds common name is the seed-cracker. Thomas Smith initiated breeding studies on the *P. ostrinus* to determine the genetic basis of the behavioral and morphological differences between bill morphs. High levels of nest predation in the wild precluded the possibility of obtaining pedigrees from the wild. Smith circumvented this problem by importing breeding pairs for his study in a collaborative effort with the Riverbanks Zoo in South Carolina. In order to study the inheritance of beak morphology which is strongly associated with feeding behaviors, he first had to get the animals to breed in the laboratory.

Finding the environmental factor that triggers reproduction in undomesticated animals is often a daunting task. Many temperate birds are triggered to initiate reproduction with a change from short photoperiods to long photoperiods. The early attempts by the Riverbank Zoo to trigger reproduction by manipulating the photoperiod met with failure. However some tropical Finch aficionados suggested that they try an unusual environmental trigger. In the wild, the finches can rely on a very time Honoured cue, they experience two dry seasons and two wet seasons every year. One of the wet seasons is a little more damp than the other. The workers at the Zoo played audio tapes with load thunder, simultaneously drenching the aviaries with water. The finches, happy with the beginning of a simulated wet season, began to breed. This example, illustrates some of the challenges of non-model systems, but also the creative solutions used to solve practical problems of science.

Over the years, Smith and his colleagues developed a relatively large pedigree from crosses: L • L, L • s, s • s. The best fit genetic model for the control of beak size based on the genetic data in the pedigree, is a single mendelian locus with an allele for the small beaked form (s) recessive to an allele for the large beaked form (L). A simple mendelian gene causes large differences in morphology in the seed crackers, and this has cascading effects on a suite of foraging behaviors. This illustrates that a single gene can influence a large number of morphological and behavioral traits. As we shall see in the next Chapter on selection, the gene for beak morphology also influences survival in the wild.

RANDOM MATING

The Hardy-Weinberg Theorem describes what happens to gene frequencies when no evolutionary forces act on phenotypes. This assumes that no selection, migration, mutation, or genetic drift alters the gene frequencies from generation to generation. Another important assumption is that individuals in the population breed randomly with respect to genotype and phenotype. In a randomly mating population gametes combine in proportion to the frequency of each gametic type taking part in the union. For a single locus with two alternative alleles, *A* which occurs at a frequency of p, and a which occurs at frequency q (or 1-p) the proportion of genotypes is given by multiplying the frequency of each gametic type: or the more familiar: $p^2 + 2 \bullet p \bullet q + q^2 = 1$.

Genotype	**AA**	**Aa**	**aA**	**aa**
Frequency	p p	p q	Q p	q q

These calculations describe how gametes pair up randomly during fertilization. Another expression of random mating occurs at the level of the phenotype. If the frequency of each phenotype is: p^2, 2pq, and q^2, and phenotypes pair up randomly we would expect to see the following crosses to occur with the following frequencies:

RANDOM MATING FINCHES

Smith observed the following frequencies of matings in *Pyrenestes ostrinus*: $S_{males} \bullet S_{females} = 34$, $S_m \bullet L_f = 14$, $L_m \bullet S_f = 14$, $L_m \bullet L_f = 6$. A few simple computations are in order to learn how the birds are breeding. The birds may mate randomly, assortatively by phenotype (like breeds with like) or perhaps by disassortative mating (birds seek out a more dissimilar partner). While we cannot distinguish between all the genotypic classes in *P. ostrinus*, we can ask whether the phenotypes are breeding randomly. What frequency of matings would we expect by chance? We need to compute the following items.

Frequency of small-billed males	(34+14)/(34+14+14+6) = 48/68 = 0.71
Frequency of small-billed females	(34+14)/(34+14+14+6) = 48/68 = 0.71
Frequency of large-billed males	(14+6)/(34+14+14+6) = 22/68 = 0.29
Frequency of large-billed females	(14+6)/(34+14+14+6) = 22/68 = 0.29

What is the probability that a small-billed male pairs randomly with a small-billed female? The probability that a small breeds with small is given by multiplying the overall frequency of each:

0.71 * 0.71 = 0.54 and, we would expect to see
a total of 68 * 0.54 = 34.3.

By the same logic, we can compute our random expectations for the other three kinds of matings to derive an expected number of matings if the birds were randomly mating. By inspection alone we can see that the observed and expected random frequencies are nearly identical. We could carry out a formal test, the Chi-square, which is based on observed versus expected frequencies. I leave this as an exercise for the reader.

Observed (Expected)		**Female of the Pair**	
		S	**L**
Male of	S	34 (34.3)	14 (14)
the Pair	L	14 (14)	6 (5.7)

MENDELIAN INHERITANCE OF THREE ALLELES

For a single locus with three alleles, the formula for frequency of each genotypic class is slightly more complicated as we add a few more heterozygous classes and one homozygous class.

<table>
<tr><td>Genotype</td><td>|</td><td>|</td><td>|</td><td>|</td><td>|</td><td>|</td></tr>
<tr><td>Frequency</td><td>p p</td><td>q q</td><td>r r</td><td>2 p q</td><td>2 q r</td><td>2 p r</td></tr>
</table>

The two allele and multi-allele Hardy-Weinberg Law really only implies that gametes achieve union randomly with respect to genotype. Given the observed *Ams* gene frequencies in isopods, what is the frequency of observed mating phenotypes that we would expect under random mating?

MOLECULAR TOOLS OF BEHAVIORAL GENETICISTS

The search for genetic factors in natural populations and laboratory stock depends largely on the pre-existence of genetically-based morphs. Examples include the case of the pedigree analysis of crosses between bill morphs in finches or in the screenings for alternative behaviors carried out in *Drosophila*.

Many human genetic disorders are uncovered by analyses of pedigrees that are obtained from isolated populations where the affliction occurs at a relatively high frequency. When we see a maladaptive genetic disorder genetic drift is the probable cause. In small isolated populations, a single copy of a genetic disorder can drift to high frequency when inbreeding occurs. Inbreeding is much more likely in a small population where consanguineous matings occur at high frequency versus those in a large population where encountering a relative is quite unlikely.

Genetic sleuths refine their search for genetic factors by screening the genetic material of key families in which the affliction is particularly well characterized. If researchers find a perfect match between transmission of a certain piece of DNA and the transmission of the malady, they can then isolate where the gene coding for the Behaviour is to be found on a genetic map of the human genome. The number of behavioral genes discovered in humans is increasing at a rapid rate using inference from pedigrees and the transmission of specific regions of DNA.

Another important technique for isolating genes that control Behaviour entails a mutant screen in which a parental generation in a colony of animals is subjected to a mutagen, and then the progeny in the colony are screened for behavioral disorders. The researchers then search for the genetic basis by molecular methods that are similar to those described above in the analysis of human pedigrees. The next set of examples illustrate how molecular methods aid in the determination of genetic factors underlying behaviors.

MUTATION ANALYSIS OF A GENE FOR PARENTAL CARE IN MICE

Brown et al (1996) have recently identified a gene in mice, fosB, that is essential for the correct expression of maternal behaviors. They induced a mutation in a single gene, *fosB*, that when disabled appears to extinguish nurturing behaviors in female mice. Evidence from their study suggests that a very simple neural pathway may be involved. Lesion experiments have shown that the hypothalmus in female mice is critical for nurturing Behaviour. By deleting a gene whose gene product is expressed in the hypothalmus, Brown et al have isolated a key genetic factor involved in nurturing. It appears that the gene products of *fosB* are expressed in the small part of the hypothalmus called the preoptic area of the brain.

Fos B deficient mothers do not show the following nurturing behaviors:

- They do not retrieve their young, despite normal maze running ability.
- They do not nurse the young, despite normal mammary gland development.

By validating that *FosB* deficient moms are not incapacitated in basic organismal functions like spatial ability or hormone physiology, Brown et al have shown that *fos B* is important to nurturing per se and not merely a pleiotropic effect of *fos B*'s effects on other non-nurturing traits.

Indeed, *fosB* deficient mothers have normal expression of the reproductive hormones Estrogen and Progesterone which change during the course of the development of nurturing behaviors of the mother. In addition, fosB deficient mothers also have normal olfactory abilities based on olfactory discrimination tests. *fosB* deficient mothers simply do not nurture their young, there is little else wrong with their phenotype. When normal female mice, intact virgins, and even males, are exposed to pups the expression of *fosB* genes are triggered in the preoptic area of the brain. Olfactory neurons in this neural pathway appear to be particularly important. The presence of *fosB* is necessary to induce normal parenting Behaviour, but it is not the only gene required — it alone is not sufficient to elicit all aspects of parental care. Nevertheless, its correct function is crucial for the expression of normal nurturing. The FosB gene represents a crucial link to the chain of proximate factors that lead to a complex suite of parenting behaviors in mice, and perhaps many other mammals.

PATERNITY ANALYSIS AND A GENE FOR ALTERNATIVE MALE STRATEGY IN THE RUFF

Molecular methods are also useful for determining pedigrees of animals which have bred in the wild. In such cases the mother is usually known with certainty as she laid the eggs. However, the female could have mated with a large number of putative sires. DNA paternity analysis has become a standard for determining the identity of an offspring's sire.

Ruffs are a shore bird that breed and nest in northern Europe and they come in two plumage and Behaviour morphs. "Independent" male ruffs have a territory and defend females against other independents. The "territories" are very small and localized in "leks", which are areas where many males aggregate and display to attract visiting females. Non-territorial "satellite" males move among the independent males and obtain copulations from females on the independent's territory. Ruffs are fixed in their plumage Colour and Behaviour throughout their lives. Eighty-four percent of ruffs are independents, and sixteen percent are satellite males.

David Lank and his colleagues used molecular probes called "mini-satellites" to determine which male sired the chicks on ruff breeding grounds in Finland. By comparing the alleles in chicks and alleles in the mother and putative sires, they were able to determine the father and reconstruct the father's morphology. They also scored the morphology of the female parent's brother and father to determine the likely phenotype that the female would have expressed had she been male. Such genetic sleuthing is particularly important with sexually dimorphic traits which have a sex-limited expression (*e.g.*, traits only seen in males or only seen in females). Females do not express the alternative male morphology, yet they might carry genes for the morphology that they pass on to their male offspring. Lank and his colleagues then reared the field caught chicks to maturity when they could score the breeding morphology of the male progeny. They also reared an additional generation of chicks in captivity that were derived from the field collected cohort of chicks. They were certain of the father's identity in captive bred birds and did not need to use molecular probes to determine paternity.

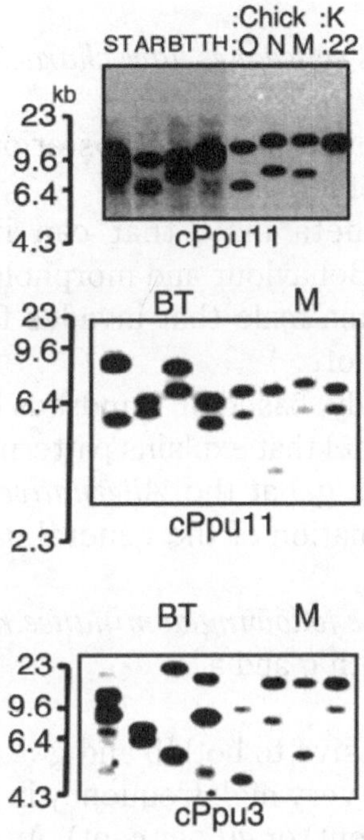

Fig. Variation in Minisatellite Alleles in Male Ruffs (ST, AR, BT, TH) that are Used to Determine the Sire of Progeny (0, N, M) of the Female (K:22). Allelic Variation at three Microsatellite Loci are Shown (cPpu11, cPpu11, cPpu3). Given the Mothers known Contribution to Offspring, Male BT is the Only Male that could be could Contribute Alleles at all Three Loci and he is the Likely Sire.

Because only two phenotypes are seen in the ruff, a single genetic locus with two alleles must be controlled by a completely dominant allele and a recessive allele, otherwise we would expect to see three phenotypes. The data on ruff pedigrees can discriminate between several simple mendelian patterns of inheritance.

Mendelian genes are either found on sex chromosome or on autosomal chromosomes. Their genetic crosses rule out the possibility that the gene is a sex-linked dominant. Their pedigree data is consistent with an autosomal gene with Satellite dominant.

While the example of male ruffs provides solid evidence of major genes which affect the phenotype, the sample size in these studies is still inadequate to test for the existence of other genetic loci that control development of male behaviors. Testing for the effects of two or more factors requires sample sizes in the thousands, particularly if those factors interact epistatically or in a non-additive fashion.

EPISTATIC GENES FOR ALTERNATIVE MALE STRATEGY AND SEX TRANSFORMATION IN MARINE ISOPODS

A final example of genes that have a major effect on Behaviour involves the genetic control of alternative male behaviors in a marine isopod, *Paracerceis sculpta*.

Stephen Shuster and his colleagues have characterized three alternative male morphs in the marine isopod:

1. A large alpha male with elongate posterior appendages that they use to defend harems of females
2. A medium-sized beta male that can invade female harems by mimicking female Behaviour and morphology, and
3. A small-sized gamma male that invades female harems by virtue of its secretive behaviors.

In a large breeding study, based on hundreds of genetic crosses, Shuster has developed a genetic model that explains patterns of inheritance of the three morphs. Three alleles (a, b, g) at the *Alternative male strategy* (*Ams*) locus provide a reasonable explanation of the general pattern of inheritance of the male morph.

The three alleles have the following dominance relations:

1. b is dominant to both g and a ,
2. g is dominant to a ,
3. a is of course recessive to both b and g .

The a allele occurs at a very high frequency in the population (93per cent) compared to either b(1per cent) or g(6per cent). An alpha male phenotype must be homozygous (e.g, a|a) at the *Ams* locus because a is recessive to the other to *Ams* alleles. Gamma males can be either a|a or g|g . However, the heterozygous form of gamma g|a is very common while the g|g form is rare in a natural populations. Likewise, beta males can be b|g , b|a , or b|b . However, as a is the most common allele in the population, the b|a is the most common genotype for a beta male. As an exercise, compute the frequency of each male genotype assuming Hardy-Weinberg gene frequencies and convince yourself that a|a , b|a , and g|a are the most common male genotypes in the isopod population.

Shuster and Sassaman found that a careful inspection of genetic crosses between females and the predominant male genotypes revealed a significant departure from a 50:50 sex ratio of the progeny. The existence of a second locus termed transformer (*Tfr-two alternative alleles, 1 and 2*) is required to adequately explain the aberrant sex ratio found only in certain genetic crosses. The *Tfr* locus causes males to transform into females, or females to transform into males.

However, the direction of sex change depends on the genotype at the Ams locus. Homozygous alpha males transform from male to female if they bear at least one copy of the *Tfr* - 2 allele (*e.g.*, genotypes 1|2 and 2|2). Alpha males with the 1|1 genotype at the *Tfr* locus are not sex-transformed. Conversely, beta males transform from female to male if they bear at least one copy of the *Tfr* - 1 allele (*e.g.*, genotypes 1|1 and 1|2). Beta males with the 2|2 genotype at the *Tfr* locus are not sex-transformed. Gamma males transform from female to male if they are homozygous for the Tfr-1 allele, but not if they carry one or more copies of the Tfr-2 allele. The epistatic interaction between two loci governs the expression of male and female behaviors in this marine isopod. The gene for *Alternative male strategy*interacts in a very non-additive fashion with the gene for *Transformer*.

Finally, we find an added complexity in this interesting genetic system. An extrachromosomal factor (ECF) accentuates the effect of the *Tfr* locus, but only in the *Tfr*heterozygotes of alpha and beta males. The exact nature of the ECF remains obscure but cytoplasmic effects on that determine phenotype are common in the animal kingdom. They could be transmitted from mother to egg (as are many cell organelles such as mitochondria).

Table. The Interaction between Three Alleles at the Alternative Male Strategy Locus (*ams*) and Two Alleles at the Transformer Locus (*tfr*) in a Marine Isopod, *Paracerceis Sculpta*.

		Tfr genotype	
Ams genotype	**Sex**	**1**	**1**
	M F		
	M F		
	M F		

3

Animal Genetic Engineering

INTRODUCTION

There are some serious issues related to genetic modification of animals using animal genetic engineering techniques. One is not sure of the consequences of these genetic modifications and the further interaction with the environment. Proper clinical trials are also necessary before one can use it for commercial purposes. In the recent past people have raised objections on some of the methods used *e.g.* the transfer of a human genes into food animals, use of organisms containing human genes as animal feed. Some religious groups have expressed their concern about the transfer of genes from animals whose flesh is forbidden for use as food into the animals that they normally eat. Transfer of animal genes into food plants that may be objectionable to the vegetarians.

Besides this, there are several other aspects of this issue have to be sorted out:

- What will be the consequences, if a modified animal will breed with other domestic or wild animals thereby transferring the introduced genes to these populations?
- What are the health risks to human on consumption of genetically modified animals and their products?
- With the production of disease resistant animals, what will be the effect on ecology?
- There is also wide spread concern about the risks of human recipients getting infected with animal viral diseases after a xenotransplantation., which might infect the population at large.
- There are also concerns about the risk that drug resistance gene markers used in genetic engineering procedures might inadvertently be transferred and expressed.

The need of the hour is to formulate clear guidelines which should be followed while using genetic engineering techniques in bio-medical research. *e.g.* products from transgenic organisms should be clearly marked to give choice to people who follow dietary restrictions due to religious beliefs. In fact all the ethical and moral issues raised by some aspects of biotechnology should be addressed by open discussion and dialogue.

MAPPING GENES

Modern molecular biology is making it possible to identify how different genes control different characteristics. Scientists are "*mapping*" the genes on the chromosomes, so that they can see where genes are located and how they function within cells and processes within the animal. This will help identify which combinations give the traits that are desired, making selection of animals carrying the best of the naturally occurring genetic variation more effective. Some traits are determined by a single gene and these are relatively easy to study. But others, including commercially important traits such as growth rate and feed conversion efficiency, are controlled by many genes working together. One important research goal is to find genes that make animals resistant to diseases. Selecting for these genes will help produce naturally healthier animals and reduce farmers' reliance on antibiotics and other veterinary drugs. In addition, knowing the processes that contribute to health and disease in animals will help us understand the function of the very similar systems that affect human health.

The full sequence of all genes (the genome sequence) has recently been derived for chickens and cattle and will soon be available for pigs. These sequences allow us to identify the similarities and differences between different livestock species and between livestock and humans. This sequence greatly aids our ability to identify important genes and understand the causes of healthy development or disease susceptibility.

GENETIC MODIFICATION

The traditional selection and mapping described above makes use of naturally occurring gene variants. However, there are a number of reasons why it might be valuable to alter existing gene variants or add new genes. As well as identifying and isolating individual genes, molecular biology enables scientists to transfer these genes from one individual to another - so called genetic modification or transgenics. This could further accelerate livestock improvement but its usefulness in the foreseeable future is most likely in the area of obtaining high-value medical products from animals rather than in agriculture.

Uses of Transgenic Animals

There are five main areas in which the production of transgenic animals offer potential benefits:

- To produce therapeutics for human medicine - In 2006, the European Medicines Agency approved the use of a therapeutic used to prevent blood clotting that is produced in goats' milk. Recently scientists have bred several generations of transgenic chickens that produce therapeutics in the egg white. This offers the potential for producing high-value therapeutics in the eggs of genetically modified chickens.

- To provide models of human disease - The function and control of genes implicated in human disease can be studied in animals.
- To provide tissue for use in human transplant surgery - The recognition molecules on animal cells may be tailored so that they are no longer rejected as foreign by humans.
- To improve the efficiency of livestock breeding and to increase the range and quality of livestock products.
- More speculatively, to improve animal health - Long term it should be possible to identify genes that confer resistance to disease and to introduce these into livestock, so reducing reliance on antibiotics and chemicals to control disease.

Genes are Introduced into Animal Cells

Newly fertilised egg cells contain two pronuclei (one originating from the egg cell and one from the sperm) each of which carries half the full complement of chromosomes. The pronuclei fuse to form the nucleus of the fertilised cell which then divides to form the embryo. Newly fertilised cells can be isolated and immobilised under a microscope. A very fine glass needle is used to inject DNA directly into one of the pronuclei. The egg is then transferred into a foster mother where it grows and divides normally to produce the embryo. A new technique called nuclear transfer enables scientists to remove the nucleus from an egg cell and replace it with a nucleus from another cell. In this way the complete genetic information of the donor cell is incorporated into the recipient. The cells may be genetically modified in the lab before they are transferred to the recipient.

GENETICS OF SCRAPIE RESISTANCE IN SHEEP

Scrapie is an infectious disease of sheep that affects the central nervous system and is always fatal. Upon necropsy, infected animals will have holes or vacuoles in the tissue of the brain. The disease is classified as a transmissible spongiform encephalopathy (TSE). Some TSE's in other species include bovine spongiform encephalopathy (BSE, Mad Cow Disease), Creutzfeldt-Jakob Disease (CJD) in man, new variant CJD (vCJD) in man thought to result from eating meat from BSE-infected cattle, transmissible mink encephalopathy (TSE), chronic wasting disease (CWD) in some U.S. populations of deer and elk, and feline spongiform ecephalopathy (FSE). Federal and State animal health authorities have considered scrapie a high priority disease for elimination for many years, but the relatively recent evidence that ties vCJD in humans to BSE in cattle has resulted in increased attention on scrapie in sheep.

INCIDENCE

Scrapie was first diagnosed in the U.S. in 1947 in a flock of Suffolk sheep in Michigan that had imported sheep from Canada of U.K. origin for several years. From 1947 through July 2001, scrapie had been diagnosed in over 1,000 flocks, and approximately 1,600 individual cases of natural sheep scrapie and 7 individual cases of natural goat scrapie have been found (Sutton, USDA/APHIS/VS, personal communication).

Through 1992, the breed distribution was as follows: 87 per cent Suffolk, 6 per cent Hampshire, and 7 per cent other breeds (Border Leicester, Cheviot, North Country Cheviot, Corriedale, Cotswold, Dorset, Finnsheep, Merino, Montadale, Rambouillet, Shropshire, and Southdown) and crosses (Detwiler et al., 1996; Wineland et al., 1998, USDA, 1999). While 1,600 appears to be a large number of cases of sheep scrapie, it actually is not since it represents an average of only about 30 cases per year. Relative to most other sheep diseases, the incidence of scrapie is quite low.

Clinical signs

Early signs of scrapie include subtle changes in behaviour and temperament. These changes may be followed by scratching and rubbing against fixed objects. Other signs are loss of coordination, weight loss despite a good appetite, biting of feet and limbs, and lip smacking.

Gait abnormalities such as high-stepping of the forelegs, hopping like a rabbit, and swaying of the back end may be seen. Signs or effects of scrapie usually do not appear until 2 to 5 years after an animal is infected. Sheep may live 1 to 6 months (sometimes longer) after the onset of clinical signs (USDA, 1999). Infected animals show no immune response to the infective agent as is the case with viral and bacterial diseases. There is currently no live-animal test to determine if an animal is infected. This is a major area of research, and some promising live-animal tests are in the development and evaluation stages.

The scrapie agent is thought to be spread most often from ewe to offspring and to other lambs in contemporary lambing groups through contact of lambs with the placenta and placental fluids of infected ewes (USDA, 1999).

Lateral transmission from infected rams to ewes and lambs, from infected ewes to other ewes, and from an infected environment to adult animals are not thought to be major routes of transmission; although there are documented cases of scrapie where such routes of transmission are thought to have occurred.

There is not agreement among scientists on the nature of the infectious agent. However, infected animals have large amounts of an abnormal type of prion protein (denoted as PrP^{Sc}), especially in the central nervous system tissues of the brain and spinal cord.

Scrapie-free animals have normal prion protein (PrP^{C}) and no PrP^{Sc}. Some feel that PrP^{Sc} itself is the infectious agent (Prusiner, 1998) while others feel that PrP^{Sc} is the result of another infectious agent. However, scientific evidence increasingly supports the theory of PrP^{Sc} itself as the infectious agent. Dr. Stanley B. Prusiner, University of California, San Francisco was awarded the

1997 Nobel Prize in Physiology or Medicine for his work on the infectious nature of prions. The infectious agent is extremely resistant to heat and to normal sterilization procedures.

Prion Protein

Prion protein is a constituent of normal mammalian cells ((Prusiner, 1998). Sheep prion protein is a protein of 256 amino acids in length. There is some variation among sheep in the particular amino acids that comprise their prion protein. However, the PrP^{Sc} found in a scrapie-infected sheep is of the same amino acid sequence as some or all the normal prion protein (PrP^{C}) found in that same sheep. The difference between PrP^{Sc} and PrP^{C} is in its molecular shape. The prion infectious agent theory contends that PrP^{Sc} infects a normal sheep and serves as a template to change the normal shape of PrP^{C} to the infective shape of PrP^{Sc}.

Of the 256 locations for amino acids along the sheep prion protein molecule, amino acid changes in at least three of these locations (amino acids number 136, 154, and 171) have been shown to confer increased or decreased susceptibility to scrapie (Goldmann et al., 1994; Westaway et al., 1994; Belt et al., 1995; Clouscard et al., 1995; Ikeda et al. 1995; Hunter et al., 1997b; Hunter et al., 1997a; Elsen et al., 1999; Thorgeirsdottir et al., 1999; Tranulis et al., 1999). Each particular amino acid is specified in the prion protein gene by a group of three nucleotide bases known as a codon. Therefore, these three important locations are referred to as codon 136, codon 154, and codon 171 on the prion protein gene.

The most common amino acids coded for at these three codons are: alanine (A) or valine (V) at codon 136, arginine (R) or histidine (H) at codon 154, and glutamine (Q) or Arginine (R) at codon 171. Since genes occur in pairs, sheep have two prion protein genes. Each gene can be the same in which case the sheep will produce only one type of prion protein. If the genes are different, then two types of prion protein are produced. Just considering these three codons, a population of sheep theoretically can contain individuals with one of 27 different prion protein genotypes. Not all 27 genotypes have been documented, so some may not occur in nature.

GENETIC SUSCEPTIBILITY OR RESISTANCE

Genetic resistance to scrapie depends not only on the prion genotype of the sheep but also on the strain of scrapie present. Some genotypes found to be resistant to one strain of scrapie have been shown to be susceptible to another strain. It appears that there are differences between the U.S. and Europe in the major strains of scrapie that are present. The fact that the U.S. is free of certain scrapie strains is good justification for the stringent regulations for importation of sheep and goats into the U.S. from countries with scrapie. Codon 136. Sheep homozygous for alanine (AA) have been shown to be more resistant to scrapie than sheep homozygous for valine (VV) or heterozygous (AV) in European studies. While a lower incidence of scrapie is anticipated in AA sheep in Europe than in the other genotypes, some AA sheep still have been diagnosed with scrapie. AA sheep are not 100 per cent resistant to scrapie. In the U.S., amino acid changes at codon 136 appear to be less important to scrapie susceptibility than in Europe. This may be due to the different strains of scrapie found in the two regions.

Codon 154. Of the three important codons, amino acid changes at 154 appear to have a slightly less dramatic effect on scrapie susceptibility than do the other two, and the susceptible genotypes are not consistent across studies. In some studies, sheep with the arginine (R) allele have a lowered incidence of scrapie, and in other studies, sheep with the histidine (H) allele have a lowered incidence. At the present time, it appears that screening sheep on the basis of codon 154 genotype has little value in increasing resistance to scrapie.

Codon 171. Amino acid changes at codon 171 have a large effect on scrapie susceptibility in sheep in both Europe and the U.S. Virtually no sheep homozygous for arginine (RR) have been identified with scrapie. The one exception is a single RR Suffolk in Japan that was diagnosed with scrapie (Ikeda et al., 1995). The frequency of heterozygous (QR) sheep also is very low among scrapie-infected sheep. However, the frequency of sheep homozygous for glutamine (QQ) is very high among scrapie-infected sheep.

The last column of Table presents a very crude scrapie resistance index for each genotype in U.S. sheep. The higher the number, the greater the expected resistance to scrapie. The index is weighted heavily in favour of RR and QR genotypes at codon 171, less heavily in favour of the AA genotype at codon 136, and neutral with respect to the genotypes at codon 154.

SUSCEPTIBILITY VERSUS SCRAPIE-INFECTED

It must be remembered that a susceptible genotype like QQ at codon 171 does not imply that the animal is scrapie-infected. Scrapie is not a genetic disease. Scrapie is not caused by a particular genotype. An infectious agent causes scrapie. In order to have scrapie, both an infectious agent and a susceptible genotype need to be present. For example, Australia and New Zealand are scrapie-free, but Suffolk, Cheviot, Merino, and Poll Dorset sheep of susceptible genotypes are found in these countries (Hunter and Cairns, 1998). There is no scrapie in these countries because the scrapie agent is not present.

Likewise, susceptible genotypes are found in many breeds of sheep in the U.S., but almost 90 per cent of the scrapie cases are in the Suffolk breed. This may be due to the major mode of transmission of the disease from an infected ewe to her newborn lambs. Scrapie was brought into the U.S. in Suffolk sheep. A possible scenario that followed was that infected Suffolk ewes infected their lambs, and infected females from infected ewes continued to pass the disease to their lambs. However, infected Suffolk rams would not tend to infect the ewes they were mated to or their lambs, so infected Suffolk rams mated to ewes of a different breed, as is common in crossbreeding systems in the U.S., would not tend to pass scrapie onto that breed.

While scrapie is primarily a disease of Suffolk sheep in the U.S., the disease is found in the Suffolk, Cheviot, Swaledale, Bleu du Maine, Herdwick, Poll Dorset, Shetland, and Soay breeds in the U.K. (Hunter et al., 1997ab), Texel breed in The Netherlands (Belt et al., 1995), Suffolk and Corriedale breeds in Japan (Ikeda et al., 1995), Romanov and Lacaune breeds in France (Clouscard et al., 1995; Elsen et al., 1999), Rygja breed in Norway (Tranulis et al., 1999), and Icelandic breed in Iceland (Thorgeirsdottir et al., 1999). This is not an exhaustive list of sheep breeds or countries with scrapie, but a sample to indicate that scrapie is found in several breeds. A discussion of theories on how scrapie initially becomes established in a breed of sheep would be interesting.

CLONING AND GENETICALLY ALTERED ANIMALS

At Eurogroup for Animals, we believe that humans have a duty to safeguard animals' welfare and health and we oppose all human activity that causes distress or suffering. In particular in the area of animal breeding, we are of the opinion that animals genetic constitution should not be changed if doing so causes suffering to themselves or to future generations.

Genetic selection must always seek to maximise the wellbeing of the resultant animal. Animal breeding should support a diverse gene pool in order that the genetic resources to maximise wellbeing are sustained for future generations. Techniques and technologies used in animal reproduction must not be used if they cause distress or suffering.

For these reasons Eurogroup is opposed to the cloning of animals and would urge the EU to introduce an immediate ban on the cloning of animals for food production, and on the sale of imported food products from cloned animals and their offspring. Eurogroup is concerned that delaying a decision will mean that food from cloned animals will enter the supply chain uncontrolled and without consideration of the opposition of EU citizens and consumers. Under Article 13 of the Lisbon Treaty, in the EU it is a legal obligation to pay full regard to the welfare requirements of animals when formulating and implementing the Union's agriculture, fisheries, transport, internal market, research and technological development and space policies.

Our concerns are based on the fact that some of the developments and applications of animal cloning and other genetic alteration techniques:

- Involve procedures that may cause animals pain, suffering or distress;
- Use a very large number of animals;
- Encourage a wider variety of applications leading to increased animal use;
- Increase the perception of animals as commodities for human use and/or gain, such as research tools or units of production;

- Are progressing at a rate that is outstripping public understanding and ethical and public debate.

CONCERNS RELATING TO HUMAN AND ANIMAL HEALTH

The main applications of cloned and other genetically altered (GA) animals (and their offspring) that are of concern at the present time are:

- The creation of genetically manipulated (GA) and cloned animals;
- The subsequent use of these GA animals (and their offspring):
  - As livestock on farms (used to standardize/maximise quantity and reduce the costs of processing/producing animal derived food products *e.g* milk, meat, eggs);
  - As disease models;
  - In fundamental research *e.g* to understand gene function;
  - In toxicity testing;
  - As bioreactors to produce biologically active compounds for experimental and/or medical purposes;
  - As sources of cells, tissues and organs for xenotransplantation;
  - In the creation of cloned pets, sports animals and 'living art'.

However, any new biotechnology, or new application of existing technologies, that is developed and/or tested in animal models, or that causes the animals pain, suffering, distress or lasting harm is of serious concern to Eurogroup. Examples include tissue engineering and the development of nanotechnology generally.

PRODUCTION OF CLONED AND OTHER GA ANIMALS

Use of animals in experiments is a matter of serious public concern and pressure to reduce numbers has contributed to a downward trend in some countries. Creation and use of GA animals is reversing this trend and there has been an exponential rise in the number of GA animals used in scientific procedures each year in some EU member states. For example, in the UK, the number of GA animals has risen from around 50,000 in 1990 to 900,000 in 2004 (Home Office Statistics 1990-2004). In Germany, Ireland, Finland, Sweden and the Netherlands, there is also increasing creation and use of GA animals (EU Stats 1999, 2002).

GA animals can be produced by a number of different techniques, all of which are inherently wasteful with respect to the number of embryos that are manipulated relative to the number of GA offspring subsequently produced. Current estimates suggest only a 3-5 per cent success rate when generating new GA animals (Nuffield Council on Bioethics, 2005). The remaining offspring are surplus to requirements and are killed. A large number of animals are therefore used to provide sufficient eggs or embryos for genetic manipulation, and to act as recipients and foster mothers for the manipulated embryos. These animals will usually be killed, either before harvesting the eggs or embryos, or, in the case of recipients, once their young are weaned. Thus GA technology is wasteful of animal life and the ethical implications of such wastage are important and need to be acknowledged.

The procedures used to produce GA and/or nuclear transfer cloned animals involve hormonal and surgical interventions that can cause pain, suffering and distress. These include; superovulation, vasectomy, semen and embryo collection and embryo transfer (JWGR, 2003; Pew Initiative, 2002).

Many of the manipulated embryos die during gestation. This may also cause suffering, although this depends on the stage of development at which death occurs and the species involved.

ADVERSE EFFECTS AS A CONSEQUENCE OF GENETIC MODIFICATION AND CLONING

Where GA animals are created as models of specific diseases, they can experience a range of adverse effects associated with the condition in question, which can have profound effects on their health and welfare. However, being genetically altered does not necessarily compromise the welfare of individual animals, and indeed some GA animals are indistinguishable from their non-GA siblings. It also needs to be acknowledged that adverse effects may only become apparent when animals are subsequently maintained in a less well defined or controlled, environment than that of the laboratory or experimental farm. Nevertheless, in many cases, genetic alteration can have a deleterious effect on animal welfare and the harms caused depend on a number of factors. Studies have shown that some *in vitro* culture procedures carried out during genetic manipulation or cloning protocols can lead to unpredictable complications in the animals subsequently produced (ECVAM, 1998).

An example of this is Large Offspring Syndrome (LOS), which affects sheep, mice and cows following nuclear transfer. As well as causing complications for the mother during birth, LOS encompasses a range of debilitating pathologies for the offspring including malformations in the liver, brain and urogenital tract, immune dysfunction, placental abnormalities, stillbirth, fetal overgrowth, respiratory failure and circulatory problems (Vajta and Gjerris, 2006).

A further concern is that there is currently no mechanism to ensure consistency in the training of personnel in specific GA technologies and procedures, or to ensure that experimental and other refinements are disseminated throughout the biotechnology community. This can have a profound influence on the levels of pain, suffering and distress experienced by animals undergoing these procedures. It can also have a negative impact on the success rates achieved and levels of animal wastage.

Concerns regarding mutagenesis programmes

Mutagenesis (using chemicals such as ENU, or physical mutagens such as radiation, to increase the natural mutation rate of DNA) is the quickest method for producing large numbers of GA mice. This technique is being widely applied both in Europe and beyond, with mutagenesis programmes aiming to explore the function of every gene in the mouse genome. This project presents a number of animal welfare concerns:

- the number of animals involved in mutagenesis programmes is vast; *e.g.* around 35-36,000 for a 3 year project, or around 50 animals per mutant line established (Mammalian Genetics Unit, Harwell, 2006);
- the process of mutagenising animals has significant animal welfare implications for the animals involved, for example males require 12-14 weeks to recover their fertility following treatment with a mutagen. Furthermore, only around 50 per cent of mutagenised males are able to go on to sire offspring with the remainder being culled (Goldwitz et al, 2004);
- the mutations induced are by their very nature unpredictable and the scientific usefulness of mutagenised animals cannot therefore be predicted, nor can the effect of any mutation on the health and welfare of the animals. This makes the justification for producing GA mice in this way highly questionable.

Concerns regarding knock-in and knock-out technology

A more targeted approach to the production of GA animals uses DNA constructs inserted into the animals genome to either completely remove a gene of interest (knock-out), or to replace a given gene with an altered version (knock-in). This is done either in the whole animal, or within specific tissues of an animal (conditional). The animal welfare concerns include:

- At the current level of efficiency, the numbers of animals used is high - at least 200 animals will be used in the production of a single GA animal (Mammalian Genetics Unit, Harwell, 2006);
- The site of insertion into the host genome and the number of copies of the DNA construct inserted cannot always be controlled unless ES cell manipulation is used. This can be a problem because the incorporation of the construct DNA at an incorrect location can result in the random inactivation of other genes or alterations in the expression of surrounding genes, both of which can impact on the health and welfare of the animals produced;
- When a knockout animal is generated the level, or pattern of expression observed for other genes can be altered to compensate for the lost gene (Okkenhaug, 2003). Thus any effects observed in the GA animal may not only reflect the loss of the gene of interest, which means that investigation of the effects is not always straightforward. The use of knock-out as an alternative to knock-in or conditional technology, therefore needs careful consideration and justification.

Special concerns for non-human primates

This submission relates to all animals used in biotechnologies, but Eurogroup has particular concerns about the application of such technologies to non-human primates. Macaques have already been produced by nuclear transfer cloning in the USA (Chan et al, 2001) and recent technical advances show that the production of GA primates is currently increasing (*e.g.*: Genetically altered primates with Huntingdon's disease produced in 2008; Yang et al, 2008). Eurogroup firmly believes that the genetic modification of non-human primates should not be allowed for any purpose, given the ethical issues raised by such developments, the large number of animals required to produce each GA animal, and the associated potential for harms.

Animal models of disease

A common justification for the creation of GA animals is that they will provide, or contribute to, 'improved', more predictive models of disease. This is an oversimplification because a) this is not necessarily always true, and b) even where a new GA model is more appropriate, it may be used alongside other, older models by different researchers. There is no mechanism for ensuring only the most relevant are used and that newer models are available to all researchers. In such cases the GA animal is not the new *definitive* model but just an *additional* one.

The motivation for the research may merely be an interest in the model for its own sake, but a medical application may be used to justify the work since this is likely to be more acceptable publicly and politically. Basic, fundamental research carried out within academic research establishments is not regulated under Directive 86/609, so it may not undergo an ethical review with appropriate assessment and weighing of harms and benefits. This directive is under review and is expected to be adopted end 2010. The revised directive may incorporate basic, fundamental research with proper review, however only if implemented and enforced properly.

There is also currently no requirement for GA animals to be cryopreserved and stored within central archive facilities or depositories. Such methods can reduce repetition and duplication of work by providing a central resource for use by the wider scientific community and protect against adverse events such as environmental disasters or genetic drift. They also reduce the need for live transportation of GA animals, with associated welfare problems, because frozen gametes or embryos could be sent instead.

GA ANIMALS IN TOXICITY TESTING

GA mice and rats are increasingly used in genotoxicity and carcinogenicity testing, within studies that are done to fulfil regulatory requirements for the marketing of chemicals and pharmaceuticals. Such animals are likely to suffer equivalent (or even more severe) levels of pain and distress to those experienced by animals in traditional tests, but it has been claimed that fewer animals will be needed.

Eurogroup believes that reducing the numbers of animals used in research and testing is an important goal, provided that this can be done without increasing the level of suffering experienced by individual animals. However, relatively severe adverse effects have been reported in some GA strains used in toxicity testing (ECVAM, 1998). For example, mortality rates are higher in *c-neu* and *c-myc* mice used in carcinogenicity testing than in conventional mice used in the same type of test. Many GA mice used in carcinogenicity tests are more susceptible to developing cancer and so will develop tumours more rapidly, which could make it more difficult to implement humane endpoints. Reducing numbers is thus not automatically a positive outcome for animals - the impact on individuals must be taken into account and it may be justifiable to use more animals who will suffer less.

In any case, the contribution that biotechnologies can make to reducing animal numbers in carcinogenicity and genotoxicity is not consistent, largely due to actual or potential regulatory requirements. In carcinogenicity testing, GA mice were introduced to try to reduce animal numbers and the time taken to obtain test results. However, there are proposals to add additional control groups (a positive control and treated and untreated controls using non-GA mice) that would decrease the magnitude of the reduction if they are adopted (OECD, 2005). It is also uncertain whether testing on one GA strain will be regarded as sufficient by regulators, in which case the reductions will be further diminished by requirements for results obtained using other strains. For genotoxicity testing, GA models have distinct scientific advantages over existing *in vivo* assays when used as second tier tests for *in vitro* genotoxins. The reduction in numbers of animals used would be fairly substantial, perhaps from 50 to 20 per substance tested, but again it is not certain that a test on one GA model would be regarded as sufficient. There are potentially much greater savings in animals if GA tests can be used in place of very large tests of heritable mutation which are currently used, albeit rarely, for chemicals of high concern.

Conversely, using GA animal models in toxicity testing could increase the use of animals by making some tests more practical, or by increasing the amount of information they produce. For example, the use of GA animals could make investigations feasible that would otherwise require very large and impractical numbers of non-GA animals. Regulators might be inclined to ask for the GA test in cases where the conventional test would not have been requested (and could therefore presumably have been done without) because it was regarded as too cumbersome and possibly uninformative. Eurogroup believes that, in toxicity testing as in the other research fields discussed in this document, the focus should be on developing *in vitro* alternatives to replace animals and not on developing different animal models.

ANIMALS AS 'BIOREACTORS'

This category of GA animal use includes:

- "pharmed" animals who produce therapeutic substances *e.g.* pigs who express the blood protein Factor IX in milk, and goats that express the anti-clotting agent Atryn in their milk (Van Cott et al, 1996; Lindsay et al, 2004; GTC Therapeutics);
- Animals producing specialist materials *e.g.* goats who produce spiders' silk proteins in their milk for use to make 'biosteel'. This has both medical and non-medical applications (*e.g.* in sutures and bullet-proof vests respectively) (Nexia Biotechnologies, Quebec);

The substance produced may have an adverse effect on the animal, either at the point of expression, or if it can enter the animal's bloodstream. For example, a strain of rabbits genetically engineered to express human erythropoietin (EPO) in the mammary glands also expresses the protein at low levels in other organs, resulting in greatly elevated numbers of red blood cells, infertility and premature death (Massoud et al, 1996). Using animals in this way, and referring to them as 'bioreactors', reinforces the perception of animals as units of production and/or biological tools, rather than as sentient beings with the ability to experience pain, suffering and distress.

Xenotransplantation

The use of GA animals to supply organs, tissues or cells for transplantation into humans is a highly controversial issue which has been the subject of a great deal of debate. There are many legal, scientific, human health, animal welfare and ethical concerns, which have been described in a number of documents (Nuffield Council on Bioethics, 1996; Kennedy Report, 1997; Council of Europe, 2002). Only the ethical and welfare issues relating to animals are addressed here. These include:

- The ethics of genetically modifying animals of any species as a source of cells, tissues and organs for human transplantation;
- The harms associated with the initial creation of GA animals as source animals (see section 2.1);
- The suffering and/or distress associated with production and maintenance systems for high health status source herds. This includes hysterotomy-derivation, early weaning practices, and barren husbandry environments, which have a serious negative impact on animal welfare because they prevent animals from satisfying their physical, social and behavioural needs.

Furthermore, development of xeno technology to a point where it can be used still requires a great deal of pre-clinical research. To date, such research has included studies of efficacy, physiology, immunology, and infection risks in a range of species including primates, goats and dogs. This research, by its very nature, causes considerable suffering. Experiments involving organ transplantation require major surgery, which in itself causes suffering that is exacerbated by tissue rejection and immunosuppressive treatment. Xenotransplantation is also an example of a biotechnology where over-optimistic claims are made to justify the approach, the funding and the use of animals. For example, in September 1995, the UK company Imutran *"envisaged the first xenotransplants of transgenic pig hearts into human patients taking place in 1996"* (Nuffield Council on Bioethics, 1996). Yet despite some progress, particularly with cell transplants, the transplant of whole organs is no closer and xenografts still rarely survive for more than a few months (Chapman, 2004).

Cloning companion animals

Eurogroup believes that, without doubt, some applications of modern biotechnology are trivial, scientifically unnecessary and ethically unjustifiable. Examples include;

- The cloning of champion racing and show jumping horses for sport (Galli et al, 2003; Cryozootech);
- The generation of a green fluorescent rabbit 'GFP Bunny' as transgenic "art" (Eduardo Kac, 2000);
- The cloning of companion animals for example cats, purely to satisfy humans' emotional requirements (Shin et al, 2002; Genetic Savings and Clone Inc).

Concerns relating to agricultural production

Gene mapping was/is the most widely used technique in agriculture to enhance selective breeding schemes, however as more and more 'super' animals exist this is being overtaken by the import/export of embryos and/or sperm from cloned 'super' animals. Examples of agricultural applications of gene mapping and other genetic altering techniques that have given rise to ethical and animal welfare concerns include:

- *Increasing productivity or changing body composition*: Gene mapping has been used primarily to select individuals for breeding with the aim of increasing productivity *i.e.* growth rates, litter sizes and production traits such as egg laying, milk volume and meat quantity and quality (including the proportion of lean meat to fat). Whilst still being practiced to create 'super' individuals for subsequent breeding, gene mapping is being/has been replaced by the import/export of embryos and/or sperm from cloned 'super' livestock (primarily from the US). Farmed species have also been genetically manipulated to alter the composition of meat and milk. For example, pigs and cows respectively have been genetically altered to have higher levels of Omega 3 in their muscle (Lai et al, 2006), and to express higher levels of casein in their milk (Brophy et al, 2003).

- *Increasing disease resistance:* Animals are genetically manipulated to be resistant to disease, for example, cattle have been engineered to express the antibiotic lysostaphin in their milk, which results in increased resistance to mastitis (Wall et al, 2005). Animals (including pigs, sheep, mice and rabbits) have also been modified to express antibodies providing immunity to specific diseases (Lo et al, 1991; Weidle et al, 1991), for example mice have been generated with protection against prion disease (Heppner et al, 2001).
- *Making animals more 'environmentally friendly':* An example of this application is the 'Enviropig™', which has been engineered to contain the enzyme phytase in the pigs' saliva so that they can digest sources of dietary phosphorus. This results in faeces with a lower phosphorus content, which in turn reduces the pollution of surface and ground water with phosphorus (Golovan et al, 2001).

ETHICAL AND ANIMAL WELFARE CONCERNS

The selective breeding of farm animals has been conducted for thousands of years, and has given rise to a number of welfare concerns. However, Eurogroup believes that the use of cloning or other GA technologies to speed this process, or to introduce genes that could never be incorporated into the genomes of farm animals by any natural process, is a serious ethical and welfare issue. Directly altering an animal's genome is viewed by many as an unacceptable assault on the integrity of the animal that is incompatible with the concept of respecting farmed animals, and minimising the harms that are caused to them for human benefit. These views are important and should be respected as a legitimate part of the debate on biotechnology and farmed animal welfare.

Eurogroup also questions the necessity of further increasing production in farm animals. In many cases, productivity is already pushing animals to their physical and metabolic limits, so with any further increase there is an enhanced likelihood of animal welfare problems. Enhancing selective breeding, by gene mapping or genetic modification, can also cause suffering if the trait that is selected for has a negative impact on the rest of the animals' physiology. For example, hens who produce high numbers of eggs suffer from osteoporosis because the majority of the calcium they ingest is used in eggshell production (J. Mench, Pew Initiative, 2002). There can also be less direct effects on welfare, in that some GA animals may receive lower standards of husbandry than conventional animals. For example, clinical mastitis is a major welfare problem in dairy systems with sub-optimal standards of hygiene, and early detection is reliant on routine inspections by parlour staff at milking. The creation of cattle resistant to mastitis may encourage the perception that mastitis is no longer a problem. This could not only compromise standards of parlour hygiene, but may reduce the level of attention paid to each animal at milking. This would increase the potential for other clinical or welfare problems to go undetected. "High productivity" animals may also be at risk if their husbandry is not appropriate. It may be possible for them to be properly managed and cared for in the controlled environment of a breeding company or experimental farm, but there are serious concerns regarding the welfare of such animals once they are released into commercial agriculture.

GENETICS OF TUBERCULOSIS IN IRISH DAIRY COWS

Tuberculosis has been an on going problem for the Irish cattle industry for many years. Prior to the introduction of the 'test and slaughter' policy in 1954 bovine tuberculosis was wide spread, with an animal infection level of 3.0 per cent (140,000 reactors per annum). Progress was rapid during the initial stages of the programme, when animal infection level was reduced to 0.4 per cent (23, 000 reactors) by 1965 when Ireland received 'official tuberculosis free status'. However progress has stalled at this point; in 2006 animal infection level was 0.4 per cent (26,000 reactors). Current control measures, which include annual testing of every bovine each year, have effectively minimised cattle-to-cattle transmission of infection.

However, there are three main factors constraining the eradication of tuberculosis in Ireland including: the continuing expansion of the national herd, the frequency of movement during the production cycle, and the presence of a wildlife reservoir, the Eurasian badger *Meles meles*. Furthermore, the current programme represents a very substantial cost to the industry and government (approximately □80 million per annum). Selection for increased resistance to tuberculosis susceptibility provides an alternative strategy that could enhance existing control programmes. However, this strategy requires additive genetic variation for resistance to tuberculosis.

There is some evidence at the genus level that *Bos indicus* is more reistant to tuberculosis than *B. taurus*. Furthermore, there is evidence that certain family lines are more resistant than others in; a heritability ranging from 0.06-0.08 has been estimated in cattle in Latvia. Similarly, a heritability estimate of 0.06 has been published for an analogous disease, paratuberculosis (Johne's) disease in Dutch dairy cattle.

Nevertheless, there is a lack of information on genetic parameters for tuberculosis resistance in cattle, which is largely a result of the lack of suitable data. Ireland is uniquely placed to address this issue, given the availability of three national databases (including the: AHCS [Animal Health Computer System], CMMS [Cattle Movement Monitoring System] and ICBF [Irish Cattle Breeding Federation database]) with linkable animal and herd level records of animal breeding and disease control. Furthermore, the sound epidemiological and immunological knowledge generated in Ireland and abroad were an invaluable resource in the development of a trait definition, data editing criteria and for the selection of herd and animal level risk factors to be included in analyses.

The objective of this study was to estimate the genetic variation in tuberculosis resistance in Irish dairy cows. A total of 15,321 animals (including 1,268 tuberculosis positive animals) from 724 herd test dates during the period of 1st November 2002 and 31st October 2005 were available for inclusion in the analysis. All animals were Holstein-Friesian and had an identified sire. Tuberculosis positive animals were defined as animals with a skin change where the *Mycobacterium bovis* purified protein derivative response was 4 mm greater than that of the *M. avium* purified protein derivative response.

Animal linear, animal threshold and maternal effects linear models were used to estimate the additive genetic and residual variance for tuberculosis susceptibility. Fixed effects in the models included, herd episode, month of test, age at test, year of calving, month of calving, proportion of Holstein genes, heterosis and recombination loss.

Heritability of tuberculosis susceptibility was 0.0317 (SE=0.0106). The animal threshold model estimated an almost four fold larger heritability (h^2=0.1189 (0.0281). Threshold models have been shown to be theoretically better, and are more accepted for all or nothing (reactor/non reactors) trait heritability estimation, indicating the higher heritability estimate should contribute to greater genetic improvement.

The heritability estimate from the maternal effects model was 0.0285 (SE=0.0110), which was less than the estimate of the animal linear model, indicating that maternal inheritance dose not influence tuberculosis resistance in cows. The low heritability estimates observed are characteristic for disease traits, and are similar the earlier estimates of 0.06-0.08 and 0.06 for tuberculosis and paratuberculosis resistance respectively. This study has demonstrated that exploitable genetic variation for tuberculosis susceptibility exists among Irish dairy cows, and that selecting for increased resistance should be possible.

The results from this study will be used to generate of breeding values for the industry, allowing breeders to identify Irish and foreign animals that are genetically resistant to infection with tuberculosis. Furthermore, while selecting for increased resistance to tuberculosis, breeders maybe indirectly selecting for general health, *i.e.* heightening resistance against a variety of pathogens that infect cattle in a similar manner.

The breeding values estimated from this study will also be used in a collaborative project in Trinity collage Dublin, to identify the major genes associated with tuberculosis resistance. This additional knowledge will provide for more (accurate) information on the genetic merit of animals, which will be used to enhance selection programmes in the future.

Increasing the resistance of our national herd to tuberculosis would prove synergistically beneficial with existing eradication efforts, through reduced animal/herd incidence, resulting in reduced animal/environmental bacterial loads, ultimately reducing cattle-cattle, cattle-environment, cattle-wildlife, and environment-wildlife transmission of the disease, a major step towards the eradication of tuberculosis, and attainment of the much coveted official bovine tuberculosis free status in Ireland.

GENETIC MANIPULATION OF ANIMALS

Experimental animals have been used in biomedical research for decades. In many cases, aspects of physiology and biochemistry have been investigated, and artificial manipulations have often been confined to examining the effect of altering the animal's environment or some aspect of its phenotype. Some animals, notably *Drosophila* and mice, have been particularly amenable to genetic analyses and traditional genetic manipulation of animals has involved carefully selected breeding experiments or exposure of animals to powerful chemical or radioisotopic mutagens. A new era in animal research was ushered in during the early 1980s when successful experiments designed to genetically modify animals by inserting foreign DNA were first reported. These new methodologies were expected to have many advantages for research but two major areas have benefited:

- *Gene function:* While the use of cultured cells and cell extracts can be extremely valuable in studying gene expression and function, the ability to insert genes into whole animals or to selectively delete or alter single predetermined genes in an animal provides enormous power in studying gene function.
- *Animal models of disease:* Nature has provided some animal models of disease and some have been generated by random mutagenesis programmes, *but not in a predetermined way*. The new technologies held the promise of altering at will even single genes within a living animal in such a way as to mimic mutations faithfully in an analogous gene in humans, thereby providing a higher chance of resembling human disease phenotypes.

In order to create genetically modified animals, it is necessary to modify the DNA of germline cells so that the modified DNA is heritable. As a result, certain cells that have the capacity to differentiate into the different cells of an adult animal (or at least to give rise to germ line cells) were considered to be the optimal targets for introducing foreign DNA. The fertilised oocyte is one such cell, being totipotent. Other target cells are cells of very early stage embryos, including *embryonic stem (ES) cells.* Although such cells are postzygotic they represent a stage in development where there has been incomplete separation of the *soma* and the *germline.* Such cells are therefore capable of giving rise to both somatic and germline cells.

When a foreign DNA molecule is artificially introduced into the cells of an animal, a transgenic animal is produced. The foreign DNA molecule is called a transgene and may contain one or many genes. By inserting a transgene into a fertilised oocyte or cells from the early embryo, the resulting transgenic animal may be able to transmit the foreign DNA stably in its germline. Many different types of transgenic animals have been created including transgenic *Drosophila,* transgenic frogs, transgenic fish and a variety of transgenic mammals including mice, rats, and various livestock animals. The technology of transgenesis and its applications. Although transgenes often integrate into the host chromosomes without affecting the expression of any endogenous genes, occasionally the integration event alters endogenous gene expression (insertional mutation), producing a recognisable phenotype. This constitutes a form of *in vivo* mutagenesis, albeit at an *unselected* target gene.

Gene targeting was developed as a method of *in vivo* mutagenesis in which the mutation is introduced into a *preselected* endogenous gene. This can be achieved in somatic cells, but gene targeting in cultured ES cells is particularly powerful because it can lead to the construction of an animal in which all nucleated cells contain a mutation at the desired locus.

In mammals, gene targeting has been possible only in mice but research on ES cells from other species may extend the capacity for gene targeting in the near future. Note that in some cases gene targeting is used to produce a subtle mutation and as a result the ES cells used for blastocyst implantation do not contain any foreign sequences. The resulting mice may be described as genetically modified but not as transgenic.

The ability to produce transgenic mice and particularly the ability to perform specific changes in a predetermined gene by gene targeting has permitted the design of many new animal models of human disease. Another experimental approach involving genetic manipulation of animals has had a major impact recently.

In 1997, a new era in mammalian genetics was heralded when the procedure of somatic cell nuclear transfer permitted "*cloning*" of an adult mammal for the first time. This involved transfer of the nucleus from an adult cell into an enucleated oocyte and the technology has subsequently been applied as an alternative route to generating transgenic animals.

THE CREATION AND APPLICATIONS OF

TRANSGENIC ANIMALS

Of the different transgenic animals that have been made thus far, transgenic *Drosophila*, transgenic frogs, and transgenic fish have been very important for understanding aspects of gene function and development in these species. Transgenic sheep and other transgenic livestock animals have been produced largely to serve as bioreactors, whole-animal expression cloning systems in which introduced genes are expressed to give large amounts of therapeutic or commercially valuable gene products.

But it has been transgenic mice that have been the most useful to biomedical research, both in providing animal models of disease and in permitting the most useful analyses of mammalian gene function.

Transgenic Animals

Transgenesis involves transfer of foreign DNA into totipotent or pluripotent embryo cells (either fertilised oocytes, cells of the very early embryo or cultured ES cells) followed by insertion of the transferred DNA into host chromosomes. If the foreign DNA integrates into the chromosomes of a fertilized oocyte, the developing animal will be *fully transgenic* since all nucleated cells in the animal should contain the transgene. If chromosomal integration occurs later, at a postzygotic stage, the animal will be amosaic, with some cells containing the transgene and some others lacking it.

If the transgene is present in germline cells it can be passed through sperm or egg cells into some of the animal's progeny, and PCR-based tests can be used to quickly screen for the presence of the transgene. Progeny that are transgene positive can be expected to be fully transgenic; all their cells should have developed from a fertilized oocyte containing the transgene.

Pronuclear Microinjection

To obtain transgenic mice by this route, females are superovulated, mated to fertile males and sacrificed the next day. Fertilized oocytes are recovered from excised oviducts.

The DNA of interest is then microinjected using a micromanipulator into the male pronucleus of individual oocytes. Surviving oocytes are reimplanted into the oviducts of foster females and allowed to develop into mature animals.

During this procedure, the microinjected DNA (transgene) randomly integrates into chromosomal DNA, usually at a single site, although rarely two sites of integration are found in a single animal. Individual insertion sites typically contain multiple copies of the transgenes integrated into chromosomal DNA as head-to-tail concatemers (it is not unusual to find 50 or more copies at a single insertion site). As a result of chromosomal integration, the transgenes can be passed on to subsequent generations in mendelian fashion: if the foreign DNA has integrated at the one-cell stage, it should be transmitted to 50 per cent of the offspring.

Transfer Into Pre- or post-implantation Embryos

Cells from very early stage embryos may be totipotent or at least pluripotent and can provide a route for foreign DNA to enter the germline. DNA can be transferred into unselected cells of very early embryos or into cell lines derived from ES cells.

One method that allows foreign DNA to integrate into the chromosomes of the target cells uses retroviruses, RNA viruses which naturally undergo an intermediate DNA form prior to integrating into cellular genomes. Infection of preimplantation mouse embryos with a retrovirus such as Moloney murine leukaemia virus or injection of the retrovirus into early postimplantation mouse embryos results in mosaic offspring. Retroviruses should integrate rarely and at random into accessible cells, and the use of replication- defective retroviruses provides heritable markers for clonal descendants of the target cell (unlike wild-type viruses which spread from cell to cell). This approach has been used, therefore, for studying cell lineage using reporter genes.

In *Drosophila,* efficient chromosomal integration is possible by using sequences from a *Drosophila* transposable element known as the P element to permit insertion of single copies of a gene at random in the genome. The gene or DNA fragment to be inserted is first manipulated so as to be flanked by the two terminal sequences of the P element. The modified DNA is then microinjected into a very young *Drosophila* embryo along with a separate plasmid containing a gene encoding a transposase. In the presence of the transposase the terminal P element sequences allow the intervening DNA fragment to transpose and as a result of the ensuing transposition events, the injected DNA often enters the germline in a single copy.

Cultured Embryonic Stem (ES) Cells Provide a Powerful Route to Genetic Modification of the Germline

The microinjection of foreign DNA into fertilized oocytes is technically difficult and not suited to large-scale production of transgenic animals or to sophisticated genetic manipulation. A popular alternative, but one which has so far been restricted to the construction of genetically modified mice, involves transferring the foreign DNA initially into cultured ES cells. Mouse ES cells are derived from 3.5–4.5 day postcoitum embryos and arise from the *inner cell mass* of the blastocyst. The ES cells can be cultured *in vitro* and retain the potential to contribute extensively to all of the tissues of a mouse, *including the germline*, when injected back into a host blastocyst and reimplanted in a pseudopregnant mouse.

The developing embryo is a chimera: it contains two populations of cells derived from different zygotes, those of the blastocyst and the implanted ES cells. If the two strains of cells are derived from mice with different coat colours, chimeric offspring can easily be identified. Use of genetically modified ES cells results in a partially transgenic mouse. Because the injected ES cells can form all or part of the functional germ cells of the chimera, it is possible to derive fully transgenic mice. This is usually accomplished by screening the offspring of matings between chimeras (usually males) and mice with a coat colour recessive to that of the strain from which the ES cells were derived.

The big advantage of ES cells is that they can be grown readily in culture. This means that a variety of genetic manipulations can be conducted in cultured ES cells. Importantly, *the desired genetic modification can be verified in tissue culture*, before injecting the genetically modified cells into a blastocyst prior to implantation. For example, the desired gene can be ligated to a markergene, such as the *neo* gene, enabling a positive selection for cells that have been successfully transfected. The presence of the desired gene can also be verified quickly by a PCR-based assay. ES cells also offer the huge advantage of *gene targeting* by homologous recombination, a method which can permit a programmed selective alteration of a single *predetermined* gene and also highly specific ways of chromosome engineering. Such approaches are extremely powerful for understanding gene function and also for creating animal models of disease.

The ES cell approach to constructing transgenic mice was made possible by the successful establishment in the early 1980s of stable cell lines from isolated mouse ES cells. ES cells were not so readily identified in other mammals, although there have been some important recent successes.

Transgenic Animals Variety of Studies Investigating Mammalian Gene Expression and Function

Transgenic animals have been extremely important for analysing human genes, and have helped greatly in our understanding of a variety of fundamental biological processes, notably in immunology, neurobiology, cancer and developmental studies. The following list is far from exhaustive but is intended to illustrate some major types of application:

- *Investigating gene expression and its regulation:* Although evidence for *cis*-acting regulatory elements is often inferred initially from studies using cultured cells, they need to be validated in whole animal studies. Transgenes consisting of the presumptive regulatory sequence(s) coupled to a reporter gene, such as *lacZ*, provide a sensitive method of detecting gene expression and a powerful way of investigating regulation of gene expression. Long-range control of gene expression is often investigated using YAC transgenes.
- *Investigating gene function by targeted gene inactivation:* Specific genes can be inactivated by a gene targeting procedure to introduce a transgene into the target gene (insertional inactivation). The effect on the phenotype of creating a null mutation in the gene of interest can provide powerful clues to gene function.
- *Investigating dosage effects and ectopic expression:* In some cases, valuable information can be gained by over-expressing a transgene or by expressing it *ectopically* (the transgene is coupled to a tissue-specific promoter which causes expression in cells; the phenotypic consequence may provide valuable clues to function).
- *Cell lineage ablation:* Transgenes can be designed consisting of a tissue-specific promoter coupled to a sequence encoding a toxin, for example diphtheria toxin subunit A or ricin. When the promoter becomes active at the appropriate stage of tissue differentiation, the toxin is produced and kills the cells. Thus, certain cell lineages in the animal can be eliminated (cell ablation) and the phenotypic consequences monitored.

- *Investigating gain of function:* In principle, any mammalian gene that produces a dominant negative effect or gain of function can be investigated by introduction of an appropriate transgene. In some cases, this can provide proof of a suspected biological function. A classical example concerns the *Sry* gene. A variety of different genetic analyses had implicated this gene as a major male-determining gene but convincing proof was obtained using a transgenic mouse approach. The experiment consisted of transferring a cloned *Sry* gene into a fertilized 46,XX mouse oocyte. As a result of this artificial intervention, the resulting mouse, which nature had intended to be female, turned out to be male.
- *Modelling human disease:* And have helped greatly in our understanding of a variety of fundamental biological processes, notably in immunology, neurobiology, cancer and developmental studies. The following list is Insertional inactivation is often used to model loss of function mutations whilst gain of function mutations can often be modelled by inserting a mutant transgene.

THE USE OF YAC TRANSGENES AND INDUCIBLE PROMOTERS

Inducible Promoters

For many applications it is desirable to have a transgene expressed under the control of a tissue-specific promoter or an inducible promoter. In the former case, genetic engineering can be used to splice known tissue-specific promoters from cloned genes to the gene of interest.

In some cases, coupled regulatory elements can confer both position-independent and tissue-specific expression as with sequence elements from the â-globin locus control region.

Various attempts have also been made to create inducible transgenic mice (*e.g.*, by using heavy metal ions to induce expression of an integrated gene which has a coupled metallothionein promoter, etc.).

Generally, the use of inducible promoters has been hampered by "*leakiness*" in gene expression and by relatively low levels of induction, and they have often been applicable to a limited range of tissues. More recently, however, more promising systems have been developed.

For example, methods employing tetracycline-regulated inducible expression have permitted construction of both highly inducible transfected cells (with much greater efficiency than the constitutive system) and transgenic mice.

In the latter case, the expression of a reporter gene, such as the luciferase gene, can be controlled by altering the concentration of tetracycline in the drinking water of the animals.

YAC Transgenics

Early studies of gene expression and regulation in transgenic animals involved transfer of small genes. However, expression of small transgenes often fails to follow the normal temporal and spatial patterns of expression or match the expression level of the endogenous homologue.

An increasing number of human genes are known to be very large. Even in the case of small genes, important regulatory elements that are required for correct expression may be located many kilobases upstream of the coding sequence. In order to be able to study the expression and regulation of a human gene under the control of its own *cis*-acting regulatory elements, it was therefore necessary to establish transfection conditions which would allow the transfer of large DNA clones.

A major breakthrough in transgenic studies was the development of so-called YAC transgenics. The first report to be published described transfer of a 670 kb YAC containing the human *HPRT* (hypoxanthine phosphoribosyl transferase) gene into mouse ES cells. This was accomplished by spheroplast fusion.

Fragments from the yeast genome can integrate at the same time, however, and so alternative methods have sought to purify an individual YAC by size-fractionation on a preparative gel using pulsed-field gel electrophoresis (assuming that the YAC migrates at a position in the gel that is different from any yeast chromosome).

The purified YAC can be inserted into a fertilized oocyte by pronuclear microinjection. This method is, however, limited to small YACs: the DNA of large YACs is more likely to fragment following microinjection with very fine micropipettes. Alternatively, purified YACs have been transferred into ES cells by using liposomes, artificial lipid vesicles that are used to transport molecules into a cell following fusion of the lipid coat with the plasma membrane of the recipient cell. YAC transgenics have permitted study of large genes such as the 400 kb human *APP* gene, a gene known to contribute to Alzheimer's disease, encoding the amyloid protein precursor. A YAC transgene containing *APP* showed tissue- and cell type-specific expression patterns closely mirroring that of the endogenous mouse gene.

Long-range gene regulation mechanisms (locus control regions, imprinting, and other chromatin domain effects) can be modelled. An interesting application has been in the production of fully human antibodies in the mouse by transfer of human YACs containing large segments of the human heavy and kappa light chain immunoglobulin loci into mouse ES cells, and thence the creation of transgenic mice able to produce human antibodies.

Transchromosomic Animals

Ultimately, even YACs have upper limits for the size of foreign inserts that can be transferred. Mammalian artificial chromosomes have also been generated, including first generation human artificial chromosomes. Such systems will have the capacity of transferring hundreds and possibly thousands of genes into transgenic animals, although a preferred route may be by using nuclear transfer technology rather than ES cells.

Recently, however, transfer of whole chromosomes or chromosome fragments into ES cells has been possible by microcell-mediated chromosome transfer. Using this approach Tomizuka *et al.* (1997) were able to transfer human chromosomes or chromosome fragments derived from normal fibroblasts into mouse ES cells. The resulting chimeric transchromosomic mice were viable, and the chromosome fragments appeared to show functional expression and could be transmitted through the germline.

USE OF MOUSE ES CELLS IN GENE TARGETING AND GENE TRAPPING

Gene Targeting by Homologous Recombination

Gene targeting involves engineering a mutation in a *preselected* gene within an intact cell. It can therefore be viewed as a form of artificial site-directed *in vivo* mutagenesis. The mutation may result in inactivation of gene expression (a "*knock-out*" mutation), or altered gene expression, and so can be useful for studying gene function. In addition, the same method can be used to "*correct*" a pathogenic mutation by restoring the normal phenotype, and so has therapeutic potential.

Gene targeting typically involves introducing a mutation by homologous recombination. A cloned gene (or gene segment) closely related in sequence to endogenous target gene is transfected into the appropriate cells. In some of the cells, homologous recombination occurs between the introduced gene and its chromosomal homologue. Gene targeting by homologous recombination has been achieved in some somatic mammalian cells, such as myoblasts. However, the most important application involves mouse ES cells: once a mutation has been engineered into a specific mouse gene within the ES cells, the modified ES cells can then be injected into the blastocyst of a foster mother and eventually a mouse can be produced with the mutation in the desired gene in all nucleated cells.

Homologous recombination in mammalian cells is a very rare occurrence (unlike in yeast cells, for example, where it occurs naturally at high frequencies, enabling sophisticated genetic manipulation). The frequency of homologous recombination is increased, however, when the degree of sequence homology between the introduced DNA and the target gene is very high. As a result, the introduced DNA clone is a *mouse* sequence which should preferably be isogenic (derived from the same mouse strain as the strain of mouse from which the ES cells were derived). Even then, the frequency of genuine homologous recombination events is very low and may be difficult to identify against a sizeable background of random integration events. To assist identification of the desired homologous recombination events, the targeting vector (often a plasmid vector) contains a marker gene, such as the *Neo*- gene, which permits selection for cells that have taken up the introduced DNA. PCR assays are used to screen for evidence of a homologous recombination event (by using a marker-specific primer plus a primer derived from a sequence present in the target gene but absent from the introduced homologous gene segment).

The targeting construct is transferred into cultured mouse ES cells by electroporation, a method in which pulses of high voltage are delivered to cells, causing temporary relaxation of the selective permeability properties of the plasma membranes. Two basic approaches have been used:

- *Insertion vectors:* Target the locus of interest by a single reciprocal recombination, causing insertion of the entire introduced DNA including the vector sequence. This is the most reliable way of causing a knock-out mutation.
- *Replacement vectors:* Are designed to replace some of the sequence in the chromosomal gene by a homologous sequence from the introduced DNA. This can occur as a result of a double reciprocal recombination or by gene conversion. The replacement method can inactivate a gene when the introduced sequence contains one or more premature termination codons or lacks critical coding sequences. It can also be used to correct a pathogenic mutation.

The replacement vector approach, as well as the insertion vector method, often leaves foreign sequences at the target locus. In some cases, however, a more subtle mutation is required. For example, it may be desirable to investigate the effect of changing a single codon. Various two-step recombination techniques can be used to accomplish this method, and the resulting mouse, although genetically modified lacks any foreign sequences and so can no longer be described as transgenic.

Gene targeting in mice is popularly used for producing artificial mouse models of human disease. In addition, it provides a powerful general method of studying gene function. The gene in question is selectively inactivated, producing a "***knock-out***" mouse, and the effect of the mutation on the development of the mouse is monitored carefully.

Sometimes there is little or no phenotypic consequence after inactivating a gene that would be expected to be crucially important, such as some genes which encode a transcription factor known to be expressed in early embryonic development. The lack of a phenotype in such cases is often thought to be due to genetic redundancy (another gene is able to carry out the function of the gene that has been knocked out). As a result, in some cases double or even triple gene knock-outs have been carried out to analyse gene function, as in the case of some of the *Hox* genes.

Tissue- and Cell Type-specific Knock-outs

Some genes are vital to early development and simple knock-out experiments are generally not helpful because death ensues at the early embryonic stage. To overcome this problem, methods have been developed to inactivate expression of the target gene in only selected, predetermined cells of the animal (conditional knock-out). The animal can therefore survive and the effect of the knock-out can be studied in a tissue or cell type of interest.

An early example of a conditional knock-out sought to study the role of DNA polymerase â (an enzyme that is essential for embryonic development) in T lymphocytes. The gene targeting procedure replaced part of the endogenous gene by an introduced homologous gene segment flanked by *lox*P sequences. Mice carrying this targeted mutation were then mated with a strain of mice which carried a *Cre* transgene gene under the control of a T cell-specific promoter. Offspring with the *lox*P-flanked pol â sequences plus the *Cre* transgene were identified and survived to adulthood. The Cre product was expressed only in T cells, leading to inactivation of the target gene in these cells by excision of DNA polymerase â gene segment between the two *lox*P sequences.

Tissue- and Cell Type-specific Gene Activation

This approach is the opposite to that described above: it involves selective *activation* of a gene in certain cells of the animal to switch on a foreign gene only in predetermined cells of the animal.

Chromosome Engineering

Another important recent development is a strategy for chromosome engineering in ES cells which relies on sequential gene targeting and Cre-*lox*P recombination. Gene targeting is used to integrate *lox*P sites at the desired chromosomal locations and, subsequently, transient expression of Cre recombinase is used to mediate a selected chromosomal rearrangement. Chromosome engineering strategies of this type offer the exciting possibility of creating novel mouse lines with specific chromosomal abnormalities for genetic studies.

The multiple targeting and selection steps in ES cells used in the above chromosome engineering methods can be avoided using the novel approach of Herault *et al.* (1998). This targeted meiotic recombination method takes advantage of the homologous chromosome pairing that occurs naturally during meiosis at the first cell division.

A transgene is designed to express Cre recombinase under the control of a *Sycp1* promoter (the *Sycp1* gene encodes the SCYP1 protein which is part of the synaptonemal complex that facilitates crossing over). As a result Cre recombinase is produced in male spermatocytes during the zygotene to pachytene stages when chromosome pairing occurs.

Gene Trapping in Mouse ES Cells Allows an Efficient Approach to Functional Analysis

Large-scale approaches to investigating animal gene function have principally relied on exposing animals to high doses of radiation, or to potent chemical mutagens such as *ethylnitrosurea* (*ENU*) or *ethyl methylsulfonate* (*EMS*). In mouse mutagenesis programmes, for example, males are typically exposed to high levels of a suitable mutagen to induce a high frequency of mutation in sperm DNA. The progeny of irradiated mice are then screened for obvious phenotypic abnormalities. Such mutagenesis screens have been very useful in producing novel mutants, but a major problem is that the mutations occur essentially at random. Identification of the structural change(s) in the DNA of a single mutant animal may therefore require a laborious positional cloning approach.

A mutation that is induced by inserting a known foreign DNA sequence (transgene) into the mouse genome has a major advantage over one induced by chemical mutagens or X rays: *it leaves a sequence tag at the locus which is mutated*. As a result, rapid molecular characterisation of the mutated locus is possible. Mouse ES cells provide a way of introducing such mutations into the germ line, and so random insertional mutagenesis in ES cells using transgenes was considered a useful way of producing a large number of mouse mutants which could quickly be characterised at the molecular level. Unlike gene targeting, the transgene should not be related to endogenous sequences, but should integrate essentially at random by non-homologous recombination. However, because large sections of the mouse genome are non-coding, many random insertion events may not result in gene disruption.

In order to improve the efficiency of recovering mutations that are likely to have a phenotypic effect (by altering a gene or its expression) the gene trap approach was devised. The underlying principle is that the transgene which is inserted into ES cells contains a defective reporter gene or marker gene *which lacks some component needed for gene expression*. The reporter or marker gene is designed to be expressed only after it inserts into a gene (within an intron or exon) or at a promoter. When it integrates at such positions it can acquire the expression element that it lacks.

Different gene trap strategies are possible. In some cases, the reporter lacks a functional promoter and so relies on chance integration next to an appropriate *cis*-acting sequence element that can activate its transcription. Other approaches have used a marker gene coupled to a suitable promoter but lacking a downstream polyadenylation signal. Here, the marker gene is designed to be expressed after integrating into a host cell gene such that a fusion RNA product is made that utilises 32 host sequences in order to acquire a poly (A) tail. Already, this type of approach has been applied on a large scale, with a recent report describing disruption and sequence identification of 2000 genes in mouse ES cells.

REQUIREMENTS FOR ANIMAL CELL CULTURE

Among the essential requirements for animal cell culture are special incubators to maintain the levels of oxygen, carbon dioxide, temperature, humidity as present in the animal's body. The synthetic media with vitamins, amino acids and fetal calf serum.

Following parameters are essential for successful animal cell culture:

- *Temperature*: In most of the mammalian cell cultures, the temperature is maintained at 37°C in the incubators as the body temperature of Homo sapiens is 37°C.
- *Culture media*: The culture media is prepared in such a way that it provides-
  - The optimum conditions of factors like pH, osmotic pressure, etc.
  - It should contain chemical constituents which the cells or tissues are incapable of synthesizing. Generally the media is the mixture of inorganic salts and other nutrients capable of sustaining cells in culture such as amino acids, fatty acids, sugars, ions, trace elements, vitamins, cofactors, and ions. Glucose is added as energy source-it's concentration varying depending on the requirement. Phenol Red is added as a pH indicator of the medium.There are two types of media used for culture of animal cells and tissues- the natural media and the synthesised media.
  - *Natural Media*: The natural media are the natural sources of nutrient sufficient for growth and proliferation of animal cells and tissues. The Natural Media used to promote cell growth fall in three categories.

  i. Coagulant, such as plasma clots. It is now commercially available in the form of liquid plasma kept in silicon ampoules or lyophilised plasma. Plasma can also be prepared in the laboratory taking out blood from male fowl and adding heparin to prevent blood coagulation.

  ii. Biological fluids such as serum. Serum is one of the very important components of animal cell culture which is the source of various amino acids, hormones, lipids, vitamins, polyamines, and salts containing ions such as calcium, ferrous, ferric, potassium etc. It also contains the growth factors which promotes cell proliferation, cell attachment and adhesion factors. Serum is obtained from human adult blood, placental, cord blood, horse blood, calf blood. The other forms of biological fluids used are coconut water, amniotic fluid, pleural fluid, insect haemolymph serum, culture filtrate, aqueous humour, from eyes etc.

  iii. Tissue extracts for example Embryo extracts- Extracts from tissues such as embryo, liver, spleen, leukocytes, tumour, bone marrow etc are also used for culture of animal cells.

Synthetic Media

Syntheic media are prepared artificially by adding several organic and inorganic nutrients, vitamins, salts, serum proteins, carbohydrates, cofactors etc. Different types of synthetic media can be prepared for a variety of cells and tissues to be cultured. Synthetic media are of two types- Serum containing media (media containing serum) and serum- free media (media with out serum). Examples of some media are: minimal essential medium (MEM), RPMI 1640 medium, CMRL 1066, F12 etc.

Advantages of serum in culture medium are:

- Serum binds and neutralises toxins,
- Serum contains a complete set of essential growth factors, hormones, attachment and spreading factors, binding and transport proteins,
- It contains the protease inhibitors,
- It increases the buffering capacity,
- It provides trace elements.

Disadvantages of serum in culture medium are:

- It is not chemically defined and therefore it's composition varies a lot,
- It is sometimes source of contamination by viruses, mycoplasma, prions etc,
- It increases the difficulties and cost of down stream processing,
- It is the most expensive component of the culture medium.
  - pH- Most media maintain the pH between 7 and 7.4. A pH below 6.8 inhibits cell growth. The optimum pH is essential to maintain the proper ion balance, optimal functioning of cellular enzymes and binding of hormones and growth factors to cell surface receptors in the cell cultures. The regulation of pH is done using a variety of buffering systems. Most media use a bicarbonate-CO_2 system as its major component.
  - *Osmolality*: A change in osmolality can affect cell growth and function. Salt, Glucose and Amino acids in the growth media determine the osmolality of the medium. All commercial media are formulated in such a way that their final osmolality is around 300 mOsm.

CELL CULTURE ENGINEERING

Animal cells are used largely in the production of proteins such as antibodies and other pharmaceutical products. With increasing demand for these products, there is likewise a parallel effort to find ways in which to inrease the culture efficiency and productivity. These efforts take form in many ways. Among these are attempts to develop better reactor control algorithms and the formulation of mathematical models that will sufficiently describe the cell culture process. These are hoped to give a better understanding of the nature and limitations of animal cell culture and will, in the process, enhance production.

Mammalian cells, under typical conditions, convert most of the glucose consumed into lactic acid, that has adverse effects on cell growth. The accumulation of lactate limits the maximum cell concentration achieved in cultures. Fed-batch cultures have been used to cultivate hybridoma cells where glucose was controlled and maintained at low levels.

Dynamic nutrient feeding based on on-line measurement of oxygen uptake rate (OUR) was employed to control the feeding rate by estimating the metabolic demand of the cells. This strategy led to a significant reduction in lactic acid production. The altered metabolic state of hybridoma cells could also be maintained in continuous cultures to give a high steady state cell concentration of 5×10^6 cells/ml. This concentration was achieved by altering the cellular metabolism in a fed-batch prior to switching to continuous mode. The strategy was used to achieve various distinct steady states with the molar ratio of lactate produced to glucose consumed (L/G) ranging from 0.03 to 0.5. In comparison, a control initiated from a batch culture had a L/G of 1.4 and yielded a cell concentration of only 1.5e6 cells/ml.

The different steady states were analysed to demonstrate the differences in glucose and amino acid metabolism under different metabolic states. The specific uptake rates of glucose and amino acids were found to reduce by an order of magnitude with the reduction in L/G. For some amino acids, the specific rates were lower even at higher extracellular concentrations, indicating a possible down-regulation of the amino acid transporters. A few amino acid transporter activities were studied using radiolabeled substrates to analyse their characteristics under differing metabolism. Intracellular fluxes were also calculated using Metabolic Flux Analysis and were found to vary over the different metabolic states.

Specifically, the fluxes around pyruvate were seen to redistribute indicating potential enzymatic regulation. To probe the energy metabolism, the activity of mitochondria was observed using a vital stain Rhodamine 123. A distinct difference was observed between the mitochondrial activity and distribution in cells of different metabolism. The specific rate of antibody production was found to be the same for all the metabolic states studied. These results demonstrate the potential of controlling cell metabolism, to reduce the metabolite production, in order to achieve high cell and product concentrations in mammalian cell cultures. The altered metabolic state could be maintained in continuous cultures, while increasing the feed nutrient concentration, to achieve 1.0e7 cells/ml.

To grow animal cells in culture, scientists once used a wide variety of animal-based components to create a conducive environment, but that is changing. "In the past 10 years," says Chuck Velnoskey, market development manager at BD Biosciences, Sparks, Md., "we've been moving away from animal-origin to animal-free. The next step is completely chemically defined cell-culture products." He adds, "This increases the consistency from lot to lot, which provides consistency to the end user." Despite the seeming simplicity of such a change, it creates some challenges. For one thing, the chemically defined products must produce the yields that researchers have come to expect from previous products. Second, the new products must be cost-effective.

Many researchers add supplements to cell-culture media to improve cell viability, growth, and productivity. For example, Elizabeth C. Dodson, PhD—manager, research and development at BD Biosciences says, "Some people use peptones for supplementation, and these can be derived from animal-origin sources." To provide a completely chemically defined option, BD developed its Recharge Media Supplement. "It's a new way to think of supplementation," Dodson says. "Here's a supplement that can be used in place of say, yeast extract, and it's completely chemically defined."

In fact, Velnoskey points out that Recharge is "ideally suited for mammalian systems, and it is recommended for CHO lines that work with yeast-based supplements." He adds, "It contains the bioactive and nutritional constituents found in yeast peptones, so that you maximize cellular activation and protein production."

Using such a precisely defined supplement allows for predictable yields in culturing cells. As Dodson says, "Recharge is designed so that mammalian cells yield protein amounts that are similar to what is produced with peptones and the quality is similar." She adds, "We have in-house and customer data that support this up to bioreactor scale."

Beyond efficient yields, Recharge is also easy to use. "Just drop it into the cell-culture medium in place of the other supplementation you might be using," Dodson says.

REGULATION OF CELLULAR STEROL CONTENT

The continual alteration of the intracellular sterol content occurs through the regulation of key sterol synthetic enzymes as well as by altering the levels of cell-surface LDL receptors. As cells need more sterol they will induce their synthesis and uptake, conversely when the need declines synthesis and uptake are decreased. Regulation of these events is brought about primarily by sterol-regulated transcription of key rate limiting enzymes and by the regulated degradation of HMGR. Activation of transcriptional control occurs through the regulated cleavage of the membrane-bound transcription factor sterol regulated element binding protein, SREBP. As discussed above, degradation of HMGR is controlled by the ubiquitin-mediated pathway for proteolysis.

Sterol control of transcription affects more than 30 genes involved in the biosynthesis of cholesterol, triacylglycerols, phospholipids and fatty acids. Transcriptional control requires the presence of an octamer sequence in the gene termed the sterol regulatory element, SRE-1. It has been shown that SREBP is the transcription factor that binds to SRE-1 elements. It turns out that there are 2 distinct SREBP gene, SREBP-1 and SREBP-2. In addition, the SREBP-1 gene encodes 2 proteins, SREBP-1a and SREBP-1c as a consequence of alternative exon usage. All 3 proteins are proteolytically regulated by sterols. Full-length SREBPs have several domains and are embedded in the membrane of the endoplasmic reticulum (ER). The N-terminal domain contains a transcription factor motif of the basic helix-loop-helix (bHLH) type that is exposed to the cytoplasmic side of the ER. There are 2 transmembrane spanning domains followed by a large C-terminal domain also exposed to the cytosolic side. When sterols are scarce, cleavage of the full-length SREBP takes place with the result being that the N-terminal bHLH motif is released into the cytosol. The bHLH domain then migrates to the nucleus to direct transcription. Conversely, when sterols are abundant, cleavage of SREBP is inhibited. To control the level of SREBP-mediated transcription, the soluble bHLH domain is itself subject to rapid proteolysis.

The cleavage of SREBP is carried out by 2 distinct enzymes, one of which is regulated by sterols. The regulated cleavage occurs in the lumenal loop between the 2 transmembrane domains. This cleavage is catalyzed by site-1 protease, S1P. High sterol content blocks the activity of S1P. The second cleavage, catalyzed by site-2 protease, S2P, occurs in the first transmembrane span, leading to release of active SREBP. In order for S2P to act on SREBP, site-1 must already have been cleaved.

Additional studies on sterol-regulated gene expression demonstrated that cleavage of SREBP by S1P is controlled by the level and action of an additional protein termed, SREBP cleavage-activating protein, SCAP. SCAP is a large protein also found in the ER membrane and contains at least 8 transmembrane spans. The C-terminal portion, which extends into the cytosol, has been shown to interact with the C-terminal domain of SREBP. This C-terminal region of SCAP contains 4 motifs called WD40 repeats. The WD40 repeats are required for interaction of SCAP with SREBP. Interestingly, the N-terminus of SCAP, including membrane spans 2-6, resembles HMGR which itself is subject to sterol-stimulated degradation.

This shared motif is called the sterol sensing domain, SSD. Several proteins whose functions involve sterols also contain the SSD. These include patched, an important development regulating receptor whose ligand, hedgehog, is modified by attachment of cholesterol and the Neimann Pick C1 (NPC1) protein which is involved in cholesterol transport in the secretory pathway. NPC1 is one of several genes whose activities, when disrupted, lead to severe neurological dysfunction.

The function of SCAP is to positively stimulate S1P-mediated cleavage of SREBP. The function of sterols is to inhibit this positive action of SCAP. The activity of SCAP involves movement from the ER to the Golgi and back. Because the C-terminus of SCAP interacts with SREBP, movement of SCAP takes SREBP along for the ride. When sterols are low, SCAP and SREBP move to the Golgi. This transit is required for SREBP cleavage as S1P is Golgi-localized. When sterols are high, movement of SCAP is halted. Thus, the overall effect of sterols is to regulate the ability of SCAP to present SREBP to S1P.

Glutathione

Reduced Glutathione, a tripeptide of Glutamic Acid, Cysteine, and Glycine, which serves as a reducing agent in many biochemical reactions being converted to oxidized Glutathione (GSSG) in which the Cysteine residues of two Glutathione molecules are connected by a disulfide bridge. Reduced Glutathione is important in protecting erythrocytes (red blood cells) from oxidation and hemolysis; deficiency causes sensitivity to oxidant drugs.

- *It is a tripeptide containing amino acid that contains*: Cysteine, Glycine, and Glutamate.
- Inhibits peroxide formations, reducing free radical damage.
- Detoxifies aromatic hydrocarbons common in air pollution, such as chlorine. Reported to have protective function against radiation therapy.
- Transports other amino acids into the interior of the cell.

Glycine

A non-essential amino acid, occurring as a constituent of proteins and functioning as an inhibitory neurotransmitter in the central nervous system; used as a gastric antacid and dietary supplement, and in the treatment of various myopathies.

- Is the simplest and sweetest of the amino acids. Can be used as a sweetener in herbal beverages.
- Reported useful in degenerative diseases such as MD (Muscular Dystrophy)
- Detoxifies benzoic acid (a common food additive) and aromatic hydrocarbons in the liver.
- Involved in synthesis of nucleic acids and bio-acids. May be useful in conditions characterized by abnormal nerve firing such as epilepsy, inhibition of tripeptides Glutathione and GTF.
- Is one of the amino acids in the tripeptides Glutathione and GTF; low brain concentrations of Glycine have recently been found in ALS. (Amyotrophic Lateral Sclerosis or "Lue Gehrig's Disease.")

Histidine

Is not listed as essential for adults, but is very essential for infants. An amino acid obtainable from many proteins by the action of sulfuric acid and water; it is ESSENTIAL for optimal growth in INFANTS. Its decarboxylation results in formation of histamine.

- Release of histamines from body stores are required for sexual arousal.

- Reported useful in alleviating pain associated with rheumatoid arthritis. (Should be taken with vitamin C; Should be used with caution with Manic Depressives with elevated histamines; Should be used with caution in woman with severe depression or suicidal tendencies due to Premenstrual Syndrome; Is considered a mild neurotransmitter)

PATENTING OF GENETIC INNOVATIONS IN ANIMAL BREEDING

While herdsmen have practiced the art of animal breeding since the beginning of domestication of livestock, the science of animal breeding is just now entering its second century. From its roots in Mendelian inheritance to the developments in quantitative genetics our field has now grown to include modern molecular sciences. The practical applications of the field have been to genetically improve the animals that produce meat, milk and other products of economic value.

Animal breeding as commerce has always "protected" its endeavors commercially by developing breed societies and using pedigrees to protect the intellectual property developed by the master breeders. The advent of molecular biology, sequencing of genomes and the development of cloning, have brought with them large additional investments. These developments have come with the expectation of increased rewards by investing companies and the reality that technical improvements must be protected in order to obtain sufficient return. The purpose of this paper is to discuss forms of intellectual property, in particular patents and patent law, and to suggest ways it might impact the field of animal breeding and genetics.

FORMS OF INTELLECTUAL PROPERTY

Intellectual property (IP) is the property resulting from the creative or "inventive" work of an individual or individuals. The protection of IP is described in a large body of law that includes copyrights, trademarks, trade secrets, and patents. Copyrights protect IP such as articles, books, web pages, computer software and music. Copyright protection applies when the work is in a publishable form. The ã symbol is not required, but recommended and enforcement of rights is made easier by federally registering the mark. In trademark law, the Ô claims common law trademark rights, while the ® indicates federal registration. Trademarks are used primarily to identify the source of the owners' goods, products or services. Examples include "Le Label Rouge", "American Berkshire Gold" and the gene marker *HAL*-1843 ä that is associated with the gene test for porcine stress syndrome. A trade secret represents IP that is not divulged by the owner and confers some competitive advantage. Trade secrets can last as long as the information is kept secret. An example of a trade secret might be specific information to make specialized synthetic lines or calculate breeding values. Trade secret law (in the US called the Uniform Trade Secret Acts) helps protect a technology from theft but the degree of protection is dependent on the type of technology and the means used to keep the information secret.

Patents represent the largest form of IP and their development is rooted in a large body of law both in the US and abroad. Such laws and their interpretation are subject to change, especially when new technology is introduced. The descriptions that follow are meant to be general in nature. Patents were developed to "promote the progress of science and useful arts, by securing for limited times, for authors and inventors, the exclusive rights to their respective writings and discoveries" (Article 1, Section 8, US Constitution) in return for making details of inventions available to the public. Patenting was developed in order to support the development of science and innovation.

A patent, once granted, represents a legal monopoly granted by the respective country's government to an inventor (specific to that country), permitting the patent owner to prohibit anyone else from making, using, or selling this invention for a specific period of time (now generally 20 years). Infringement of a patent is a civil wrong and the owner may sue for economic damages or injunction to stop infringement. Patents are intended to protect but not withhold technical information. Patents are very different from published papers. Inventors, unlike coauthors, must take part in the inventive process, not just participate in the research. Patents are not peer-reviewed but instead are assessed by patent examiners to determine the application's patentability.

A patent application must disclose sufficient detail to enable "one of reasonable skill in the art" to duplicate the invention. The application must be unobvious, *i.e.*, not a simple, obvious extension of existing technology. Novelty is the next criteria and refers to something created that is new. Lastly, the invention must be useful.

The scope of patent rights is determined by the claims. Some patents, like gene marker patents, may enjoy very wide coverage across species while others may be limited to single polymorphisms in one breed. Patents may be related to a process, a product produced by a process or dependent on another patent. Patents are open to interpretation by both examiners and lawyers. In the US, patents can be applied for up to one year after publication or public disclosure but this is not allowed in other countries. Patents in and of themselves do not ensure income.

They must be promoted and protected and can be viewed as tradable assets and licensed or assigned to third parties. Confusion often exists relative to "international" patents (see http://www.cambiaip.org), which actually do not exist. Under the Paris Convention Treaty, a group of countries agreed to work with each other to provide a unified examination process and to make filing in its member countries easier by forming the World Intellectual Property Office (WIPO). In unison they adopted the one-year period in which to file an application in one of the other countries without losing the benefit of the original filing date. Using this procedure avoids the citation against the applicant of any "art" that became known after the original filing date in the country of origin, but before the filing date in another country. Under this Patent Cooperation Treaty (PCT), an applicant is allowed one year to file at the WIPO office and, by designating member countries, holds the legal rights and original filing date in those designated countries without having to file in each country or pay the national expense up front. Although there are fees associated with the PCT application, they are less than one would pay for the foreign national filings at a later date. To obtain a patent in these countries, the application does eventually need to be filed in the national patent offices, fees paid, translations done and the laws and regulations of each country office followed. The applicant may have additional time from the national filing date to request examination, depending on particular country requirements. Delaying examination may provide the time desired to continue development and/or commercialization, however, during this period, fees called "annuities" must generally be paid. Due to the expense involved, applications are generally filed only in countries where protection is really needed.

A BRIEF HISTORY OF PATENTS

The early protection of intellectual property rights can be traced back to Venice in the Middle Ages when master craftsmen in guilds prohibited competition from former apprentices for a period of 20 years (Waltersheid, 1994). Such laws had considerably different economic effects for the master craftsman, the apprentice and the general public. For livestock, early breed societies were developed and monitored pedigrees to protect the IP of the master breeders.

The first modern patent act is often thought of as The US Patent Act of 1790. There was similar legislation in France in 1791. Patents related to living matter are relatively new. One of the earliest patents for living matter was granted to L. Pasteur for a yeast strain but this was done under the belief that it was an inanimate object and not living (Lesser, 2002). The first specialized patent law applied to living organisms was that of the Plant Protection Act of 1930 in the US and provided what is commonly referred to as Plant Breeders Rights (PBR) to propagate new varieties by asexual methods. In 1961, a similar law was passed in France called the UPOV (International Union for the Protection of New Varieties of Plants - Union Internationale pour la Protection des Obtentions Végétales; http://www.upov.int). Protection in the US was expanded in 1970 with the Plant Variety Protection Act to include sexually reproduced plants. The UPOV was revised in Europe in 1991. Under these laws two principles, "breeder's rights" which allows breeders to use protected varieties without permission of the owner and "farmer's privilege" which allows farmers to collect seeds from their crops and use them, were developed (Lesser, 2002). For years many seed companies have attempted to halt this latter practice by asking farmers to sign contracts prohibiting it. Recent technology, like the "terminator" technology (Kaiser, 2000), biologically prohibits the practice. Santaniello *et al.* (2000) provides an excellent review on all aspects of patent protection for plants.

The dawning and blossoming of the molecular age presented real problems for the protection of IP related to living organisms other than plants. In 1975, a French company failed to patent a "dwarf, egg-laying chicken hen produced by a process" involving a sex-linked recessive gene (Bent, 1989). In 1980, the United States Supreme Court in a 5-4 decision *(Diamond v. Chakrabarty)* declared that "anything under the sun that is made by man" is patentable. This case concerned the patenting of genetically engineered bacteria that destroyed oil sludge. In 1987, the US Patent trade office issued a pronouncement of the patentability, in principle, of non-human multicellular organisms that were not naturally occurring (Bent, 1989). This was quickly followed in 1988 by the landmark patent on the so-called "Harvard mouse" which was engineered to be susceptible to cancer. In Europe, similar laws were passed allowing patenting of animals and animal-related inventions. From 1995 to 2001, a total of 45 animal patents were granted in the US (Lesser, 2002). While these changes in patent law had large economic implications, the social and moral ramifications were also enormous.

ANIMAL BREEDING AND GENETICS – IP PROTECTION

The broad area of biotechnology encompasses many of the patent applications in our field and the US Supreme Court has established guidelines that apply to this technology (Nebel *et al.,* 2002). The Court made it clear in *Brenner v. Manson* that patent utility implied usefulness and not just "any invention not positively harmful to society." The Court expressed reservations regarding a monopoly on compounds with unknown functions, and that utility must extend beyond proving that the product is a result of scientific research (Nebel *et al.,* 2002).

Technologies in the field of animal breeding that may require (or qualify for) IP protection include, but are not limited to: statistical methods for genetic improvement, DNA markers for genetic improvement, transgenic and cloned animals and methods to produce them, new methods to measure traits, methods to identify animals, computer software and other written materials.

Patent coverage in not just confined to genetic markers. Lines of pigs or chickens have been patented and can be viewed as a specialized extension of early breed development or trademarking for protection of this IP. Other patents exist for methods involving cellular and animal manipulation and involve processes like stem cell development, transgenic production (*i.e.*, US 6,271,436) and cloning (*i.e.*, US 6,215,041 or US 6,258,998). Several advances related to mechanical or electronic devices have been made and include new A.I. or embryo transfer tools, advanced ultra-sound equipment, formulas and methods to measure backfat and other traits in livestock (*i.e.*, see early patent US 4,359,055 and more recent US 5,717,142). The increasing need for traceability of animals and animal products has spawned a number of inventions including electronic ID tags and retinal scanning methods and devices.

Considerable discussion has ensued recently from a patent entitled "Method of Bovine Herd Management" granted in the U.S. in 1994 and later in Canada (Schaeffer, 2002). The invention is for the "test-day model" and includes the gathering, mathematical treatment, and the use of the modified data by dairy producers. The novelty and unobviousness of the patent has been seriously questioned. It was pointed out that the practices of gathering, manipulating and using data by dairy producers have existed for nearly 100 years. The patent therefore claims rights to a practice that has been public knowledge for a long time. The novel idea within the patent was the specific mathematical model and procedures that Everett and co-workers developed for the analysis of test day yields.

Everett was also not the first researcher to apply a model to test day records and as has been demonstrated, the model as described in the patent, is not necessarily the best model that could be applied (Schaeffer, 2002). This patent generated considerable discussion. It has been argued "what would the field of animal breeding be if the selection index or Henderson's BLUP had been patented?" Yet while quantitative geneticists see the thought of such protection as sacrilege, molecular scientists accept (but may not like) that in a similar way the foundation patent for PCR exists and royalties must be paid for its use.

OTHER LEGAL CONSIDERATIONS

In the US and some other countries, there is no clear exception to infringement for research use of patented inventions, even in non-profit or educational institutions. Researchers are required to obtain a license or permission from the patent owner to use a protected invention for research purposes. In other countries, there is either an exemption for educational or non-commercial research or research on improving the invention. Research using the invention, however, is not allowed. Researchers should also be concerned with possible claims for "contributory infringement," *i.e.*, assisting someone else to infringe on a patent. For example, due to a claim of possible contributory infringement, the US Genome Coordinators were asked to halt the practice of supplying primers for microsatellite analysis to other researchers. To resolve the issue the Coordinators began buying the primers from a company with a research license.

ETHICAL, SOCIAL AND ECONOMIC ISSUES

Many ethical and social issues have been raised related to patenting of animals and genes (Bent, 1989; Brody, 1989; Dresser, 1988; Evans, 2002). These include: 1) patenting of animals or genes will be destructive to nature and allows man to play "God"; 2) patenting will devalue animal life and hence human life; 3) patenting will increase animal suffering; 4) patenting will lead to a decline in genetic diversity of animals and threaten species; 5) patenting speeds the trend towards commercialization of academic research and 6) patenting will undermine conventional farming and lead to increased industrial farming systems. Early humans domesticated animals and master breeders and geneticists have transformed them into productive species. Was this playing "God" or interfering with nature? The use of transgenics for making specialized animal lines for disease research is certainly adapting nature but does it devalue life and is it unethical or immoral? These are value judgements that most in society have decided are acceptable (Brody, 1989). Certainly some lines have been (and need to be) drawn to delineate what is acceptable and unacceptable. For example, most people and governments have concluded that while cloning of animals is acceptable (at least for research purposes), cloning of humans, for whatever reason, is immoral, unethical and to be avoided.

Animal welfare and animal rights issues continue to be at the forefront of livestock production and biomedical research (Evans, 2002). Individuals who believe that animals have "rights" will likely be opposed to patenting any invention derived from animal research. The most commonly cited examples are those relating to transgenic animals (*i.e.*, early transgenic pigs) in which some animals had health problems. Production and patenting of specialized lines of rodents for biomedical research that have a tendency to develop specific diseases is also considered unethical by animal rights activists (Brody, 1989). If, however, individuals believe that animal rights are subordinate to those of humans, but that they deserve proper care and welfare then the issue of patenting is much less of a concern.

It has been suggested that through the use of gene markers and highly selective breeding, or through the use of transgenics and cloning, that genetic diversity will be greatly reduced. Certainly these methods have the potential to remove within-line variation but they are likely to increase between-line variability. However, it may be argued that the patent system in fact encourages diversity as it promotes and helps establish, via patent-related deposits of biological materials, a broader genetic diversity (Bent, 1989). Related to this issue is the issue of patenting of products from animals or plants from developing countries. Should companies be allowed to sample and alter wild stocks and to subsequently earn great sums of money by then selling them to both developed countries and back to the countries from where they were obtained? Some individuals say this so-called biopiracy is encouraged by patenting and some countries are moving to address these concerns (Evans, 2002). An extension of this is the issue of species integrity. The development of transgenic pigs for xenotransplantation and other species with human genes for human protein production threatens the genetic lines that determine species. While consideration of xenotransplantation has slowed due to fears of retroviruses and diseases like mad cow disease and AIDS, as Evans (2002) points out this dispute over species integrity "is not empirical" but it is tied up in the individual and societal views on nature and "the notion of awe and wonder." It is likely, no matter how repugnant this process is to some, there will be development of transgenic lines of animals for biomedical research (not food production) and applications that do encompass genes from other species.

In a perfect world, public support of research would be 100 per cent of required funding and all IP would be publicly available in the country of origin. Many US public institutions are now funded at about 40 per cent of total budgets and pressures to obtain outside funding are growing. Companies supplying funding expect to "own", through licensing agreements or otherwise, the IP that results. Protection of IP can provide research partners a greater basis for trust and exchange of ideas and help insure that public institutions focus research on areas of high relevance and payoff. Commercial relationships also aid in technology transfer, employment opportunities for students and may allow for research opportunities not available in the public sector. Fretz and MacKenzie (2002) have suggested that: 1) such activities serve the public good; 2) short term, low quality research should not be favoured in order to obtain funding; 3) managing of IP must be a balance of serving the institution, funding agency and public; 4) independence of the researcher and institution must be maintained; 5) the mission of the public institution is not altered and 6) conflicts of interest should be avoided. Institutions must be proactive in balancing funding, IP protection and commercialization.

Does patenting increase the likelihood of "industrial farming?" Economic and governmental issues certainly play a role in the size of the average farm and the level of commercialization of farming. It can be argued that if large companies have exclusive licensing arrangements for genetic tools then small breeders will be disadvantaged. Market pressures related to size of operation and efficiency of production are much larger influences on the industrialization of farming and livestock production than patenting.

Other issues regarding patenting that have an economic basis can be discussed. Langinier and Moschini (2002) provide an excellent review of the economics of patenting and the benefits and costs that are derived. While inventors would prefer broader claims, limited claims encourage competition and further innovation. Complex products often require building on other patents. Unfortunately, development does become blocked by other patents (especially at the leading edge) and this has negative effects on both the developer as well as the public. Licensing of patents exclusively to one company may benefit that company and segments of the public but might also limit further innovation and not be in all the public's best interest. Langinier and Moschini (2002) concluded "that continued efforts are required to improve the workings of the patent system" to improve the economic performance of the system.

SELECTIVE BREEDING

Selective breeding in domesticated animals is the process of a breeder developing a cultivated breed over time, and selecting qualities within individuals of the breed that will be best to pass on to the next generation. Breeding techniques such as inbreeding, linebreeding and outcrossing are utilized by breeders in the maintenance and improvement of their chosen breeds.

Breeding Stock

"Breeding stock" is a term used to describe a group of animals used for purpose of planned breeding. When an individual is looking to breed animals, he or she is looking for certain valuable traits in purebred stock for a certain purpose, or may intend to use some type of crossbreeding to produce a new type of stock with different, and presumably superior abilities in a given area of endeavor.

To breed chickens, a typical breeder intends to receive eggs, meat, and new, young birds for further reproduction. Thus the breeder has to study different breeds and types of chickens and analyse what can be expected from a certain set of characteristics before he or she starts breeding them. Accordingly, when purchasing initial breeding stock, the breeder seeks a group of birds that will most closely fit the purpose intended.

Purebred Breeding

The idea of *breed purity* may strike an unpleasant chord with some individuals because it is reminiscent of nineteenth-century eugenics notions of a "superior strain" in humans, supposedly exemplified by members of an aristocratic social class or of certain races. The application of theories of eugenics had far-reaching consequences for human beings, particularly in cases such as The Holocaust. However, the idea of a superior strain in animals, such as that incorporated in the breeding of Thoroughbred race horses, was less controversial. By "breeding the best to the best," employing a certain degree of careful inbreeding, considerable culling, and selection for "superior" qualities, one could develop a bloodline or "breed" superior in certain respects to the original base stock or landrace which had been produced primarily by natural selection.

This process forms the basis for a "true-breeding" line of animals, which today is generally called a "purebred" line. Once created, a purified line is "closed" and preserved from dilution and possible "debasement" by unrelated stock. However, this theory is not without some negative consequences. When the ideal of purified lineage or purely aesthetic type is seen as an end in itself, without consideration of usefulness or vigour, a breed can suffer over time. Also, if a breed registry is closed with too small a gene pool, over time, degradation due to inbreeding depression may occur. For many purebred breeders, the need for a periodic outcross is recognized and incorporated into an overall plan. The observable phenomenon of hybrid vigour stands in contrast to the notion of breed purity. However, on the other hand, indiscriminate breeding of crossbred or hybrid animals may also result in degradation of quality, particularly when animals are mass-produced without concern for overall health or longevity.

Backyard Breeding

The term backyard breeder is a general term, sometimes considered derisive, used to describe people who are "home breeders", and breed only a few animals for personal use or occasional sale. However, while some backyard breeders pay strict attention to quality, the term usually describes those who allow animals, particularly dogs or horses, to procreate regardless of physical, genetic, and/or emotional health. It is also used to describe unreputable commercial breeders. In the process of careless breeding, many backyard breeders produce genetically weak animals that can be predisposed to genetic disorders or debilitating physical deformities, particularly those that result from inbreeding.

In dog breeding, these individuals focus on only a limited aspect of the dog, while ignoring the original function and temperament of the breed. When such breeding is carried out on a large scale, the venue is called a *puppy mill* or *puppy farm*. Backyard breeding is popularly blamed for the proliferation of aggressive dogs for the sports of dog baiting and dog fighting. Dog fanciers generally believe that such ill-bred dogs are the reason for the bad public reputation of certain dog breeds, and the increasingly frequent enactment of breed-specific legislation. Because of the time and expense of feeding and caring for horses, which produce one foal per year, there are fewer horse breeders who produce animals *en masse,* though some individuals do engage in animal hoarding and breed far more animals than they can support.

In the horse world, over breeding of grade animals that cannot be sold except for very low prices raises concerns that such animals will be slaughtered for horsemeat. Some, though not all, colour breeders make errors similar to puppy mill breeders by focusing on only one characteristic to the detriment of health, soundness, and quality conformation. In addition, industrial breeders, such as those who produce *pregnant mare urine* for the manufacture of premarin, allow many mares to be bred indiscriminately, simply so that they are able to produce the urine laced with reproductive hormones that forms the base for the drug. Their foals are generally considered an unwanted byproduct, particularly males (colts), and sold off, frequently to the slaughter market.

4

Method of Genetics Disorders and Behaviour

GENETICS DISORDERS

Schizophrenia is one of the more complex and less understood of the psychological disorders. For many years, studies have explored the genetic aspect of this syndrome in an effort to provide a better understanding of its etiology, symptoms, course, and development. Much of what is involved in schizophrenia still remains unclear, although great strides have been accomplished by research in the area.

In trying to understand the etiology of psychiatric disorders, one cannot avoid studying gene—environment interactions. Just as genes influence the susceptibility of an individual to the development of almost any disorder, environmental factors are also seen as important etiological contributors to disturbance. Consequently, whether or not genetic or environmental variation is involved in the development of schizophrenia is no longer considered to be an important point of contention in this area of research.

Instead, most of the current investigations appear to be concerned with acquiring a more thorough understanding of the magnitude and relevance of both environmental and genetic variation and the ways in which these interact to establish the various manifestations of schizophrenia.

Today there seems to be little doubt that a genetic basis for schizophrenia exists. However, the specific schizophrenia-producing genetic components and the method by which they are transmitted remain unclear. Genetic investigations have involved primarily three methodologies: family studies, twin studies, and adoption studies. Despite some difficulties, these methods have obtained strong evidence pointing Towards the involvement of genetic factors in the development of schizophrenia. Currently, the amount of interest and speculation in this area of research is high, and on the basis of findings, sophisticated models and new, more innovative research techniques are being developed.

The major conceptual, diagnostic, and methodological difficulties inherent in this area of investigation, the evidence for the existence of genetic factors in the etiology of schizophrenia, and the models for a genetic mode of transmission that have arisen from these lines of research.

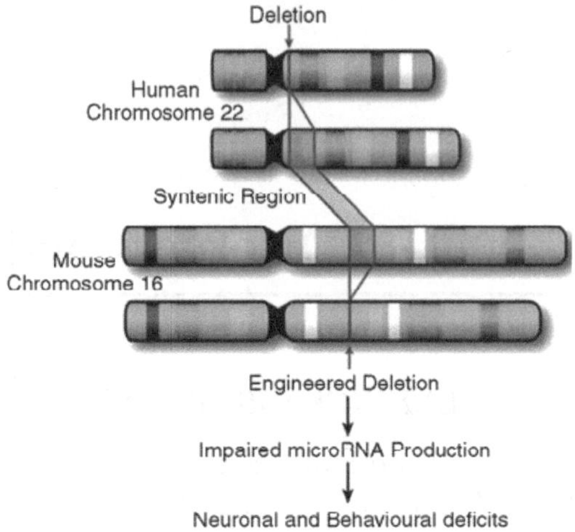

Fig. Genetics of Schizophrenia

One of the more difficult aspects of this field remains the lack of knowledge concerning the nature of schizophrenia. To date, no homogeneous group of schizophrenics has been isolated and no unitary etiology for the disorder has been found.

The possibility exists that schizophrenia is not a single, unitary disorder, but may consist of a number of different entities. Schizophrenic heterogeneity could be present at several levels. Seemingly similar phenotypic expressions may be due to quite different etiologies. As a result, the reliance upon symptomatology for the differentiation of schizophrenic subtypes may be grossly misleading. To complicate matters further, the complexity of gene operations makes it difficult to delineate the various interacting etiological components of a schizophrenic disorder.

The nature of the activating environment and the timing of that activation in addition to the gene could all combine to produce a single phenotype. Therefore, a range of possible phenotypic outcomes exists for every gene as a function of the activating environment and the time of activation. Depending on the environmental circumstances, there are a number of different genes that can produce a single phenotype. Consequently, several genetic mechanisms may underlie the expression of a single characteristic.

If schizophrenia is in fact heterogeneous, many of the present research strategies may prove inadequate because most rely on the implicit assumption of homogeneity among patients studied. A schizophrenic sample based on an overinclusive diagnostic category would result in variable schizophrenic performance. As a result, common findings among schizophrenics would be difficult to detect and valid hypotheses may be prematurely discarded on the basis that the data fail to meet statistical significance. The failure to reach a clear consensus on the nature of schizophrenia has resulted in conflicting viewpoints. To date, genetic studies of schizophrenia have relied almost exclusively on the assumption of a model in which schizophrenia consists of a complex disorder amenable to the same type of study and displaying the same pathogenic characteristics as an inheritable disease.

Other points of view, especially those among the psychodynamic and contextualist schools of thought, hold that the conceptualization of schizophrenia as a disease entity is inadequate and unjustified. For example, Sarbin and Mancuso claim that schizophrenia is a label and the product of social processes resulting in the imposition of a moral verdict in response to certain undesirable Behaviours. Others have proposed that certain forms of familial interaction predispose an individual to schizophrenia, whereas still others have preferred to focus on the schizophrenic's intrapsychic world and dynamics. An interesting model postulated by Meehl incorporates both genetic and environmental factors to explain the occurrence of schizophrenia. Meehl suggests that all that is inherited in schizophrenia is an integrative neural defect that he labels *schizotaxia*.

The imposition of a social learning history on schizotaxic individuals results in a schizotypic personality organization. That is, all schizotaxic individuals obtain a schizotypic personality organization regardless of environmental conditions. Most of these individuals will remain compensated. However, those who encounter certain detrimental influences (*e.g.*, schizophrenogenic mothering) will decompensate into schizophrenics. In light of this model, only a few of those individuals who have inherited the neural defect and are thereby predisposed to schizophrenia, will manifest the disorder.

Despite the wide disparity among theoretical models, most can be classified as elabourate extensions of the nature versus nurture controversy. It is apparent that this debate cannot be adequately resolved until a broader base of knowledge is developed.Until that time, it may be wise to bear in mind that genetic research does not necessarily negate endeavors supporting contextualist and psychodynamic conceptions. Instead, genetic research addresses both environmental and genetic factors, and is interested in delineating the effects of each separately and in interaction as they pertain to schizophrenia.

METHOD OF DIAGNOSIS

Another research difficulty related to that of schizophrenic heterogeneity pertains to clinical diagnosis. The new DSM-III allows for considerable heterogeneity as to who will be diagnosed schizophrenic; persons need to fulfill only a certain number of criteria out of a standard set of possibilities in order to receive the diagnosis. Schizophrenia cannot be diagnosed unless signs of the disorder have been present for at least 6 months. Illnesses of shorter duration, or illnesses that have resulted in the face of a psychosocial stressor and appear to be transient in nature are no longer included. In fact, these disorders are not even considered to be subtypes of schizophrenia. These changes in diagnostic criteria have left schizophrenia researchers in a peculiar bind since they now must try to define the onset of a schizophrenic disorder. In the past, the onset of schizophrenia was demarcated by the appearance of psychotic symptoms. However, it is well known that the onset of schizophrenia is often quite insidious, and that prodromal features commonly predate psychotic symptoms.

Logically, it would be best if we could define onset as the point in time when one could predict with virtual certainty that a schizophrenic psychosis would occur. Unfortunately, we still lack the necessary knowledge to make such a prediction, and the definition of onset remains strictly arbitrary. Consequently, the conceptual problem of trying to determine criteria for delineating signs along a continuum marking transitions from a schizophrenia-prone personality structure to a prodomal syndrome to a full-blown psychosis still exists.

Because of these diagnostic difficulties, a standard definition for what constitutes schizophrenia remains to be determined. The absence of a standard definition remains a major hindrance to the progression of this field. The severity of the problem is due to the fact that every research project must of necessity concern itself with identifying those people who have the disorder.

Therefore, the homogeneity of the sample studied and the Rigour of the investigation carried out on the sample depend on the accuracy of the diagnosis. Unfortunately, there is no independent validation and only moderate reliability for diagnosing a schizophrenic disorder. Even the combination of moderate interrater reliability and independent validation appears to be inadequate. Consequently, there is no truly homogeneous sample in schizophrenia research. Instead, investigators are forced to contend with the heterogeneous groupings that fit the schizophrenic label, the abundance of contradictory data, and the numerous disagreements over research findings. As long as this state of affairs exists, studies are limited to statistical rather than clinical significance.

At present, many diagnostic difficulties result because little is known about the unity or diversity of schizophrenia. In order to gain a clearer understanding of the nature of the disorder, many investigations have been conducted to determine the degree to which the range of phenotypic expression in schizophrenia is paralleled by genetic heterogeneity. These included certain studies that were designed to investigate whether these subtypes were fundamentally alike or different.

One such study by Winokur, Morrison, Clancy, and Crowe found that hebephrenic probands had more than three times as many schizophrenic relatives as paranoid probands. These authors determined that both children and other secondary cases of schizophrenia in the families of hebephrenics were more likely to be diagnosed hebephrenic as opposed to paranoid schizophrenic.

The same type of pattern was established for paranoid schizophrenics. The children and other secondary cases of schizophrenia in the paranoid probands were more likely to manifest paranoid rather than hebephrenic symptoms. These familial data led the au thors to conclude that process schizophrenia is not likely to be a unitary illness. Instead, they postulated the existence of two types of schizophrenia in the population studied, one of which may manifest itself mostly as hebephrenia but also occasionally as paranoid schizophrenia.

The second type would manifest itself exclusively as paranoid schizophrenia in which there are relatively few schizophrenic relatives. Together, these findings suggest that there may be only one type of hebephrenic schizophrenia, but two types of paranoid schizophrenia, one related genetically to the hebephrenic subtype.

Another study by Kety et al. in 1975 demonstrated that the diagnoses of chronic, latent or borderline, and uncertain schizophrenia tend to cluster in the biological relatives of schizophrenic adoptees. Similarly, a study by Slater and Cowie examined the classical pre-Kraepelinian categories and found a significant association between schizophrenic probands and their schizophrenic offspring with respect to type of category.

Kringlen in 1968 corroborated these results and also demonstrated that the entire set of Kraepelinian subtypes is represented in the children of every type of schizophrenic proband.

These studies are typical of investigations in this area. Their results seem to indicate that some groups of disorders are genetically related, whereas the relationships among others remain unclear. Despite some contradictory findings, the most common results from these studies provide evidence for a common genetic basis among various forms of schizophrenic disorders.

In general, the findings indicate that cases of all subtypes occur in the families of each schizophrenic subtype. They also suggest that only a modest tendency Towards subtype resemblance exists within families.

In addition to considering the unity or diversity of schizophrenia, one must also consider the relationship between schizophrenia and other psychiatric disturbances when making clinical diagnoses. This is especially important if one wants to formulate diagnostic criteria that are successful at differentiating among various psychoses.

At present little is known about the specificity of the schizophrenic disorder. However, there is some evidence suggesting a genetic relationship between schizophrenia and other psychiatric disorders. In a study by Reed, Hartley, Anderson, Phillips, and Johnson, a high risk of psychoses for the siblings of psychotics was found.

When the investigators looked at specific disorders, they discovered that the risk for first degree relatives of affective probands to develop a similar disorder was less than their risk of becoming schizophrenic. There is also some indication that the risk of schizophrenia for the offspring of paranoid and hebephrenic schizophrenics is lower than the risk for children of manic—depressives. Even some of the very earliest studies discovered that schizophrenic and affectively ill individuals were about equally prevalent in the families of schizophrenics. Today as well, despite efforts to select clinically homoge neous probands, the occurrence of different types of psychoses in the same family is difficult to avoid.

The suggestion that there may be a similar genetic basis to schizophrenia and other forms of psychoses raises the possibility of overlap in the symptomatology of various psychiatric disturbances; and indeed, this has been found to be the case. Another problem exists in the possibility that symptomatology is not directly related to genetic inheritance. Bleuler speculated that secondary symptoms such as delusions and hallucinations are not inherited. He also suggested that the primary symptoms do not correspond to the transmitted anomaly itself.

As a result of these possibilies, there are some major difficulties involved in basing diagnostic criteria on clinical symptoms. However, there are now some indications that future diagnostic categories will be based on biochemical indicators rather than Behavioural symptoms in order to decrease the heterogeneity of individuals in specified subgroups of psychopathology.

Using biological markers could be advantageous in several ways. First, it could avoid the genetically heterogeneous diagnostic groupings based on standard methods of symptom classification. It is well known that even relatively well formulated clinical subtypes of schizophrenia present diverse symptomatology.

As a result, it is impossible to differentiate between genotypic and phenotypic variations under the present diagnostic systems. The use of biologically independent variables in formulating diagnoses may promote the detection of biological etiologies and the formation of more homogeneous psychiatric diagnostic categories. Individuals who exhibit divergent or less severe clinical symptoms and are thereby excluded from present investigations could be included if they met a biological criterion.This in turn would allow the high risk population (those individuals with the salient factor) to be identified and studied. The use of a biologically based strategy would also permit the study of Non- hospitalized and unmedicated individuals, thereby avoiding the inclusion of these confounds within investigations.

In addition to developing biologically based criteria, many other efforts are being made to improve the diagnostic system. For example, the United States—United Kingdom Diagnostic Project, the International Pilot Study of Schizophrenia, and the World Health Organization have provided standardized input on patients for formulating diagnoses, experts to address diagnostic issues, and cross national reference points in order to establish criteria that will improve the reliability and validity of the diagnosis of schizophrenia on an international basis. Additional reasons for optimism stem from the new trends in schizophrenia nosology.

These include an increased reliance on affective symptoms as diagnostically useful, as well as a shift Towards incorporating the schizo affective disorders within the affective diagnoses. In addition, the broad use of the schizophrenic label seems to be decreasing as a function of less emphasis being placed on classic schizophrenic symptoms, and more effort being devoted to the use of operationally defined criteria tested for reliability.

Although at present it is still difficult to reach clear and definite conclusions in research dealing with schizophrenia because of the many diagnostic difficulties associated with this disorder, the increasing refinement of diagnostic issues and the development of more sophisticated research techniques makes it possible to conceive of removing the current obstacles to gathering clinically significant data and conducting therapeutic—preventive research.

DIAGNOSTIC DIFFICULTIES

Despite the numerous conceptual and diagnostic difficulties inherent in schizophrenia research, much evidence has been gathered to substantiate the existence of genetic components in the transmission of the disorder. The idea that various forms of mental illness are heritable has existed for many centuries. However, it was not until early in this century that researchers were able to provide actual documentation of this fact.

The first consanguinity study that found a significant increase in the prevalence rate of schizophrenia among relatives of index cases in comparison to the rate of the general population. More specifically, he discovered an age corrected risk for schizophrenia of 4.48per cent and 4.12per cent for other psychoses in the siblings of schizophrenics whose parents were unaffected.

Many other studies have confirmed Rudin's basic finding of a higher rate of schizophrenia in the relatives of samples of schizophrenics than in the samples from the general population. From these studies the rate of schizophrenia in the general population has been calculated to range from.35 to 2.85per cent, with a mean around.85per cent.

The findings from the consanguinity studies also reveal a clear correlation between the degree of blood relationship in relatives of an individual affected with schizophrenia and the risk of those relatives for becoming schizophrenic. More specifically, those relatives who share fewer genes in common with an affected individual show a lower expectancy rate for the disorder than those who share more genes.

In addition, these studies have demonstrated that there are no significant differences between prevalence rates for siblings and other first degree relatives. Although the difference found in the rate of schizophrenia between parents (5.5 per cent) and siblings and children of schizophrenic probands (10.2per cent and 13.9per cent respectively) appears to be large and therefore contradictory to the foregoing statement, this difference has been explained by the fact that these parents are largely selected for a degree of mental health through having been married and having children. When a correction is made for this factor, the risk in parents of schizophrenics becomes close to that of siblings and children. The somewhat higher risk in children than in siblings of schizophrenics is thought to be due to sampling error. Additional evidence for a genetic factor in the transmission of schizophrenia is found by comparing the different rates of prevalence for the disease among offspring of matings of various psychotic combinations of parents. Estimates have been obtained indicating that the risk for schizophrenia increases as the genetic loading increases. The best estimate of the risk of schizophrenia in the sibling of a schizophrenic is 9.7per cent when neither parent has the disorder.

This risk increases to 17.2per cent for the siblings of the proband when one parent also has schizophrenia. When both parents are schizophrenic, the risk to their children ranges between 36.6per cent and 46.3 per cent. Although it can be argued that this raise in risk from 17.2per cent to 46.3per cent for the siblings of a schizophrenic may be due to the extreme environmental conditions created by the mating of two schizophrenics, a study by Kallman reported that a comparable disturbed setting created by the mating of a schizophrenic and a psychopath yielded a sibling risk for schizophrenia of only 15per cent in Gottesman and Shields. The basic findings of these studies have been further substantiated by the results of more recent investigations. For example, a study by Reed et al. reported that the risk to the offspring when neither parent was psychotic increased relative to the number of aunts and uncles who were psychiatrically disturbed.

The difference among estimates of risk is a fact significant for genetic theories. Not only does this difference demonstrate that the risk of mental illness increases as a function of increased genetic loading, but the large amount of variation in the offspring undermines the prediction of environmental theories that virtually all children from dual matings of schizophrenics would be emotionally disturbed.

Another line of evidence for the role of genetic factors in schizophrenia was provided by Manfred Bleuler, who applied the methodology of consanguinity studies to clinical symptomatology and course over time. He discovered that the biological relatives of schizophrenics who also had the disorder tended to have similar ages of onset, presenting clinical pictures, lengths of duration of illness, and outcomes. These findings can be interpreted in either an environmental or genetic fashion. However, a genetic explanation appears to carry more weight because many of the relatives in the study were geographically separated and in some cases did not even know each other. Therefore, the findings of the study indicate that genetic factors play a role in several clinical parametres of schizophrenia.

Although these family studies have contributed much Towards the understanding of schizophrenia and have pointed to the role of genetic factors in the development of this disorder, some major difficulties are inherent in their methodology. The earliest family studies did not question the assumption that schizophrenia is heritable and therefore failed to consider the possibility that environmental factors such as family interactions also contribute to the etiology of the disorder.

As a result, these studies provided sources of data on the empirical risk of the occurrence of schizophrenia in different kinds of relatives, but did not elucidate the specific etiology involved in creating those risks. In other words, these studies could indicate if a trait ran in families, but could not specify what aspect of the family determined the trait. Obviously, families share more than genetic similarity, they also share a variety of social, psychological, and cultural experiences. Consequently, the results of these studies can just as validly be interpreted as supporting an environmental position.

Recently, some studies taking an environmental approach have investigated family interactions on the premise that certain conditions of family life can predispose an individual to schizophrenia. The majority of these investigations have focused on deviant role relationships and disordered communication processes among family members.

The findings of these studies generally indicate that deficits in family communication are highly correlated with schizophrenia and high risks for schizophrenia in offspring. In addition, there are some indications that families of schizophrenics contain poorly articulated and rigid role structures in the areas of dominance, autonomy, and sexuality. Despite the success of this type of research, it is unable to establish the occurrence of schizophrenia as a strictly environmental outcome.

In fact, the evidence that is presently available tends to suggest that both genetics and family process play a role in the development of schizophrenia. In light of this evidence, the current mode of investigation has changed to incorporate biological, genetic, psychological, and social factors in the context of complex, multifactorial models of causation when approaching family variables in relation to schizophrenia.

TWIN STUDIES

It was not until the advent of twin studies that the role of genetic and environmental factors in schizophrenia could be more clearly defined. The power of the twin method resides in the environmental similarities and genetic differences between monozygotic and dizygotic twins.

Both types of twins share the same intrauterine environment simultaneously with their co- twin. However, monozygotic twins share the same fertilized ovum initially, whereas dizygotic twins are no more genetically alike than ordinary siblings. As a result, one can assume the presence of a genetic factor in determining a trait if the concordance rate for that particular trait is significantly higher for monozygotic as opposed to dizygotic twins. The methodology of twin studies also provides certain advantages. For example, through the comparison of intrapair resemblance, it becomes possible to identify subtypes or components of schizophrenia that may be more under genetic or environmental control than others. Also, by noting the variability of abnormal personality traits in the monozygotic co-twins of typical schizophrenics, it becomes feasible to identify schizophrenic "equivalents" or schizotypes.

The comparison of monozygotic twins who differ in outcome will permit the formulation of inference about the relationship between life experiences and outcome. It must be kept in mind that the twin study method relies on certain assumptions. These specify that monozygotic twins as such are not specifically predisposed to schizophrenia, and that the environments of monozygotic twins are not systematically more similar than those of dizygotic twins in such a way as to foster the disorder.

Also, when assessing the outcome of these studies it is important to be aware of the usage of statistical age corrections and their effects on research findings. Age corrections incorporate estimates of a lifetime risk for developing schizophrenia in order to compensate for a current sample while the disorder exhibits a variable age of onset. Although several methods have been developed, none is entirely satisfactory, and all present only varying degrees of approximation to the true concordance rate.

It is essential to realise that the absolute size of the concordance rate is not predictive of the relative importance of the genetic contribution to the trait in question. Concordance rates can be expressed in several fashions. The basic difference in methods depends on whether concordance is expressed in terms of the pair or the individual.

The casewise or proband method refers to the proportion of cases with an affected partner, whereas the pairwise rate considers the proportion of pairs in which both twins are affected. The individual or proband concordance method yields a higher per centage than a pairwise concordance. Therefore, only those studies that use the same method are comparable.

The most influential early twin studies were conducted by Kallman and Slater. Even though the two studies contained important differences and suffered from methodological problems, both found a significantly higher concordance rate for schizophrenia among monozygotic twins.

By the end of Kallman's study and with some diagnostic revision, but without statistical age correction, 59per cent of 174 monozygotic pairs and 9per cent of 517 dizygotic pairs were concordant for what Kallman labeled "definite schizophrenia." In Slater's study, 65per cent of 37 monozygotic pairs and 14per cent of 58 same sexed dizygotic pairs were found to be concordant. Both studies used the pairwise method for calculating the concordance rates. Findings from other earlier studies were basically consistent with these results.

More recently, several studies have been conducted that are methodologically powerful, and that provide highly significant results. One of these studies, reported by Tienari, was conducted on an exclusively male sample in Finland. Initially, this investigator found a zero concordance rate for monozygotic twins. However, by 1976 the pairwise concordance rate for the sample had risen to 16 per cent. The use of the proband method, the concordance rate for the sample was found to be between 35per cent and 53per cent depending upon which borderline cases were included. As a result, even though it took some time, a significant monozygotic—dizygotic difference was demonstrated.

A study by Kringlen was conducted in Norway and included both male and female twins. His study was especially important for several reasons. His sample was well constructed and contained the largest number of psychotic twins personally investigated among the more recent studies. His investigation was very comprehensive and included case history, pedigree, clinical analysis, rating scale data, and all functional psychoses in addition to schizophrenia. Using the more conservative pairwise method, Kringlen found a significant difference between monozygotic and dizygotic twins with concordance rates ranging from 60per cent for monozygotic probands rated as severe to 25per cent for the most benign.

His results indicated no difference between the concordance rates of monozygotic twins as a function of gender. He reported that whether the sexes of the dizygotic twins were the same or different, the concordance rates did not differ, but were equal to those found in ordinary siblings. These results counter findings from some earlier studies and undermine the psychodynamic thesis that concordance for schizophrenia will increase without regard to genetic similarity, but relative to the amount of likeness and resulting identity confusion among siblings. The findings of another important twin study were reported by Fischer, Harvald, and Hauge in 1969. This study was conducted in Denmark upon a sample that was very representative with respect to twin types and sexes of probands. Only those probands who met strict criteria for chronic or process schizophrenia were included.

Even so, schizophreniform, paranoid, and atypical psychoses in co-twins led to data analysis for quantitative concordance for other than process schizophrenia. As a result of this study, Fischer et al. discovered a pairwise concordance rate of 24per cent for monozygotic twins and 10per cent for dizygotic twins.

In addition, the authors calculated the rates according to the proband method and obtained a 36per cent concordance rate for monozygotic and 18per cent concordance rate for dizygotic pairs. When those co-twins who exhibited a form of schizophrenic disorder different from process schizophrenia were included in their calculations, the investigators obtained proband concordance rates of 56per cent and 26per cent for monozygotic and dizygotic twins respectively. These figures appear to be good approximations of the true incidence rate, as the twins used in this study were followed long enough to virtually eliminate the need for any statistical age correction.

Fisher went on to report that monozygotic twin pairs concordant for schizophrenia were not at an increased risk for the disorder when compared to discordant monozygotic twins. She did not discover a relationship between severity of disturbance in the proband and the risk of developing psychosis in the co-twin.

The offspring of concordant twins were not at an increased risk for schizophrenia when compared to the offspring of discordant twins. In addition, the likelihood of a discordant twin member producing a schizophrenic child was equal to that of the schizophrenic member of the twin pair. These findings point Towards a genetic hypothesis and indicate that the genetic transmission of schizophrenia is not always manifested directly.

The Maudsley Hospital based study of Gottesman and Shields conducted in England obtained similar results to those of the studies reported earlier. These investigators also obtained a significantly increased monozygotic concordance rate over that found in dizygotic twins. Similarly, Pollin, Allen, Hoffer, Stabenau, and Hrubec, in a study based on an American sample, found a higher concordance rate for monozygotic as opposed to dizygotic twins. More specifically, the concordance rate for schizophrenia in monozygotic twins was found to be 3.3 times greater than the rate in dizygotic twins.

GRADE TONNAGE MODELS

Estimated premining tonnages and grades from over 3,900 well explored, well characterized deposits were used to construct 60 grade tonnage models. Where several different estimates were available for a deposit, the estimated tonnages associated with the lowest cutoff grades were used. Grades not available (always for by products) were treated as zero. Except for a few instances, the data base is so large as to preclude specific references. Several compilations of data were particularly useful sources for multiple deposit. The grade tonnage models are presented in graphical format to make it easy to compare deposit types and to display the data.

All plots show either grade or tonnage on the horizontal axis, while the vertical axis is always the cumulative proportion of deposits. Plots of the same commodity or tonnages are presented on the same scale; a logarithmic scale is used for tonnage and most grades. Each dot represents an individual deposit (or, rarely, a district), cumulated in ascending grade or tonnage.

Where a large number of deposits is plotted, individual digits represent the number of deposits. Smoothed curves are plotted through arrays of points, and intercepts for the 90th, 50th, and 10th per centiles are constructed. For tonnages and most grades, the smoothed curves represent per centiles of a lognormal distribution that has the same mean and standard deviation as the observed data; exceptions are plots where only a small per centage of deposits had reported grades and grade plots that are presented on an arithmetic scale, such as iron or manganese, for which the smoothed curve was fit by eye.

The number of deposits in each type is indicated at the upper right of each diagram. The deposits used to construct each model are listed with the model and cross indexed to model types. Correlations among grades and between tonnage and each grade are indicated only when significant at the 1 per cent level.There are important limitations inherent in the data base used for all grade tonnage models. Estimates of cutoff grades within individual deposit types can vary because of regional, national, or operator differences.

All too commonly there is no mention of the actual cutoff grades or mining widths that are incorporated; nevertheless, the grade tonnage figures given do represent material that the company or the government believed might someday be economic to mine. Stratiform deposits of large areal extent, such as phosphate or sedimentary manganese, are special problems because of differences in opinion and practice regarding how closely drilled they must be to "prove" ore tonnages and regarding the thicknesses and depths of what may be considered for eventual mining.

Effects of another source of variation, mining methods, are recognized in some of the placer models; typically, however, mining methods are fairly consistent within a deposit type. In a few instances, irregular cumulative frequency plots reflect mixing of economic and scientific data sources, such as in the plot of gold in porphyry copper deposits. In spite of the current difficulty of quantifying variation of grades and tonnages with respect to changes in cutoff grades or mining methods, the models presented here are believed to account for the main source of variation in grades and tonnages of mineral deposits variation due to differences among types of deposits.

The question of whether one counts deposits within a cluster of related deposits as individuals or as a total will probably never be resolved to everyone's satisfaction. Some geostatisticians would separate each ore body (and then argue about whether two operations on the same body should be counted separately), whereas some economic geologists would lump everything from a single district (and then argue about district boundaries). For the most part the entities summarized are individual deposits, but in some instances such data are mixed with data representing entire districts. Because of these inconsistencies, some care is necessary in comparing grade tonnage models between deposit types or in comparing this summary with those prepared using alternative methods.

Care is also warranted in interpreting the grade distributions for which data are missing; this concerns principally by product grades. In some instances, such as the platinum group element (PGE) contents in podiform chromite and the cobalt content of laterites, the fragmentary information given probably represents the entire class.

In other instances, such as the lead content of Cyprus massive sulfide deposits, the missing grades probably represent values below the lowest reported grades. The grades derived from studies of trace elements in ores more probably represent the former situation rather than the latter. Deposits strongly suspected to be small or very low grade are seldom sampled well enough to be characterized in terms of grade and tonnage, thus the sample of many deposit classes is truncated by economics.

Non-etheless, probably 40 per cent of the deposits used in these models are, in fact, Non- economic today; and a perusal of the figures will discover examples of both small deposits and low grade deposits.

Potential metal supply is dominated by the very few largest tonnage deposits, as shown by Singer and DeYoung (1980), who also pointed out that inverse correlations between grade and tonnage are surprisingly rare.

Thus the fact that a deposit is large does not necessarily mean that it will prove to be of low grade. This means that most low grade deposits are not likely to have huge resources and also that the omission of a few low grade or small tonnage deposits will not seriously degrade the predictions of potential national supplies for most commodities.

In contrast, the missing low grade and small deposits suggest that the grade tonnage models represent a biased sample of the large number of low grade or small tonnage occurrences and prospects found by exploration.

This fact must be considered in cases where the number of undiscovered deposits is estimated. In order for the estimated number of deposits to be consistent with a grade tonnage model, approximately half of the deposits estimated should have greater than the model's median tonnage or grade. Thus the probability that an untested prospect represents a significant deposit can too easily be overestimated.

OTHER TYPES OF MODELS

The bulk of this report deals with descriptive mineral deposit models and their grade tonnage counterparts, but there are other useful aspects which we wish to discuss even though we have not yet had the opportunity to develop or exploit them. They are the genetic, occurrence probability, and quantitative process models.

Many authors prefer to keep a clear distinction between descriptive and genetic models, apparently feeling that the descriptive models somehow represent "pure truth" whereas the genetic constitute a less objective philosophical position (or at least make the investigator "skate on thin ice"). It is altogether desirable to avoid confusing interpretation with fact; but it is well to remember, for example, that each time a field geoscientist extrapolates geology across a covered area he or she adds an element of "interpretation" to a "factual" map, and that this interpretation is not necessarily any more "real" (or "unreal") than, for example, an isotope geologist's conclusion that a given oxygen and hydrogen isotopic signature extracted from fluid inclusions points to a meteoric origin for the fluid.

The point is that the whole of our professional knowledge rests on a broad continuum of interpretations; many of them are so commonly accepted that they are no longer questioned, but many others still evoke challenges. Thus we suggest that a combination descriptive genetic model is not inconsistent with professional practice. The model begins as a description, but various aspects of the model become genetic as they acquire satisfactory genetic explanations. Eventually much of the model becomes genetic, as has happened, for example, with the Cyprus type massive sulfide deposits or the sandstone uranium deposits of the Colourado Plateau.

As the attributes of a model become understood in a genetic sense, the descriptive model evolves to a genetic model:

- F models are compilations of the properties of a group of related deposits in which the reasons for certain attributes being favourable are identified. Descriptive models evolve into genetic models, and as such they become far more flexible and powerful.

  We have presented the three model subtypes above as if they constituted a linear logical sequence leading Towards the "final" model, but in fact there must be an iterative relationship among descriptive, genetic, and grade/tonnage models. The consequence of examining any of these three may be a reassessment of the groupings of deposits chosen to be represented by a model type and the redesignation of the attributes diagnostic for that type.

  With a dominantly genetic model in hand, two more model types can be generated:

- Occurrence probability models are models that predict the probability of a deposit (of a size and grade indicated by the appropriate grade tonnage models) occurring within a given area. As with the descriptive and genetic models, probability models that are tied to lithic or structural geologic entities (that is, they are genetic) are far more focused; in fact, it is probably impossible to generate a useful probability model before the establishment of a genetic model. Accurate probability models are very difficult to construct because although the technical community has very complete data on mineral producers (mines), the data on Non- producing mineral deposits (prospects and mineral shows) are much less well documented, a point also covered in the discussion of grade tonnage models. Even more importantly, data on barren areas are sparse.

  We must extrapolate from a very fragmentary base Towards a completely unseen target.

  There is much to learn before the probability model can be made a dependable tool; yet the successful targeting of exploration programmes by industry demonstrates that, at least on a qualitative basis, areas with better than average probabilities can be identified. It is worth noting, also, that mineral fuels are much more predictable and now can have realistic probability of occurrence values attached to specific volumes of sediments provided that the initial character and post depositional histories of the sediments are well known. It is a distant but not unreasonable dream to anticipate that some day we shall approach that level of certainty for some types of Non-fuel mineral deposits.

- Quantitative process models are models that describe quantitatively some process related to mineral deposit formation; they are offshoots of the genetic model. Examples would be models of heat or fluid flow around a cooling pluton; rates of crystal growth as functions of supersaturation, impurities, and temperature; or sequences and amounts of minerals deposited from evapourating seawater.

All five of these model subtypes can be parts of the "final" model, and recycling of the model back to the original groupings stage helps refine the selection process. The persons engaged in research guidance and especially exploration and development have broad ranging needs, whereas those dealing with the availability of minerals or of land use allocation have less use for genetic or quantitative process models. Overall there is a need for a comprehensive array of mineral deposit models to meet these individual

MATURITY OF DESCRIPTIVE GENETIC MODELS

The rate at which we gain understanding and the current levels of genetic knowledge vary considerably from one deposit type to another. Such types as placers and evapourites are well known genetically and the problems in their exploration and utilization concern local site specific geologic issues rather than mineral genesis or the degree of maturation of the model.

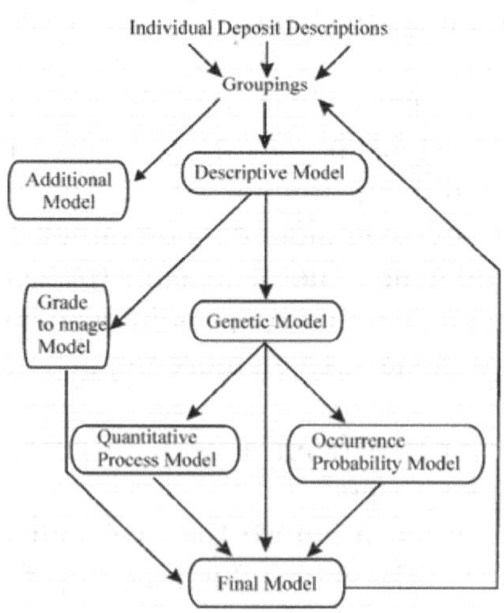

Fig. Flow Sheet Showing Evolution of Model Types.

In contrast, others such as the Coeur d'Alene Ag-Pb-Zn veins, or the massive Zn-Mn-Fe oxide/silicate bodies at Franklin and Sterling Hill, or the

Cu-U-Au at Olympic Dam, or the Cu-Zn-Pb-Ge ores of Kipushi and the Tsumeb pipe remain genetic enigmas despite, in the instances of the first two, extensive research spanning many years.

Still others, such as the diamond bearing kimberlite pipes, are geologically well understood regarding their origin yet very poorly understood in terms of the reasons for their existing at any particular site.

Our rate of acquisition of information is very irregular. The several scarps between plateaus in the knowledge curve for the marine phosphate model might mark, successively, the recognition that the phosphate was a chemical precipitate, that it occurred on continental shelves where upwelling of deep marine waters occurred, and that the upwelling regions were related to wind and current patterns that were tied to the global configuration of the continents and ocean basins.

A second example from the Mississippi Valley type ores might involve scarps marking the recognition (from fluid inclusion evidence) that the ores were deposited from warm (about 100°C) highly saline solutions that could represent neither simple surface nor marine waters. A second scarp might be associated with the recognition that the deposits were integral parts of a regional hydrologic regime whose distribution and character was susceptible to interpretation.

Bring out another point: some aspects of any model always remain to be determined, thus we never acquire a "complete" model. Indeed, the approach to "complete" understanding is asymptotic, and a lot of additional effort to clear up the "last" uncertainty in a nearly perfect model is probably unwarranted.

But, new ideas and new technologies can provide the impetus for new spurts in knowledge for heretofore incomplete models. Note that the horizontal axis is simply "years of effort" devoted to fundamental geologic investigation.

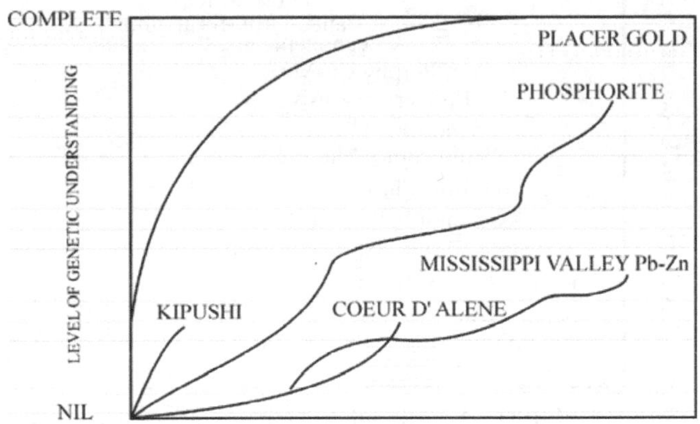

Fig. Growth Patterns of some Typical Genetic Models

The scale certainly needs to be exponential in order to fit the intensively studied and sparsely studied deposit types, but this figure is strictly schematic, there being no source of documentation for either coordinate. The figure also indicates that different deposit types may require different amounts of effort to achieve a similar level of genetic understanding.A hypothetical growth curve along which different types of deposits have been schematically arrayed.

Because some deposits (such as volcanogenic massive sulfides) are so much more difficult to understand than others (gold placers), the horizontal axis has been "normalized" by plotting a ratio of effort done to effort needed thereby permitting a smooth, although admittedly subjective and schematic, curve. There is no documentation to support this diagram, although the general concept meets with agreement among most contributors to this volume.

DeFries and Hegmann report on an activity selection study they completed. As a foundation population, they used the F_3 generation from the C_{57}BL/6J and BALB/cJ mice used in the study reported earlier.

Using an open field activity measure similar to that used in their earlier study, they were able to obtain two highly active, two low active, and two control lines, with each selection (and its replication) being successful within six generations.

Of particular interest for a genetic model for activity, however, was their finding that albino rats were less active than those with any other coat colouring (agouti, cinnamon, black, and brown).

These data together with their findings that the albino BALB/c mice are less active than the pigmented C_{57}BL/6 mice, led DeFries and Hegmann to suggest that: "Although open field Behaviour is clearly influenced by genes at many loci, a major gene effect was found: albino Ss have lower activity and higher defecation scores than pigmented animals."

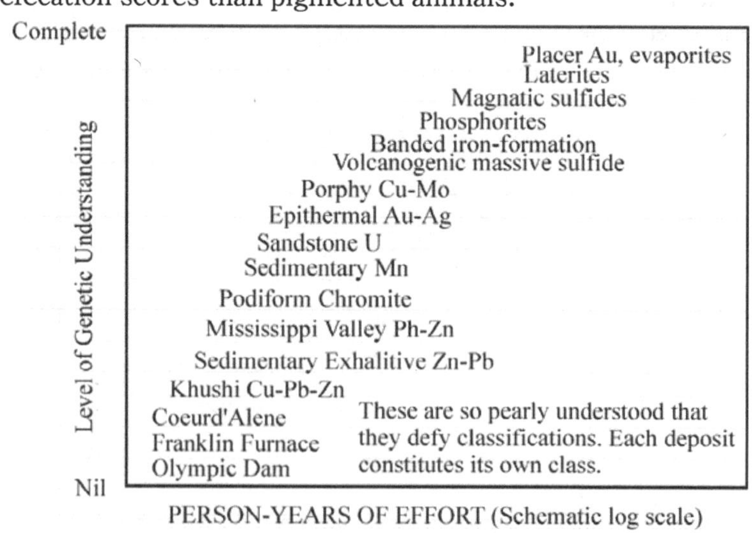

Fig. Comparison of Relative Levels of Understanding of some Important Model Types

Table. Open Field Activity Scores of $C_{57}BL/6J$, BALB/cJ and F_1 and F_2 Hybrid Mice

		Means [b]		*Variances*	
Generation	*Genotype* [a]	*Males*	*Females*	*Males*	*Females*
Parental	C	4.67	4.25	6.33	4.27
	B6	15.73	16.40	4.86	6.56
F 1	B6C	15.09	15.04	6.70	10.37
	CB6	13.98	13.66	15.80	21.06
F 2	CB6 x CB6	11.88	12.63	17.96	9.23
	CB6 x B6C	11.76	12.30	16.56	13.68
	B6C x CB6	12.66	12.86	15.87	16.44
	B6C x B6C	11.58	11.84	14.22	16.28

Table. Mean and Variance of Activity and Defecation Scores of Albino and Pigmented Animals in Generations $F_2 - F_5$, Pooled Across Generations

	Activity		*Defecation*		
	Mean ± S.E.	*Variance*	*Mean ± S.E.*	*Variance*	*N*
Pigmented	12.72 ± 0.18	18.65	1.78 ± 0.03	0.55	1729
Albino	10.44 ± 0.10	17.71	2.13 ± 0.02	0.49	557

This is plausible from the data presented, although it is possible that the effect found may be due to genes linked to the c locus in certain strains, with activity not actually moderated by the c locus itself. The finding that some albino (cc) mice do not exhibit a reduction in activity warrants caution in making generalizations about the "effects" of albinism.

It has been found, for example, that albino LG/J mice are highly active and that albino BALB/cBy mice do not show reduced activity when tested on an activity metre. DeFries and Hegmann used BALB/cJ mice, as did several of the other studies supporting the "albino effect."

Perhaps there is linkage of the c locus with other genes that reduces activity in this strain.

There could also be a different sort of linkage with the reverse effect in highly active albino strains.

At this point, however, attributing reduced activity to the lack of pigmentation seems premature.

GENETICS BEHAVIOUR

Behaviour genetics from where do diverse scientific disciplines emerge, develop, and finally coalesce? Psychology, which attempts the description, prediction, and ultimately the understanding of Behaviour, began in the universities of Europe. One tradition emerged from the 18th and 19th-century German philosophers as exemplified by Leibniz, and was given an experimental basis by Wundt. The other, emanating from England, was based on the principles of John Locke and the *philosophes* of the French enlightenment. This Anglo French influence came to America in 1882 with Herbert Spencer, who was enamored of Charles Darwin's theory of natural selection. Although these traditions are diffuse they help provide an orientation for understanding the development of psychology and its attempts to provide a firm biological basis for the study of Behaviour. The Lockean prescription offered a more comfortable cultural fit to our society, and its view of mental process as the product of nothing but sensation furnished a comfortable ambience for the growth.

MENDELIAN GENETICS

THE STUDY OF HEREDITY Z

When two gametes fuse, a zygote is formed. The haploid number of each gamete combine to form a diploid number zygote. The offspring may have some similarities with one or both of the parents. Genetics is the study of hereditary information as it is passed on from parents to offspring. Gregor Mendel was the first to study genetics in the 1800's.

MENDEL'S EXPERIMENTS

Gregor Mendel was a monk who used pea plants to study genetics. The advantages to using pea plants as his model organism were; ease of growing them, matured relatively quickly, and traits or characteristics were easily visible to the naked eye. He kept records of the following characteristics, which came in two forms;

- *Seed shape*: round or wrinkled
- *Seed colour*: yellow or green
- *Seed coat colour*:grayish brown or white
- *Pod colour*: green or yellow
- *Pod shape*: inflated or wrinkled
- *Stem length*: long or short
- *Flower position*: lateral or terminal

He used mathematics to explain his results and he published his work in a journal. His ideas were not widely accepted at the time and he died without recognition.

THE LAW OF DOMINANCE

Mendel discovered that he could create purebred plants. For example, when he allowed short plants to self-pollinate, eventually 100per cent of the offspring would be short. The same was true for the tall plants. Then he wondered what size offspring would result from crossing a short plant with a tall plant. The short and tall plants are called the parent generation or the P generation. The offspring are called the first filial generation or the F_1 generation. When Mendel crossed a purebred tall and a purebred short P generation, the F_1 generation was 100per cent tall. From this cross Mendel concluded that the tall trait was the dominant one. More accurately, he concluded that when an organism is hybrid for a pair of contrasting traits, only the dominant trait can be seen in the hybrid.

This is the law of dominance. Then, to satisfy his curiosity, Mendel crossed two of the tall offspring (no longer purebred) from the F_1 generation and found something interesting. The F_2 generation of the second filial generation was 75per cent tall and 25per cent short.

LAW OF SEGREGATION

Taking the F_1 and F_2 data into account, Mendel had to explain why the short or recessive trait was not seen in the F_1 generation and was seen in the F_2 generation. Rather than using the word gene, which did not exist to Mendel, he used the word factor to explain his results. He said there were two factors for a trait and each factor could be one of two forms. In this case the factors were tall or short and each plant could have a tall or short factor, two tall factors, or two short factors.

Mendel theorized that in the hybrid F_1 cross each plant had one tall and one short factor. He went on to theorize that the factors separate from each other during the formation of gametes or meiosis and then recombine together during fertilization. This is Mendel's Law of Segregation.

THE GENE-CHROMOSOME THEORY

Long after Mendel's death, a graduate student, W.S. Sutton, observed homologous chromosomes in diploid cells separate during spermatogenesis. Sutton realised that the separated chromosomes would be recombining in the fertilization process.

After looking at Mendel's work, Sutton concluded that Mendel's factors are carried on the chromosomes, since the chromosomes were doing the same thing Mendel's factors were theorized to do, that is separate and recombine. In Sutton's day, the word gene was used instead of factors and the Gene-Chromosome Theory was developed.

Fundamentals of Genetics and Genotypes and Phenotypes Vocabulary Review:

- *Alleles*: different forms for one gene (for example, regarding the gene that determines plant height; the alleles are short (t) or tall (T))
- *Homozygous*: the alleles are the same (for example, in a purebred tall plant, both alleles are tall (TT))
- *Heterozygous*: the alleles are different (for example, in a hybrid tall plant, one allele is tall the other is short (Tt))
- *Genotype*: the genetic makeup of an organism (for example, (Tt))
- *Phenotype*: the physical trait of an organism (for example, tall)

Probability in Genetics

The law of probability, or chance, says if there are several possible events that might happen, and no one of them is more likely to happen than any other, then they will all happen in equal numbers over a large number of trials.

THE PUNNETT SQUARE

The Punnett square diagram is an easy way to show the results of any cross. The Punnett square results for a purebred cross is always 100per cent hybrid genotype showing the dominant phenotype. The Punnett square results for a hybrid cross is 25per cent pure dominant and 25per cent pure recessive genotypes, 50per cent hybrid genotypes, 75per cent tall phenotypes, and 25per cent short phenotypes.

THE TEST CROSS

When you are trying to find out the genotype of an organism, you simply cross it with a known pure recessive mate. If 100per cent of the F_1 generation show the dominant phenotype, the unknown organism was purebred for the dominant trait. If 50per cent of the F_1 generation was dominant and 50per cent was recessive, then the unknown organism was a hybrid.

Other Concepts in Genetics: The Law of Independent Assortment states that during meiosis, genes for different traits are separated and distributed to gametes independently of one another (note: there are exceptions to this rule). Using this law in a dihybrid cross (TtRr x TtRr, for example) always produces a 9:3:3:1 phenotype ratio.

Incomplete Dominance – When organisms have this type of inheritance, the hybrid produces a trait that is a blend of each trait. Examples of this are the four o'clock flower (RW is pink), and Andalusian chickens (BW appear blue, which is actually a steel-gray)

Codominance – When organisms have this type of inheritance, the hybrid exhibits both traits equally. Examples of this include AB blood type and roan coat in cattle and horses. Multiple Alleles – For some traits there are more than two alleles for that trait. In humans, examples of traits that have multiple alleles are blood type (A,B,O alleles), Rh blood factors, and eye colour.

- Reproduction of bacteria;
- Horizontal gene transfer in bacteria;
- Cell cycles of eukarya;
- Meiosis;
- Chromosomal abnormalities;
- Life cycles of eukarya;
- Reproduction of viruses;
- Genetic processes in development;
- Sex determination and dosage compensation;

- Cancer;
- Cutting, sorting, and copying DNA;
- Genotyping by DNA analysis;
- Genetically engineered organisms;
- Genomics;
- Behaviour of genes and alleles;
- Probability and statistics toolkit;
- Genes, environment, and interactions;
- Locating genes; 35. Finding and detecting mutations;
- Cytoplasmic inheritance;
- Genetic variation in populations;
- Mutation, migration, and genetic drift;
- Natural selection;
- Quantitative genetics;
- Speciation;
- Molecular evolution and phylogeny; Glossary.

PRINCIPLES OF MENDELIAN GENETICS

Genetics formal beginnings are found in the work of Gregor Mendel whose elegant experiments with garden peas and mice provided the foundation for the science of genetics. In the garden of a monastery in Moravia (now Czechoslovakia), of which he eventually became abbot, Mendel crossbred varieties of the garden pea that had maintained relatively constant differences in physical characteristics over generations.

This hereditary constancy occurred despite the occasional, now fully scientifically explained, disappearance of a given trait during several generations and its subsequent reemergence.

Some of the factors Mendel analysed were tall versus dwarf plants, red or white coloured blossoms, seed colour and shape, pod forms, and the position of flowers on stems. He theorized that each plant contained two factors for a particular characteristic such as colour (one of each of these factors contributed by each parental plant) and that these units were transmitted to descendants in adherence to statistical predictions.

These two factors (now called genes) might be identical, in which case the organism was said to be homozygous for that gene, or they might differ and the organism would be heterozygous.

From his plant breeding experiments and observations, Mendel derived two laws that serve as the foundation, both historically and functionally, of the science of genetics. The first law concerns segregation. Genes are particulate units of inheritance that may exist in two or more variant forms (alleles) at the same position (locus) on a chromosome. Members of a chromosome pair separate when sex cells are produced. If an individual is heterozygous at the *A* locus, the gametes will be of two types—A or *a*.

The second law deals with independent assortment of alleles on different chromosomes. When the chromosomes separate to form eggs and sperm, these variant alleles (*Aa* and *Bb*, for example) are transmitted independently of one another to the next generation. When one allele is always expressed phenotypically in the organism bearing it, it is called the dominant allele (*A*); a recessive allele, expressed only when the dominant allele is absent, is designated *a*. A dominant trait is expressed when one or two dominant alleles are present (*AA* or *Aa*) but one can see only the recessive phenotype when both recessives (*aa*) are present. Some alleles are not expressed as either dominant or recessive but are intermediate, and they produce a phenotype somewhere between the phenotypes of the parents (red, pink, white).

Suppose that two individuals differing in genes at only three different loci (genetically active sites on chromosomes) are crossed: *AABBCC* x *aabbcc*. The sex cells (gametes) are carrying *ABC* or *abc*, whereas the first filial generation ($F_{1)}$ or offspring, is entirely *AaBbCc*. Each F_1 offspring can produce eight different types of gametes: *ABC, ABc, AbC, aBC, Abc, aBc, abC,* and *abc* the number of possible genotypes when three genes are involved. For an idea of what occurs when more than three gemes are involved. Inasmuch as each chromosome has many thousands of genes the number of possible combinations is astronomical. The word chromosome comes from the Greek *chroma*, which means colour, and *soma*—body, because these gene carrying units can be routinely observed microscopically when stained with basic dyes. Genes are arranged in linear order on these chromosomes, which are found in the nucleus of eucaryotic cells.

G (3/4) < W (3/4) = 9/16 GW (yellow, round)
G (3/4) < W (1/4) = 3/16 Gw (yellow, wrinkled)

g (1/4) < W (3/4) = 3/16 gW (green, round)
g (1/4) < W (1/4) = 1/16 gw (green, wrinkled)

Fig. A Tri hybrid Cross

Organisms are classified as eucaryotes if their genetic material is contained in the nucleus of their cells. This includes all life above the level of bacteria and blue green algae. The latter—called procaryotes—are considered to be earlier stages in the evolution of life on this earth. Procaryotes have no nuclei and their genetic material is distributed throughout their cells. In eucaryotes, each individual of a species has a characteristic chromosome number—46 in humans, 40 in the house mouse, 20 in corn, 48 in the potato,and 8 in the fruit fly, (*Drosophila melanogaster*). Sexually reproducing individuals often have two different alleles (are heterozygous) for many of their genes. Thus, the possible combinations of parental genotypes greatly outnumber the actually ever realised genotypes. The probability of siblings acquiring the same genetic endowment from both or even one of their parents is infinitely small.

GENOTYPE AND PHENOTYPE

The observable physical and Behavioural aspects of an organism constitute its *phenotype*. The phenotype is the joint product of the genes that an organism possesses (its genotype) plus the effects of the environments in which it developed and now lives. It has become increasingly obvious with the predominance of the Behaviourist school of psychology throughout the 1950s that the environment is of particular significance in the development of Behavioural phenotypes. A prime example of how Behaviour may be altered via the environment is phenylketonuria a genetic disease, arising from an inborn error of metabolism.

The enzyme that normally breaks down phenylalanine an essential amino acid—is lowered or missing due to gene malfunction. Unmetabolized phenylalanine accumulates in the brain, and one of the effects is a lowered intelligence quotient (IQ). If the metabolic error is circumvented by a special phenylalanine free diet, IQ is ostensibly improved.

GENETICS OF CHROMOSOMES

It might be best to consider the nature of the genotype at this point so that genetic principles are understood before we consider the complexities of the effects due to environmental fluctuations.

As noted before, the physical bases of heredity are the genes that are borne on the chromosomes carried in cell nuclei. These chromosomes can be observed during mitosis (when the cell divides and reproduces an exact replica of its genetic content, and in meiosis (including reduction division), where the chromosome number is halved to produce the gametes (spermatozoan or ovum) involved in sexual reproduction.

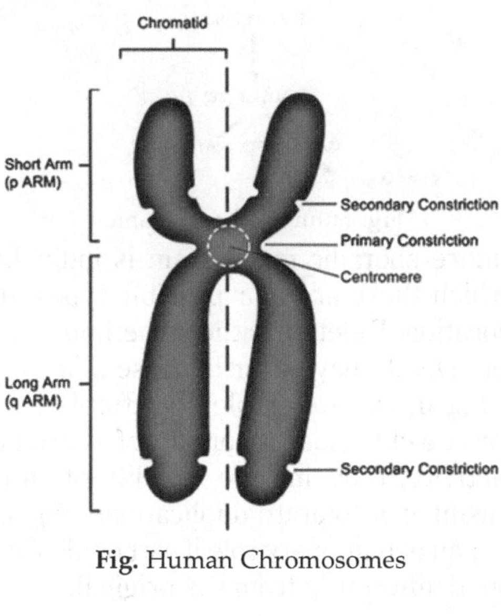

Fig. Human Chromosomes

The human species' 46 chromosomes are arranged in 23 pairs; in females all 23 pairs are homologous (correspond structurally and derivatively); in males, there are 22 homologous pairs plus a Non-homologous *X* and *Y* chromosome. The chromosome pairs, when properly stained, differ in length and appearance so that they can be individually recognized, much as one might recognize the familiar face of a friend. During meiosis, when the gametes are formed and the chromosome number is halved, all the chromosomes in a single gamete differ because each gamete has only one member of each chromosome pair. At fertilization, two gametes unite, each containing a set of 23 chromosomes to form the fertilized cell (zygote) with 23 pairs or 46 chromosomes once again. The number of chromosomes contained in gametes (23) is referred to as the haploid number and that of zygotes ($2 \times 23 = 46$) is the diploid number written as n and 2n, respectively. The process of meiosis affords a significant opportunity for increasing genetic variability. The most important consistent mechanism of change is crossing over, which takes place during the early phases of meiosis. Homologous chromosomes, lying next to each other, break, exchange equivalent portions of their genes, and reanneal to produce combinations that differ in arrangement from the genetic endowment provided by parents. This phenomenon is omnipresent in the normal meiotic process; when it fails to occur in at least one parent, the resulting organism is often sterile. Crossing over is the most consistent of a number of processes that provide the raw material upon which natural selection may act.

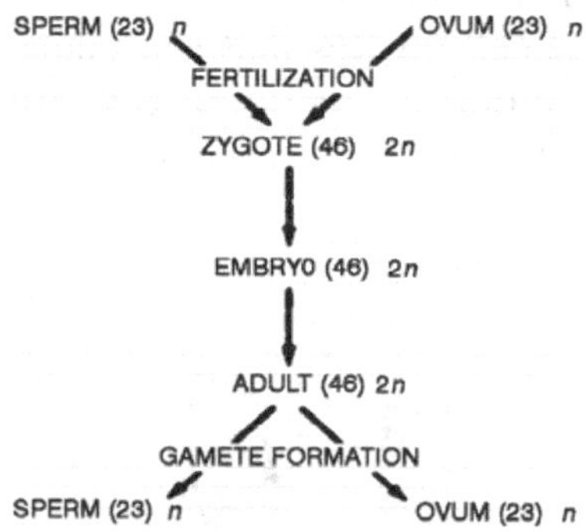

Fig. Human Chromosomes

A coarser and more sporadic mechanism is individualized chromosome rearrangement, of which there are four possible types: deletion, duplication, inversion, and translocation. Deletion is the sometimes lethal removal of a gene or sequence of genes. Death may occur because a biochemical sequence has been excised, resulting in an intolerable structural or functional gap in the organism. Duplication of a chromosome or part of a chromosome may cause an imbalance of gene activity, reducing the viability of an organism. However, because some organisms can tolerate duplications of genetic material, these duplications might play an evolutionary role if part or all of one of the duplications mutated and functioned differently from the original.

Indeed, this is one way in which evolutionary change is postulated to occur. Greater genetic variability gives rise (in minute quanta, to be sure) to physiologic and morphologic differences that endow the organism with an enlarged Behavioural repertoire so that it can more successfully cope with an ever changing environment.

This can be generalized for sexually reproducing organisms, which will be designated by n, where n is the haploid number and $2n$ the diploid number. An inversion occurs when a chromosome breaks in one or two places and the segment between the break rotates 180° leading to a reversal of gene order and a resulting change in biochemical product. A translocation occurs when two chromosomes from different pairs break simultaneously and exchange segments.

According to current knowledge, not all these chromo-somal and gene changes are equally important in their effects on Behaviour. To date, inversions and chromosome number changes appear to form the main categories of importance. But, crucial to the meiotic process (which reproductive systems evolved to support) is the shuffling and mingling of genetic material into infinite numbers of combinations.

MUTATION AND MUTANTS

Until now this discussion has centred on the shuffling of genetic material in sexually reproducing species. Genetic variability within a species has been discussed with emphasis on the variety of gene combinations attributable to meiosis, crossing over, duplication, and translocation. All these simply involve the reassortment of a set of variable genes indigenous to a population.

The essence of creating novelty is the process of mutation. These are alterations in the chemical structure of a gene that lead to changes in gene products that may ramify throughout the organism. Mutations at some loci occur spontaneously on the average of one in a million newly arisen gametes. The frequency is increased by radiation and by some chemicals known as mutagens.

A veritable storehouse of genetic mutations in mice is found at the Jackson Memorial Laboratory in Bar Harbour, Maine. Through selective inbreeding, when a mutant trait of interest is found, a strain of mice each of whose members possess the trait is established and bred for research uses. Many of the known mutant genes affect the nervous system during its development and produce a range of deficits in Behaviour from minor disturbances to severe defects. One of these found in the mouse is a mutation on chromosome five, an altered gene called *fidget*.

When this gene is present on both members of the fifth chromosome, it produces a Behavioural phenotype in which the mouse's head moves from side to side continuously. The gene is written *fi* and the mouse is said to have an *fift* genotype. In most mice the fidget gene is not present at this locus but the normal alternative gene, written as +, is. If the mouse is genotypically either *fi*+ or ++, the Behavioural alteration is not seen. Only if there are two *fi* genes present does the fidget phenotype become visible, as the wild type (+) gene is dominant to the recessive *fi*). However, dominance and recessivity are not necessarily complete, because heterozygotes (individuals with non-identical alleles) are often distinguishable from both homozygotes (individuals with identical alleles). Even if at first sight dominance appears to be complete, detailed biochemical tests may reveal differences between the heterozygote and the homozygote.

Untreated phenylketonuria in man provides one such example. Superficially viewed, the disease is controlled by a recessive gene *p*. with afflicted individuals being *pp* and normals *p*+ or ++. However, at the biochemical level, *p*+ and ++ may be distinguishable because the *p*+ individual tends to have more phenylalanine in blood serum. Dominance, in this case is incomplete. So, depending on which phenotype component is being examined, different conclusions as to the dominance may be drawn.

If a simple dominant—recessive relationship is assumed with two pairs of alleles at two loci (*i.e.*, *A* and *B* are dominant to *a* and *b*), what might ensue? If a double heterozygote (*AaBb*) is crossed with a double homozygote (*aabb*) what are the statistical expectations? The double recessive homozygote produces only *ab* gametes. From the double heterozygote each separate locus yields equal numbers of *AB, Ab, aB* and *ab* gametes according to Mendel's principles of segregation and of independent assortment of alleles during gamete formation. If these gametes are fertilized with the recessive *ab* gametes, the results, assuming two independent pairs of alleles and simple dominant recessive relationships, will yield a 1:1:1:1 phenotype ratio.

The backcross to a recessive parent (one with an *aabb* genotype) is often used as a testcross in practical breeding programmes because it can determine the actual genotype of an organism whose recessive alleles may be obscured by dominant ones. The use of this wholly recessive dihybrid parent unmasks recessive alleles in the alternate parental genotype. If we return to Mendel's garden peas and specify physical traits in a dihybrid cross (*A* is the gene for yellow and *a* the gene for green; *B* is the gene for a smooth surface and *b* the gene for a wrinkled one), we emerge with four types of gametes (*AB*, *Ab, aB*, *ab*) produced in approximately equal numbers, which yield the basic Mendelian ratio of 9:3:3:1.

RECOMBINATION AND GENETIC MAPS

There is an important exception to Mendel's second law. Assortment of genes is not generally independent if gene loci are on the same chromosome. The closer genes are to each other along a given chromosome, the more closely linked they are during gamete formation. Although genes on the same chromosome may recombine with each other during meiosis, the frequency of the recombination depends on the spatial distance between the relevant loci.

Generally, the further apart the loci are, the greater the possibility of recombination.From the frequency of these recombinational events, chromosome maps have been constructed for individual chromosomes. Genes located on the same chromosome are said to belong to the same linkage group. Thus, in humans we have 23 linkage groups (one for each pair of chromosomes, *i.e.*, the haploid number of chromosomes [*n*] corresponds to the number of linkage groups).

The house mouse, possessing 40 chromosomes, has 20 linkage groups; in *Drosophila melanogaster,* an organism of historical importance in Behaviour genetics, the numbers are 8 and 4 respectively. During the formation of human eggs and sperm (gametes), the chromosome number is halved (meiosis), leaving the egg with 22 autosomes and an X chromosome, and the sperm with 22 autosomes and an X or a Y sex chromosome. Males produce four sperm from each specialized cell in the testes and X- and Y-bearing sperm occur in about equal numbers.

GAMETES AND FERTILIZATION

In ovaries, meiosis also produces four products, but three of these become polar bodies, whereas the fourth becomes the chromosome containing nucleus of the developing egg. Human females are born with their entire reproductive quotient already present and the limits are set for the number of eggs they can produce (about 400 during a reproductive life span). Males, however, continue to produce sperm from sexual maturation until senescence.

Differences in gametic size and function are extreme. The head of the sperm is about 3 to 5 microns long and 2 to 3 microns wide. The tail, which propels it, is 10 times the length of the head. Essentially, it is a pared down packet of miraculously organized DNA (genetic material) with a launching and traveling apparatus to respond to and fertilize ova.

The human egg, in contrast, is one of the largest cells in the body—a sphere about 130 microns in diametre. It moves only passively when it is essentially shoved along, taking a few days to progress from ovary to uterus. As noted by Hartl (1977, cited in Probber and Ehrman,: "All the sperm that gave rise to all the people that ever lived could be carried in a teaspoon the eggs, would require a small bucket." The egg is comparatively enormous because in addition to its complement of chromosomes, its cytoplasm provides the nourishment for, and biochemical assistance to the developing embryo for translation of the information carried in its genes.Unlike the female gamete the function of the sperm as a cell is complete when the sperm contacts and penetrates the egg (having shed its tail), and sperm and egg fuse. With fertilization of the egg, formation of the embryo begins, but it is not until 6 weeks Afterwards that sex chromosomes become operative. At this point the *Y* chromosome, if present, will cause the inner layer of the gonads to develop into testes. If, however, there are two X chromosomes, at about 12 weeks the outer layer of the gonads will elaborate into ovaries.

X-CHROMOSOME INACTIVATION

A further complication relates to sex. Although every human embryo is predisposed to develop as a female, its eventual sex phenotype will be determined by the types of chromosomes borne by the sperm. Our species has 23 pairs of chromosomes, each of which carries thousands of genes that compose the genotype that develops and unfolds to become the visible organism, the phenotype.

In humans, 22 of the total number of chromosomes are autosomes; that is, two copies of these chromosomes are present in both sexes. The 23rd pair is made up of the unique sex chromosomes; the female has two of the X chromosomes; the male has one X and one Y chromosome. These were originally called X from X the unknown factor, and Y because it follows X alphabetically.

The X chromosome was first described in 1891 from insect material as "a peculiar chromatin body" but its function was not known then or for a long time Afterwards. Although the female had two doses of all the genes on the X chromosome and the male only one, there seemed to be no significant differences in X chromosome activity between the sexes. It was not until 1962 that a basis for dosage compensation emerged. This has been named the Lyon hypothesis after Mary F. Lyon, who first presented detailed studies on the subject. She hypothesized that in the normal female, one X chromosome is inactivated in each somatic body cell early in embryonic development. Whether the maternal or paternal X is rendered inert is a matter of chance and the result is that patches of cells with X-chromosomal lineages emerge.

The deactivated X becomes a tightly packed mass usually found close to the inner surface of the nuclear membrane of somatic cells in normal female mammals. It is called a Barr body after its discoverer. Accordingly, each cell in a male or a female has only one operative X chromosome. Because a female's cells are of two types (maternal X and paternal X), she is a genetic mosaic. A male, however, has only his maternal X chromosome to express.

What are the implications of this double X-chromosome dose for the female of our species? At the molecular level the X chromosome is a much more information packed and therefore biochemically active chromosome than the Y, which is much smaller and, except for male sex determination and fertility, inert. Recent work by McKusick and Ruddle cited more than 100 genes in the human X-chromosome map. The Y chromosome bears only two known genes, one controlling the TDF (Testis Determining Factor) and the other coding for the H—Y antigen.

The H—Y antigen plays an important role in the expression of male physiological and morphological characteristics, specifically for the differentiation of the testis. "Further male differentiation is imposed on the embryo by the action of testicular hormones against the inherent tendency

Towards the female phenotype." It is also known that males with their X and Y chromosomal complement are relatively unprotected from the expression of unpaired deleterious genes on their single X chromosome.

Every thing on their maternal X chromosome is expressed if not modified by genes on the autosomes. This is demonstrated by the much greater percentage of males afflicted by X-linked defects such as colour blindness and hemophilia. Genes on the X chromosome are said to be sex linked. In females the principles of heterozygosity and homozygosity apply to genes on the sex chromosomes just as they do for the genes on autosomes. In males, loci on the X are not matched by loci on the nearly inert Y chromosome; this condition is known as hemizygosity.

EPIGENESIS

The process of development involves constant interaction and coaction of an organism's genes and its environment. Given identical genotypes as in monozygotic human twins or highly inbred mice and *Drosophila,* phenotypes are usually very similar. Even under optimal circumstances however, phenotypic variation occurs because of perturbations in the environment. In a given situation one can speak of the average phenotype of individuals with the same genotype as a *norm of reaction*.

The *range of reaction* covers the spread of phenotypes that could potentially develop from a given fertilized ovum. Clearly, the range of reaction increases if the environment is not carefully controlled. The *epigenetic* view of development postulates that genotype and environment are involved at every stage from fertilization to birth and beyond. At fertilization, these actions begin at a molecular level.

Enzyme activities and changes in the intracellular environment are initiated. The egg provides a source of energy for the newly formed zygote and a well primed machinery for protein synthesis, which initiates the primary series of cleavages and invaginations that produce the embryo, fetus, and finally, the neonate.

The control mechanisms fostering maturation are encoded within the genes, in the internal and external environment of the DNA, and in the interrelations and interactions between them. At every level molecular, cellular, and organismic there are functional gene environment interactions and products.

5

Applications of Biotechnology

Recombinant DNA and cell fusion techniques may soon allow man to change food crops from inside out with great speed and precision, shaping life to fit the environment. The image presented of biotechnology is that the new knowledge of life processes will enable humans to live in a sustainable fashion where economic and ecological efficiency are optimally met.

For this to work, the production sector cannot treat nature as a great and infinite wellspring of resources that provides without being provided for. Have we finally reached a stage in the evolution of technology where new processes of production can be harmonious with natural processes?

This expectation carried in the rhetoric of its promoters set biotechnology apart from other industrial revolutions where the criteria for success almost always explicitly favoured economic growth and efficiency over the ideas of sustainable production, biocentrism, and ecological balance between humans and the natural habitat. Can biotechnology achieve such goals? Does this revolution embody the seeds of a gentler technology? The four chapters of this part of the book explore the applications of applied genetics to the environment and the social response to several key innovations.

Coming as they did in the wake of a new global environmental awareness, the possibilities of releasing genetically engineered organisms into the environment were met with high hopes, skepticism on issues of ecological safety, and some ideological opposition that questioned the wisdom of genetically redesigning plants and animals to increase the human harvest of nature.

THE BACONIAN LEGACY

The desire to control and rearrange species in nature is deeply rooted in the Western scientific tradition. Centuries before genes were understood, scientists contemplated a future in which the natural species were merely starting materials for the conscious rearrangement of the biotic world. For these scientists, the architecture of living forms was likened to the natural minerals of the earth. They were there to be cast into whatever form the human imagination could muster and the life sciences could render. Francis Bacon, the English scientist-philosopher, unleashed his imagination in a 1622 work titled *The New Atlantis* that describes a biological utopia. Bacon's writings heralded a new philosophy of nature and a new epistemology of science.

To have knowledge about the world is tantamount to having power over it. Control over nature was both the goal and the test of knowledge. Bacon provided a literary vision of how scientific knowledge will transform agriculture. We have also large and various orchards and gardens, wherein we do not so much respect beauty as variety of ground and soil, proper for divers trees and herbs, and some very spacious, where trees and berries are set, whereas we made divers kinds of drinks, besides the vineyards. In these we practice likewise all conclusions of grafting and inoculating, as well of wild trees as fruit-trees, which produce many effects.

And we make by art, in the same orchards and gardens, trees and flowers, to come earlier and later in their seasons, and to come up and bear more speedily than by their natural course they do. We make them also by art greater much than their nature, and their fruit greater and sweeter, and of differing taste, smell, colour, and figure from their nature. And many of them we so order, as that they become of medicinal use. We have also means to make divers plants rise by mixtures of earths without seeds, and likewise to make divers new plants, differing from the vulgar, and to make one tree or plant turn into another. Bacon's "biotopia" included an experimental programme for redesigning animals.

He described great parks with beasts and birds of all varieties. They were the natural substrate for altering phenotypes to meet human intellectual, practical, or aesthetic needs. Bacon's utopian prophecy is remarkable to the extent that it operated on the same paradigm that exists in the scientific community today, namely, that there are no unalterable properties of animals or plants. He rejected the idea of species barriers or the notion of integrated wholes. Furthermore, he recognized no moral considerations in the remaking of life-forms. The new secular science was not burdened with the ecclesiastical sanction that humans should not seek to improve upon God's creatures.

JACQUES LOEB AND THE CREATION OF LIFE

In the latter part of the nineteenth century, German-born physiologist Jacques Loeb advanced the field of biological engineering and believed its fundamental task was to produce living matter artificially. Like Bacon, Loeb refused to accept species boundaries or evolutionary constraints. He actively sought to transcend natural barriers if for nothing more than the challenge of it. Loeb first received international acclaim for his experiments in circumventing biological fertilization. In 1899, while working with sea urchins, Loeb discovered what he called "artificial parthenogenesis." He devised a method of fertilizing the eggs of sea urchins without male sperm by using inorganic salts. In Philip Pauly's *scientific biography of Loeb*, he notes that "artificial parthenogenesis was a vindication of Loeb's hopes and a model for science to come in which biologists would consciously work to reconstruct the natural order to make it more rational, efficient, and responsive to the ongoing development of engineering science."

Loeb demonstrated that what was believed to be a pure biological function-fertilizing an egg–was reducible to physical chemistry. His discovery brought banner headlines such as "Science Nears the Secret of Life," as well as a host of satirical cartoons. Like Bacon, Loeb rejected the Greek view of science as wisdom and understanding of the world as given.

In its place he substituted conquest and transformation as the goals of science, "utilizing the forces of nature to bring about new combinations, creating things which have never been created outside of nature's workshop." For example, Loeb studied the possibilities of redesigning animals that regenerate certain body parts.

Have succeeded in finding animals in which it is possible to produce at desire a head in place of a foot at the aboral end, without injuring the vitality of the animal. now have animals that have heads on each end of their bodies; thus have animals with bilateral symmetry that in nature have different oral and aboral poles. This idea is now hovering before me that man himself can act as creator even in living nature, forming it eventually according to his will.

Loeb developed a specialized vocabulary to describe his research programme in biological engineering. "Abiogenesis" meant the creation of living things from simple substances, whereas "heteromorphosis" described how a physical process could be used to create biological mutants (two-headed tubularia). Through a series of experiments on lower animals, Loeb sought a "technology of living substance" *(einer Technik der lebenden Wesen.)*

After the late nineteenth century, practical efforts to fulfill Bacon's utopian prophecy were initiated by plant breeders who sought perfection first through the process of breeding, screening, and selection, and subsequently through techniques of cross-fertilization and crop hybridization.

Plant breeders have altered the genetic strains of crops and flowers by methods developed over centuries, and hybrids have been disseminated widely in the environment. The genetic architecture of the hybrid plant is modified by human intervention, but not by human design, since there is considerable randomness and uncertainty to the process. Millions of plants may be screened in search for a few favourable genotypes. Modern genetics goes a step further than its classical counterpart.

No longer is the plant breeder constrained by species boundaries. As Bacon had imagined, plant species may be transformed according to some archetype or blueprint. Seeds, once thought to be the basic units of plants, are displaced from their privileged status by the cellular and subcellular units. A similar change has taken place with respect to animal life. The breeding of animals, which can be traced back to the earliest period of domestication, contributes an artificial form of environmental pressure. Evolution is guided by human intention.

Thus a sluggish, overweight sow is the desired phenotype for the feedlot, but not, perhaps, for survival in the wild. Industrial and biological evolution became one and the same. The results are animals that are selected for compatibility with humancentered systems–domestic, agricultural, or food processing. As a method to direct the evolution of animals, breeding is a slow, limited, and ponderous technology in comparison to what may be achieved by the application of molecular genetics to animal reproduction. Since genes can be inserted directly into the fertilized egg, obstacles of species incompatibility are easily overcome.

This was dramatized by several experiments on rodents. The first transgenic experiment on an animal embryo took place in 1981 when a rabbit globin gene was inserted into a mouse. In 1982 rat genes and subsequently human genes were implanted into the fertilized eggs of mice. The result was a mouse twice its normal size.

Also, with the methodology of germline manipulation of animals it is possible to create identical clones either by splitting the early embryo or by nucleus transplantation. In the latter case the desired full complement of genes is inserted into a fertilized, denucleated egg.

These and other applications of cell mechanics and genetic technology contributed to the commercialization of a new class of environmental products. The following section provides a framework for classifying these research programmes in biotechnology as a prelude to an examination of early public policy responses.

HISTORY OF BIOTECHNOLOGY

Biotechnology was being used even in ancient times although in it's more primitive and nascent form. A lot of evidence is available to suggest that making of curd, bread, wine and beer was known to pre-historic man. These processes were used in the kitchen by using the process of fermentation. Other methods of preserving the food items like-drying, smoking, curing , pickling in salt and brine etc were widely used by the prehistoric civilization residing around Near East and Europe. The traditional biotechnology refers to these conventional technologies which have been used for many centuries.

We are also familiar with the well documented process of mummification of dead bodies in Egypt using the technique of dehydration with a mixture of salts, largely sodium carbonate. Even before 2000 B.C. Egyptians also knew how to make vinegar from the crushed dates stored over a long period of time. Around 1500 B.C. people in Mesopotamia, Palestine and Egypt, were widely using the technique of making wine from crushed grapes and beer from bread leaven using yeast and germinated cereals. According to some scientists, Louis Pasteur should be considered as the "father of biotechnology" as it is he who identified the role of microorganisms in the process of fermentation. Later the process of Fermentation was properly documented in 18th century.

With the development of new apparatus such as Barometer and air pump, the Fermentation process was classified into three groups:

- That which accompanied the evolution of gas.
- That which accompanied the formation of alcohol
- That which accompanied the production of acid.

THE VERY FIRST BIOTECHNOLOGISTS

A number of scientists contributed towards the better understanding of the process. Joseph Black showed that upon heating magnesium carbonate with acid, a gas is released which is quickly absorbed by quick lime in water. Henry Cavendish, showed that same gas is released when the brown sugar in water was treated with yeast. In 1772, Daniel Rutherford reported that when an animal is kept in a closed container and then died the residual air after passing through a solution of alkali was noxious air. This air proved to be Nitrogen as experiments done by Karl Wilhelm Scheele.Later on Antoine Lavoisier provided the chemical basis for the nature of alcoholic Fermentation by using the analytical techniques for the quantitative estimation of carbon.

A French man, Nichola Appert (1810) described the method of food preservation. Peter Durand (1810) gave the use of tin container for food preservation by putting an air tight vessel containing food material in the boiling water.In 1830, Charles B. Astier gave the concept that air is the carrier of all kinds of germs.Later experiments proved microorganisms to be the cause of fermentation.

It was Charles Cagniard-Latour (1838) who gave the proof that alcoholic fermentation is cause due to the multiplication of brewer's yeast. In 1857, Louis Pasteur first showed the formation of lactic acid from sugar through fermentation due to the presence of lactic yeast using a polarimeter.

In the early 19th century, the contributions of Eduard Buchner, Hans Buchner and Martin Hahn proved that fermentation process does not require the presence of living organism as it was possible to achieve the same results by using the cell-free press juice or the proteins obtained by disrupting the microbial cells.In 1884, Charles Chamberland constructed a porcelain bacterial filter which helped in the discovery of viruses and their role as disease causing agents.

With the understanding of the genetic constitution of various living organisms, the biotechnological revolution began in 1970s and 1980s.This is called the era of modern biotechnology.

The major differences between the modern biotechnology and traditional / classical biotechnology are:

- The modern biotechnology uses the capability of science to change the genetic material for getting new products for specific requirement through recombinant DNA technology.
- The modern biotechnology specifies the ownership of technology and evaluates it's socio-political impact.

A strong foundation of genetic engineering and modern biotechnology was laid down by Cohen and Boyer in 1973. They successfully introduced the desired genes of one organism into another and cloned the new genes. The development of recombinant DNA technology also greatly influenced the advancements in the field of biotechnology. In 1961, the first journal of Biotechnology and Bioengineering was launched. The first guidelines for rDNA research were issued by National Institute of Health, USA in 1976. In 1978, the European Federation of Biotechnology was set up.

In 1980, the U.S. Supreme court gave the ruling that genetically engineered microorganisms can be patented. The U.S. approved the Humulin (human insulin) – the first pharmaceutical product of rDNA technology, for human use way back in 1982. Encouraged by the rapid progress made in this field, the human genome project was officially launched in 1990 and by 2001 the human genome was sequenced. Another milestone was achieved in 1997 when the first mammalian sheep, Dolly was created by nuclear cloning. In 1993, the first genetically engineered tomatoes, FlavrSavr, were sold in the market.

In the 19th century a lot of Industries came up in the area of the production of wine, beer, whisky etc. and a lot of growth also took place in the canned food Industry which helped to preserve the food for a longer time and made them available during off seasons.The 20th century witnessed the transformation of small scale Industrial fermentation in to large-scale manufacturing in Industrial sector and commercial application of the techniques.Another aspect of the scenario came into fore, during the Gulf war in 1991, when the work on microorganisms dominated in the preparation of biological warfare, antibiotics and fermentation process. Suspected preparation of biological and chemical warfare led to US attack on Iraq in 2003. As a result of interaction between various disciplines, lots of new branches came up like bioprocess engineering, food science engineering, metabolic engineering, artificial neural networks etc.The development in this exciting field is still going on rapidly which will solve some of the mysteries of life processes and at the same time open up many more new questions.

PRESENT SCENARIO

Due to multidisciplinary nature of the field of biotechnology, a wide range of different branches of science have made significant contributions to the fast development of this field. Some of these disciplines are- biochemical engineering, physiology, biochemistry, food science, material science, bioinformatics immunology, molecular biology, chemical engineering etc.

Biotechnology is also improving the lives of people around the world.The most famous example is the production of insulin. Before 1982, insulin required for the treatment of diabetes was obtained from pig and cow pancreas. The procedure used was very costly, tedious and often lead to complications while using the animal insulin. Using biotechnological tools, the human gene for insulin was isolated, cloned and expressed in microorganisms and large scale production of insulin started. Insulin was the first pharmaceutical product of recombinant DNA technology that was approved for human use. Millions of diabetic patients world over benefited by the biotechnological production of insulin.

Biotechnology also has affected economy in a positive way due to the creation and growth of small business, generation of new jobs. Agricultural biotechnology has reduced our dependency on pesticides. Bioremediation technologies are being used to clean our environment by removing toxic substances from contaminated ground water and soils.

The biotechnology has contributed in health services (diagnostics, vaccines, therapeutics, foods etc), agricultural sciences (improved crop yield, food quality, improved health of livestock and farm animals), and in environmental monitoring, pollution control. Modern biotechnology products started coming to the market in the late 1980s. Today, about 60per cent of the biotechnology products in the market are Health care products and 21per cent are products used in agriculture and animal husbandry. A considerable amount of efforts in research are on, to use and extract benefit from this interesting and upcoming field for the betterment of human life and the environment.

Many biochemical companies such as National Pituitary Agency (U.S.A), GeneTech Co. (U.S.A), Biogen (Switzerland), E. Lilly (U.S.A) are involved in the production of biotechnological products using genetic engineering techniques. Many countries have developed collaborative networks and relationships on various aspects of biotechnology such as" International Cell Research Organisation "(ICRO), " Regional Microbiology Network for South-East Asia" (supported by Japan and UNESCO), Microbiological Resource Centres (MICRCENS, supported by UNESCO).

ONGOING BIOTECHNOLOGICAL PROJECTS AND AREAS OF RESEARCH

The biotechnology is going to play a major role in the following areas of research and technology:

- *Production of bio-pesticides and bio-fertilizers*: This will reduce the dependency on chemical pesticides and will safeguard the environment.
- *Production of photosynthetically efficient plants*: This will solve the relevant energy related issues.
- *Production of transgenic animals as bioreactors for producing valuable drugs*: This area has generated a lot of debate as there are concerns

regarding the interaction of these transgenic animals with the environment and the consequences under certain conditions e.g if these undergo spontaneous mutations.

- *Human Gene Therapy*: Excellent alternative in medicine to cure certain diseases at genetic level.
- *Production of plants resistant to herbicides, viruses, insects and other pests*: This approach will not only reduce the indiscriminate use of chemical pesticides but will increase the crop productivity as well.
- *Production of diagnosis kits*: Production of diagnosis kits for toxoplasmosis identification.
- *Production of monoclonal antibodies*: Production of monoclonal antibodies for organ transplant tissue typing- This will help to reduce the failure rates of graft/tissue rejection.
- Production of human interferons microbiologically.
- *Production of human insulin microbiologically*: An alternative treatment for diabetics.
- Production of xanthan gum in oil fields for enhanced crude oil recovery.
- Microbiological upgradation of hydrocarbons.
- Production of ethanol by continuous fermentation.
- Bioprocessing alkenes to valuable oxides and glycols.
- Production of fructose using inexpensive forms of glucose.
- Improved production of Vit B12.
- Developing immobilized cell and enzyme systems to use in chemical process industries.
- To use genetic engineering for constructing organisms with desirable traits for their use in specific industries.

EVOLVING POLICY OF BIOTECHNOLOGY

FROM THE LABOURATORY TO THE FIELD

The timely development and rational introduction of R-DNA modified organisms into the environment depend on the formulation of sound regulatory policy that stimulates innovation without compromising good environmental management. Ecologists are unable to predict which introduced species will become established and which will not, and it is often not possible to explain successes or failure after the fact.

The weighty publicity over biotech products and research directions helped bring about a new configuration of public advocacy. Transgenic animals were an issue that linked animal rights organizations, environmentalists, and alternative agriculture groups. The development of more refined genetic screening techniques brought warnings from civil liberties and disability rights advocates. Disclosures of rising expenditures in the Department of Defense's biological Defence Programmes stimulated interest from activists in the disarmament community. Feminist-health advocacy groups began exploring the impacts of genetic techniques on reproductive technologies. Bioethicists, clinicians, and religious leaders began tackling the thorny problems of human gene therapy. The prospect of major pharmaceutical advances through rDNA research provided the grist for debates in the international health community on the priorities for developing vaccines. Food and agricultural organizations questioned the impact of biotechnology and the new patent provisions on control over plant genetic resources.

It appeared that every major industrial innovation in applied genetics tapped a wellspring of new issues that were brought to the social agenda. Of these, the issue that ignited the strongest public reaction during the early stages of the biotechnology revolution was the introduction of genetically engineered organisms (GEOs) into the environment. This chapter examines the origins of regulatory oversight over deliberate releases and the federal efforts at creating an orderly transition from small-scale labouratory applications of gene-splicing to large-scale releases of GEOs into the environment.

SOCIETAL CONCERN OVER GEOS

Why has there been so much concern over the risks of releasing plants and microorganisms that have been modified by genetic engineering techniques? Shouldn't the emphasis be on the product and not the particular way the product is created? There are three plausible explanations for the cultural selection of genetically engineered products as a special area of concern. First, the perception of risk associated with deliberate release has largely been formed from prior concerns about recombinant DNA research. In other words, the environmental problems of genetically modified organisms were inherited from earlier stages of the genetics debates. Recombinant DNA-produced organisms are what R. E. Kasperson calls a "social amplifier" in the public's perception of risk.

A second explanation is based upon the notion that genetic engineering provides far greater specificity and control over the product than one could achieve by plant hybridization or breeding of animals. As a result of the specificity of rDNA techniques and their capability of joining quite distant forms of life, the novel life-forms span wider species boundaries and are subject to fewer natural constraints.

By inserting a single foreign gene, a phenotypic property of a bacterium may be radically altered. Resistance to antibiotics is such a property. The ability to change phenotypes in the labouratory with such ease has heightened concerns about deliberate release. Will nature have an opportunity to accommodate to these sudden changes?

To the contrary, some scientists have argued that the specificity of genetic modification makes modern gene-splicing safer than conventional genetics, where genes get mixed randomly and in large clusters. Following this line of reasoning, the precision of gene-splicing means that the resulting properties of the organism will be easier to predict. According to plant geneticist Winston Brill writing in *Science,* in conventional breeding it is impossible to predict the properties of the progeny from most of the crosses. Genetically engineered plants have greater specificity: "If we compare plants derived from breeding Programmes with those derived through genetic engineering, it is clear that, in the latter case, the addition of a few characterized genes to the plant results in properties that are relatively easy to predict."

From another perspective, despite the specificity of rDNA techniques, with them one might be capable of producing more substantial changes in organisms with fewer genetic alterations than with classical genetic techniques. The issue of whether modern genetic engineering techniques are capable of producing varieties of plants, microbes, and animals that could never have arisen from the natural rearrangement of genes remains unresolved. It is widely acknowledged that we humans can at least accelerate or redirect the evolutionary process even if we cannot create qualitatively new life forms.

A third reason why genetically engineered plants and organisms designed for environmental release have attracted more concern than the release of similar products prepared through conventional genetics is related to the reputed power of the new technology. Gene-splicing has been the raison d'être of a technological revolution. This is not simply another discovery in the slow, incremental growth of science. This discovery has given birth to a new industrial process for radically reconfiguring biological matter. The disclosure that there is a new power to transform nature is one of the sources of public and scientific anxiety.

It might be argued that if rDNA technology embodies a power to stimulate the growth of a multibillion dollar industry, why should its risks be considered comparable to those of conventional genetics? What is the likelihood that the industrial potential of gene-splicing (gentechnics), which is, let us say, a thousand times greater than that of conventional biotechnology, will be unleashed without any increases in environmental risks? It is certainly a question worth considering.

Setting aside for the moment the body of scientific argument about the potential risks of new biotechnologies, there is an undisputed equation between technological power and risk anxiety that must be considered in fully understanding the public reaction to biotechnology. The simultaneous pronouncements about power and safety seem incongruous to a popular culture that has been sensitized to technological failure.

In trying to comprehend the risks of releasing genetically altered species into the environment, inevitably we are drawn to comparisons. Two technologies of commensurate transforming quality to gene-splicing are synthetic organic chemistry and nuclear physics. Both of these technologies are capable of creating new arrangements of matter in a fashion analogous to the creation of novel species through biogenetic engineering.

Risk assessment for synthetic chemicals has been in progress for several decades. There have been some important breakthroughs as well as notable impediments. The identification of a chemical substance is a well-defined process. It is, therefore, possible to construct a precise inventory of chemical compounds. The same is not true for biological agents, at least in some practical sense. Microorganisms and plants are classified by phenotype, and therefore the addition or deletion of a few genes will not necessarily warrant a change in the classification.

It is estimated that 60 to 80 thousand distinct chemicals are used in industry out of a pool of several million that have been synthesized. If a genetic identification system was used for biological organisms, the number of extant chemicals would pale against the number of distinct life-forms since, for the latter, a single nucleotide change would be a differentiating factor. Keeping track of novel organisms and establishing an identification system is a problem of enormous complexity, and probably unrealistic since genetic mutation is a constant occurrence. Yet any serious regulatory effort in biotechnology must address the identity question.

An obvious difference between inert chemicals and life-forms is that the latter are self-reproducing. Throughout the history of the chemical industry there have been countless cases where toxic carcinogenic chemicals were disposed of in lagoons and landfills. These chemicals saturate the soil and eventually migrate to subsurface water supplies where they contaminate drinking water. Once embedded in the earth, many industrial chemicals are difficult to remove. Entire Neighbourhoods in areas such as Times Beach in Missouri and Love Canal in New York have been evacuated because of toxic waste. In other, more manageable situations, contaminated soil is removed or filtration methods are applied to poisoned wells.

The mistaken release of a nuisance biological agent cannot be handled by techniques developed for chemical contamination. At the worst, the released organisms are beyond recall and will grow to population orders of a magnitude beyond the density of the inoculation. Moreover, if a novel organism were introduced and subsequently found to be dangerous, geographical isolation and community evacuation would simply not work.

Considerable progress has been made in standardizing toxicological testing for chemicals. The use of accepted methodologies, standardized target species such as germ-free mice or rats, and microbial assays such as the Ames Test have contributed to uniform standards of risk analysis. Notwithstanding the progress, there are still many areas of uncertainty and scientific debate.

Among them are questions about dosage and extrapolation from animal to human effects. Also, human epidemiological studies are frequently too insensitive to pick up small increases in cancer incidence over a lifetime exposure. While there are many effects of chemical exposure that are not well understood, there is at least a basic methodology for gathering the data.

There is no commensurate methodology for assessing the risks of released organisms. Moreover, the risks associated with certain chemical releases are real. Their effects on humans and the biotic world have been observed. In contrast, the potential risks of genetically altered life-forms are currently speculative. As a result, the social demand for evaluating the risks of bioengineered products designed for environmental use may not evoke the same urgency as if the hazards were confirmed.

When chemicals enter the environment, it is not always obvious what effects the breakdown products (metabolites) will have on the ecology. Chlorine has been used extensively to purify drinking water. Its use has been associated with the appearance of chloroform, a potent carcinogen, and other troublesome compounds called tribalomethanes. While there may be risks in the continued use of chlorine, the alternatives are not good. No safer method for purifying water is available. We can never be sure how released chemicals will reconfigure themselves in ecosystems, but compared to the possibilities of biological entities, the range of unexpected outcomes for inert chemicals is probably much narrower since the biological entities mutate.

A subclass of all possible mutations affects the phenotype of the entity. Whether it is more complicated to predict the mutational possibilities of a novel organism than to predict the synthetic pathways and metabolites of a new chemical is the subject of debate. Organisms are certainly more complex than inert chemical compounds. That does not mean that they are inherently riskier, but it does portend a high level of complexity in analyzing environmental risks of genetically novel entities in contrast to newly synthesized chemicals.

There are also some comparisons to be made between biotechnology and nuclear technology. The number of radio nuclides is relatively small, surely in comparison to industrial chemicals. They are generally used in well-defined and controlled settings. There are laws regulating the release of radioactive materials. Also, the health effect of radionuclides in high and moderate doses has been studied and is reasonably well understood. This is not the case, however, for low-dose, long-term exposure.

Radioactive materials are detectible in minute quantities with a sensitive monitoring device. To improve the confidence of residents living in the vicinity of a nuclear power plant, some communities have been provided with radiation detection counters.

There is nothing analogous in biotechnology. In theory, one can identify and track bacteria that have been released into the environment. The organisms would have to be tagged in a special way. Even then the identification can be a difficult task depending upon the Behaviour of the microbes and the conditions of the environment. Each case is unique. At this time, there are no standard methods of detection and no cannonical procedures for distinguishing safe from deleterious organisms. In comparison to bacteria, it is much simpler to detect the spread of an unwanted plant (a weed). But once released, plants, like bacteria, may be impossible to recall. We have seen how, within a period of a decade, a single critical discovery was the progenitor of an industrial revolution. The investment, scientific, and corporate communities moved expeditiously to capitalize on the commercial opportunities of the new genetics. To whom was this new industry accountable? How were the public policy issues handled? What social guidance was imposed upon the new technological direction? The next section discusses the early regulatory response to environmental applications of biotechnology.

Nih's Early Role

Beginning in the 1980s, industry and university proposals for field-testing genetically modified plants and organisms triggered a major science policy debate in the United States that spilled over to the European community. Those who have followed the recombinant DNA controversy from its inception will recognize that the current configuration of policy alternatives is the result of a ten year historical process. Initially, molecular geneticists cast the problem of genetic engineering in technological terms. Gradually, public perception of the problems associated with gene-splicing focused attention on the ethical and ecological issues.

The emergence of a second generation of genetics policy debates brought participation from new disciplines, new communities, new public interest groups, and new federal agencies. Public concerns slowly shifted from the singular issue of labouratory safety to a much broader range of problems. And while these changes were taking place, regulatory oversight of biotechnology also shifted from the National Institutes of Health to other governmental bodies.

For nearly a decade, the agency that assumed primary responsibility for the safe uses of genetic engineering was the National Institutes of Health (NIH). Essentially a science-funding agency under the Department of Health and Human Services, the NIH established the Recombinant DNA Advisory Committee (RAC) in 1974 at the recommendation of Paul Berg, the Stanford University biologist who provided leadership in the early efforts to assess the risks of rDNA research.

According to Berg's plan, an international meeting of biologists would result in a set of principles for safe handling of rDNA molecules. Those principles were then to be used by the NIH's newly formed scientific advisory committee to establish guidelines for all genetic experiments involving the cutting and splicing of foreign genetic material. Berg, along with other scientists who organized the Asilomar meeting, surmised that if the NIH did not act with dispatch in responding to the potential risks of gene-splicing, Congress might pass restrictive legislation. They viewed the passage of such legislation as detrimental to the interests of biology; particularly at stake was the legitimacy and progress of molecular genetics. First, biologists could lose influence over the risk assessment and risk management process. Second, the field of molecular genetics might become stigmatized as the only discipline whose principal research method was regulated. Third, there was considerable concern that rDNA legislation would be inflexible and difficult to amend. Asilomar organizers feared that biology would be saddled with irrational requirements.

Written by scientists for the general use of scientists, the NIH guidelines made their debut in June 1976. No explicit references were made to industrial processes or non-NIH supported uses of rDNA. The guiding principle behind the development of the guidelines was containment. Since one could not predict at the outset with any reasonable degree of certitude that a foreign gene introduced into an organism could not inadvertently transform it into an epidemic pathogen, the consensus at Asilomar was to construct a set of containment provisions consisting of physical barriers, safety operations, and carefully selected host organisms chosen because they do not survive well outside of the labouratory setting.

Each containment level was matched with a class of experiments that was permissible under the stipulated conditions. As scientists on the RAC and elsewhere became more confident in the safety of rDNA techniques, labouratory containment requirements were substantially relaxed.

The early NIH guidelines contained a provision that restricted industrial applications of rDNA technology. Cultures of rDNA organisms produced or handled in volumes greater than ten liters (classified as large-scale) were prohibited for NIH grant recipients. The large-scale prohibition was not based upon a scientific assessment of risk. It was a convenient threshold introduced by academic scientists to protect the use of standard labouratory beakers in basic research.

A second provision of the NIH guidelines restricted industrial activity by explicitly proscribing the intentional release of an rDNA organism into the environment. Since absolute containment was nothing more than an idealization, unintentional releases were considered unavoidable. However, by limiting the volume of rDNA culture, the probability of escape could be minimized. Also, since the volume of agent released is correlated to survival and propagation, the large-scale prohibition also supported the general containment strategy.

Commercial interest in rDNA techniques grew rapidly in the mid- 1970s. Estimates of industry growth vary. My own count indicates that a minimum of 14 new biotechnology enterprises (NBEs) were formed in 1976, the year the NIH guidelines were introduced. During 1979, NBEs grew by at least 26; and in 1981, at least 66 NBEs entered the biotechnology industry.

The NIH had no legal jurisdiction over the research in private companies. In practice, however, firms were not willing to risk the negative publicity that might arise if they violated the NIH guidelines. The nascent biotechnology industry was comprised predominantly of small firms started by relatively young scientists, many of whom retained their university affiliation. The close link between academe and industry may help explain the high degree of compliance among new firms with the standards adopted by the NIH.Since these scientists were groomed on the NIH guidelines, the complications of compliance that might have beset a new industry were minimized.

Despite a watchful media, there is no evidence that the NIH guidelines were flaunted by the biotechnology industry. Ironically, the few cases where violations of the guidelines were reported took place at universities. Geoffrey Karny cited two factors responsible for industry's compliance with the voluntary guidelines: "First, the possibility of tort lawsuits provides monetary inducement to comply with the *Guidelines,* which would probably be accepted as the standard of care against which alleged negligence would be evaluated. Second, the threat of statutory regulations, which the companies have sought to avoid, always exists."

Between 1976 and 1979 the NIH process for overseeing rDNA research was put to its severest test. First, the city of Cambridge, Massachusetts, issued a moratorium on rDNA experiments requiring moderately high physical and biological containment. After a widely publicized citizen review process, the city passed the country's first rDNA law in 1977. The law departed from the NIH guidelines in a few minor respects. More importantly, it symbolized the right of local government to exercise control over where the research gets sited and the safety conditions of its performance. Moreover, it established uniform and legislatively mandated guidelines for both publicly and privately funded research.

After the Cambridge rDNA law was passed, nearly two dozen states and local communities debated the issues. Legislation was enacted in about half the jurisdictions. In response to local events and a national mood of concern Towards gene-splicing, fifteen distinct bills were filed in Congress between 1977 and 1978 to regulate rDNA research. Some of these bills would have shifted the regulatory authority from NIH to a national commission. These bills also varied in the degree to which local laws were subject to federal Pre-emption. Congress spent two years debating the issue of an rDNA law. A compromise bill was finally voted out of committee early in 1978, but for lack of strong congressional leadership and interest it failed to reach the House or Senate floor for a vote.

Congressional failure to enact legislation strengthened the NIH's position as the sole agency overseeing rDNA activities. Responding to continuing public concern over the research, the Department of Health, Education, and Welfare (HEW, currently the Department of Health and Human Services) rewrote the RAC's charter and increased the size of the committee from sixteen to twentyfive. The new charter, issued in 1978, stipulated that one third of the committee was to consist of individuals with expertise and interest in public health and the environment. The change in the composition of the RAC drew sharp criticism from prominent scientists who argued that rationality was being compromised by including Non-scientists in what was essentially a technical process. When the expanded RAC met in early 1979, its agenda was filled with petitions for relaxing containment requirements and approving additional host-vector systems.

Over the next few years several important changes were made in the rDNA guidelines that established a role for the NIH in the review of field tests. A voluntary compliance Programmes was established that gave the private sector access to the NIH review process, while prohibitions against large-scale rDNA activities and the intentional release of genetically altered strains into the environment were removed.

After the decision by Congress not to enact rDNA legislation, there was a strong residue of criticism that the NIH guidelines could not protect society from the potential adverse consequences of commercial gene-splicing. In response, NIH developed a voluntary compliance Programmes for institutions that did not fall under the agency's purview. This initiative gave the fledgling biotechnology industry the opportunity to gain the imprimatur from the NIH for both labouratory and commercial-scale rDNA work. A firm wishing to participate in the Programmes first submitted the composition of its institutional biosafety committee (IBC) to the NIH's Office of Recombinant DNA Activities for approval. Once IBC approval came, a firm could file requests with the RAC following procedures similar to those of university petitioners.

One difference between the NIH's handling of academic and industry proposals is that, on the occasion of the latter, the RAC went into closed session. Members were required to sign confidentiality pledges for the protection of information deemed proprietary by the firm. Some RAC members were opposed to having the committee review proposals in closed session. One public interest member refused to participate in the review of industry proposals. He argued that oversight of the private sector was the responsibility of those agencies of government with statutory authority to protect workers, public health, and the environment.

Since the NIH lacked authority to carry out these functions, he felt his participation would give legitimacy to this extension of the NIH's role. Another RAC member expressed the following sentiment: "Voluntary compliance is the worst of all possible worlds.... You achieve none of the objectives of regulation and none of the benefits of being unregulated. All you're saying is 'I give a stamp of approval to what I see before me without any authority to do anything.' "

The Programmes was also the target of mainstream critiques. As an example, in its 1981 report on biotechnology, the Office of Technology Assessment wrote: "The most significant limitation in the scope of the Guidelines is their Non-applicability to industrial research or production on other than a voluntary basis. This lack of legal authority raises concerns not only about compliance but also about NIH's ability to implement a voluntary Programmes effectively." In May 1979, the NIH's advisory committee went on record opposing voluntary compliance by a vote of nine to six, with six abstentions. The RAC voted that non-NIH funded institutions should be required to comply with the guidelines. This recommendation notwithstanding, the voluntary compliance Programmes became a permanent part of the guidelines in January 1980.

Many biotechnology companies considered it in their long-term interest to secure RAC approval for their rDNA activities. Regardless of how the firms' management felt about the potential risks of gene-splicing, submitting experiments for NIH approval was excellent public relations. The voluntary compliance Programmes helped the biotechnology industry respond to the criticism that the private sector was operating without regulations. When pressed by local communities to demonstrate the safety of genetic experiments, a company's most compelling response was that it complied with the NIH guidelines.

In the wake of sporadic episodes of local opposition to genetic engineering research, the biotechnology industry sought a predictable and stable regulatory climate, but one that would not impede research and development. The NIH contributed to this goal, but with limited success. The voluntary compliance Programmes proved functional to the industry during its early years of development when there was intense competition, investment instability, and the uncertainties of local regulation. As the commercial activities progressed from labouratory research to product development, the NIH policies accommodated to the new stage of industrial activity, particularly in their response to large-scale work and the release of genetically modified plants and organisms into the environment.

Large-scale Prohibitions

Provisions were built into the early NIH guidelines for waiving the large-scale prohibition in cases where minimal risks were balanced against important societal benefits. The wording of the prohibition was clarified in the 1978 version of the NIH guidelines. No exceptions to the prohibition against the production of large scale cultures were permitted "unless the recombinant DNAs are rigorously characterized and the absence of harmful sequences established." Under NIH's leadership the risk assessment paradigm was in large measure still under the primary influence of geneticists.

For a short period of time, the NIH gave serious attention to all phases of large-scale work with rDNA molecules. Proposals submitted to the RAC were required to include a description of labouratory practices, specifications on physical and biological containment, risk data, characterization of genes, and the design of the fermentation equipment in conjunction with the physical description of the facility. In 1980, the RAC published a standard that it planned to use for reviewing large-scale proposals. The committee's role in evaluating plant design and operations for large-scale fermentation drew criticism from some members. They argued that the RAC should confine itself to advising the NIH on the nature of biological procedures and not plant operations. The committee, after all, had no special expertise in bioprocess safety engineering. It depended on outside consultants for guidance in this area.

Also, it was a matter of some significance that molecular biologists on the RAC found the review of plant design boring and uninformative. Other RAC members contended that if the voluntary compliance Programmes was to mean anything, a comprehensive review of large-scale operations was essential. Since no other federal agency was engaging in this review, they believed that it was the NIH's responsibility to fill the regulatory void.

The internal debates over this issue were a poignant reminder of the NIH's ambivalence in serving as overseer of commercial genetic engineering. These debates also cast doubt on the logic of having a biomedical funding agency guide a nascent industry exclusively through a system of voluntary measures. The peculiar nature of this regulatory Programmes began to reveal its contradictions. For example, within the RAC opposing factions interchanged positions.

Initially, the group most skeptical about the safety of genetic engineering expressed opposition to the RAC's review of industrial proposals, particularly large-scale ones. They reasoned that the NIH was acting as de facto regulator without enforcement powers or congressionally derived authority. If the RAC refused to serve this function, members of this faction believed that agencies like the Occupational Safety and Health Administration (OSHA) and the Environmental Protection Agency (EPA) would enter the field.

Concurrently, there were other members of the RAC who opposed broader federal involvement in biotechnology, particularly legislative or rule-making actions. They viewed NIH's continued role in overseeing commercial rDNA work as essential to Pre-empting such initiatives.

The unexpected reversal took place among committee members around 1980. The pro-regulatory group grew less sanguine about the involvement of other agencies in regulating biotechnology. As a consequence, this faction began supporting a stronger role for the RAC. Meanwhile, the committee's regulatory minimalists, also confident that broader agency involvement was unlikely, backed an NIH role over commercial rDNA activities that were limited to issues of pure genetics, namely, sequence characterization and approval of host-vector systems.

The regulatory minimalists succeeded in so limiting the RAC's review of the safety of rDNA products. By 1981, firms taking part in the NIH's voluntary compliance Programmes were no longer obligated to submit data on sterilization of their fermentation system or their procedures for disposal of rDNA cultures. This policy change was expressed in a proposition adopted at the September 1980 RAC meeting: "The RAC will determine if a given recombinant

DNAcontaining strain is rigorously characterized and the absence of harmful sequences is established. Such a determination shall include specification of a containment level. These determinations should not in any way be construed as RAC certification of safe labouratory procedures for industrial scale-up."

The potential release of rDNA organisms through the effluent of bioreactors did not attract any public attention. While the EPA had jurisdiction over such biological releases under the Clean Water Acts, a constituency for agency action did not develop. The newly formed Committee (later Council) for Responsible Genetics published an article on biogenetic waste in its public interest bulletin *GeneWatch.*

However, the issue of biogenetic waste was not a rallying point among environmentalists or the general public. It seemed to lack some important features that influence risk selection. First, it did not impart a clear and present environmental danger. Second, the waste stream, even if it carried genetically engineered organisms, did not excite the media. Bioeffluent was not intentionally designed to alter nature. Since science writers are captivated by the frontiers of science, there was not much of a story in fermentation sludge.

Quite a different media and public reaction took place when proposals appeared before the RAC requesting approval to field-test genetically modified plants and bacteria. There is an informative distinction to be made in the public reaction between intentional and unintentional releases of genetically modified life-forms. To fully grasp the distinction, we have to look at the kinds of experiments proposed, the types of media coverage they were given, the role of scientists in raising safety concerns, and the development of a public interest constituency.

Deliberate Release of GEMs

As the RAC reduced its oversight over industrial bioprocesses, it became more active in reviewing environmental releases of genetically engineered microor ganisms (GEMs). Consequently, field-testing was given greater visibility in the media. From 1980 to 1984, when companies and university scientists were preparing to field-test their products, the NIH was the only federal agency with active responsibility over deliberate release experiments involving genetically altered plants or microorganisms.

Progressive relaxation of the rDNA rules finally led to the removal of barriers to field-testing. The first revisions of the guidelines in 1978 still prohibited "deliberate release into the environment of any organism containing recombinant DNA," but individual waivers were permitted after proper public notification, RAC review, and approval by the director of the NIH. The revised NIH guidelines of 1982 eliminated the entire list of proscribed experiments. By June 1983, the prohibition against intentional release of rDNA organisms was replaced by a multitiered review process. Submissions for deliberate release required approval by the RAC, the institution's biosafety committee (IBC) and the NIH director, in addition to various subcommittees. Agricultural applications of biotechnology, widely publicized in the media, were being readied for field trials. By December 1983, the RAC had reviewed and approved three proposals for releasing genetically altered life-forms. In each case, the committee concluded that the tests posed no significant risk to health or the environment.

The first three proposals for deliberate release came from university scientists and therefore did not involve proprietary information. In March 1980, a Stanford University scientist requested approval from the RAC to field-test a corn plant into which had been inserted the corn storage protein gene with modified sequences. The genetically altered strain included a new corn protein gene that encoded all the amino acids essential to humans (including an enhancement of lysine and methionine, in which corn is deficient).

By genetically engineering the new corn protein with the full complement of the essential amino acids, its value to the human diet would be improved. The RAC failed to come up with a hazard scenario for the genetically modified corn; nevertheless, the investigators were required to detassel the plant to prevent pollen dispersion during the field trials. Permission for the test was granted in August 1981. There was no significant public reaction to the decision.

A second proposal for field-testing was brought to the RAC by a Cornell University scientist in June 1982. Tomato and tobacco plants were transformed with DNA from yeast and *E. coli* to provide them with antibiotic resistance. In reviewing the experiment, the RAC raised concerns about the possible spread of antibiotic resistance and the effect of an itinerant plant on the ecology. Neither of these concerns delayed the committee's approval, which it gave on October 25, 1982. When the recommendation reached the director of the NIH, he referred the proposal to the U.S. Department of Agriculture (USDA) for a reading. After USDA approval was given in February 1983, NIH accepted the field test in April 1983, nearly a year after the proposal was first brought to the RAC.

The third of the early field test proposals for genetically modified life-forms proved to be the most controversial. Ironically, it was viewed by some experts as among the safest deliberate release experiments one could perform in the environment. There was one important distinction between the first two proposals and the third: the latter consisted of genetically altering a soil bacterium. The difference between plants and bacteria proved to be a critical factor in the public perception of risk. Also, there was an established tradition of introducing hybridized plants into the environment. However, no analogous tradition existed for microorganisms. The relative novelty of this field test was reflected in the RAC's review.

In September 1982, scientists at the University of California at Berkeley proposed to field-test two soil organisms (*Pseudomonas syringae* and *Erwinia herbicola*) from which about a thousand base pairs of DNA sequences had been deleted. In their natural state these organisms synthesize a protein that provides a nucleation point for ice crystallization, and are known as ice-nucleating agents (INA). By excising the genes responsible for ice nucleation and establishing the genetically modified organism in the environment, scientists believed they could reduce the temperature at which frost begins to form on crops.

When the proposal for field-testing the microbes with the deleted ice-crystallization gene (denoted INA^- or ice minus) was first brought before the RAC, one member expressed concern about the role the organisms might play in the environment. Questions raised about the effect of INA^- on precipitation patterns were based upon some theories that INA^+ (normal ice-nucleating agents) played a role in atmospheric weather. The precipitation issue was not resolved by the committee. Nevertheless, the RAC approved the field test, although by a small plurality (seven yes; five no; two abstentions).

When the recommendation was transmitted to the director of the NIH, more data along with additional safeguards were requested. Several months later on the second round of review, the RAC approved the proposal unanimously. The NIH gave the final go-ahead for the field test in June 1983, some nine months after it had initially received the proposal. Investigators agreed to mark the strains of the ice minus organism with antibiotic resistance, which allowed them to monitor the dispersal of the organism.

They also agreed to limit field-testing to a single location, the University of California agricultural field station at Tulelake in northern California, a site isolated from the major fruit tree and citrus-growing regions of the state. Despite these precautions, the RAC's decision on the field test for ice minus was met with citizen opposition and litigation from the Foundation on Economic Trends, Jeremy Rifkin's organization based in Washington, D.C. Between 1980 and 1985 the RAC reviewed five proposals for field-testing GEMs. Four were approved and one was voluntarily withdrawn.

GENETICS AND THE ENVIRONMENTALISTS

In the mid- 1970s, when rDNA molecule research was still an exotic technique, the public entry into the debate over its safe uses was determined by several factors. First and foremost, a group of distinguished biologists called attention to the new technique by publishing letters in science journals that caught the attention of the popular press.

The international meeting at Asilomar was attended by science writers, whose coverage of the events dramatized the disagreements among biologists at the meeting. In drawing attention to the new discovery, the Asilomar organizers did not intend for the fate of the new technology to be decided by popular acclamation or democratic process.

However, some biologists, frustrated by the NIH's role in setting safety standards, brought their concerns to university campuses and local communities. To the public, it appeared that scientists were polarized on the dangers of rDNA research.

Second, the issues were dramatized through the world press by the controversy that erupted in Cambridge, Massachusetts, which pitted community values against the interests of national science. The 1976 Cambridge rDNA controversy represented the birth of public involvement in genetic engineering. Prior to that, the issues were debated exclusively in professional groups and on university campuses, notably the University of Michigan.

Third, national environmental groups such as Friends of the Earth, the Environmental Defence Fund, and the Sierra Club devoted some attention to genetics issues between 1976 and 1979. But the commitment by these groups was limited usually to a single staff person. Their activities were met with skepticism, and in some cases, disapproval by some board members who had ties to the biomed ical community. Fourth, but to a lesser extent, litigation had drawn public attention to the issues. Two suits were filed against HEW/NIH in 1977. A resident of Frederick, Maryland, petitioned the court to enjoin the NIH from performing a risk assessment experiment until an environmental impact statement was prepared. A second suit filed by Friends of the Earth alleged that HEW failed to comply with the requirements of the National Environmental Policy Act when it established policies for rDNA research.

Fifth, congressional interest in the safety of rDNA research intensified in the mid-to-late 1970s. House and Senate hearings provided a forum for activist scientists and environmental lobbyists. The accumulated effect of these factors resulted in heavy media coverage for both local and national events related to gene-splicing experiments.

The subsequent public debate over the specific issue of the deliberate release of genetically engineered organisms has been fostered by a similar configuration of factors, namely, scientist-critics, litigation, environmental coalitions, local opposition to experiments, and congressional interest. However, the old and new rDNA debates differ in the weighting of these factors. For example, Jeremy Rifkin has proven particularly effective in drawing public attention to the environmental and ethical issues associated with the deliberate release of GEOs. Operating with the modest resources of his Non-profit organization, Rifkin has utilized a variety of techniques, including confrontational protests, litigation, and short-term coalition building, to dramatize breaking issues in genetics and public policy.

Rifkin's philosophical opposition to genetic engineering has been expressed in his writings and media accounts that feature his work: "Genetic engineering represents the ultimate negation of nature." On genetics policy, he has had more impact on the media that any single group or individual in the United States. His notable success as a publicist and genetics critic can be explained by several factors. Rifkin can act quickly since he is not accountable to a board of overseers or a mass organization. He is able to identify and capitalize on the weakest link of any policy process. With properly timed and targeted litigation, Rifkin has been able to strike at the jugular of the bureaucracy.

He has proven time and again that a well-placed lawsuit is a magnet for media attention. Rifkin made his national debut as a genetics activist in 1977 when he and a group of supporters invaded a meeting at the National Academy of Sciences devoted to rDNA science and policy. His strategy at the NAS is best described by the 1960s term "guerrilla theater," a dramatically staged political confrontation. Fearing major disruption of its conference, academy officials allotted Rifkin time at the start of the Programmes for a statement. Meanwhile, his associates, donning silk stocking face masks, stood like motionless icons, projecting a grotesque apparition of human mutants. While Rifkin took over the podium of the academy auditorium, his colleagues stretched a banner across the stage with a quote from Adolf Hitler that stated: "We will create the perfect race."

On another occasion, Rifkin organized a signature campaign of scientists and religious leaders who were asked to support a statement against genetic manipulation of the human germline. The list of supporters included individuals of vastly different ideological persuasion. To have such a politically diverse group on the same side of an issue was itself a news event. It proved worthy of a front-page story in the *New York Times.*

In the fall of 1983, when the RAC was reviewing a proposal for field-testing ice minus, Rifkin brought national attention to the problem of deliberate release by filing suit against NIH on the grounds that its approval was given prior to a full study of the environmental consequences. His suit was supported by two organizations, Environmental Action and the Environmental Task Force. To strengthen his case, Rifkin received sworn affidavits from leading ecologists who opposed the release of the ice minus strains, at least until more could be learned about the environmental consequences.

Rifkin's lawsuits have been rallying points for critical debate on genetics issues. Philosophically, he personifies the modern genetics Luddite. But on a practical level, he has made effective use of environmental law to impose a greater burden of responsibility for assessing the risks and eliciting the ethical consequences of new technologies. Ironically, he is secretly applauded by many who, while they might differ with him philosophically, find a certain degree of relief in slowing the pace of regulatory approval. Environmental groups, skeptical about deliberate release, have been hesitant to act more aggressively because the hazards of genetic engineering are still hypothetical. For Rifkin, it was less important to prove the hazards exist than to rally support against the hubris of science and reexamine its assumptions about progress.

There is a striking comparison between the old and new genetics controversies on the role of scientific participation. In 1976, the molecular geneticists were taken to task for not calling upon infectious disease epidemiologists to evaluate the risks of transforming *E. coli* into an epidemic pathogen. Subsequently, the infectious disease community did participate in the risk evaluation. Similarly, ecologists played no role in the rDNA debate until the controversy over deliberate release erupted. Their entry into the discussion over deliberate release came when the social and political context was ripe for their participation. Ecologists are viewed as natural allies to environmentalists; when the latter became involved in the debate over field tests, they sought advice from the former.

Also, the Committee on Science and Technology of the House of Representatives made specific recommendations to NIH and USDA that they revise the memberships of their advisory committees to include individuals specifically trained in ecology and environmental sciences.

Finally, the EPA began developing its own policy on deliberate release in the early 1980s. The agency is accustomed to working with ecologists, for example, in evaluating pest control methods. The first genetically engineered organisms planned for environmental release were classified as biological control agents. The EPA began hiring consultants to study the potential environmental problems of releasing such agents. A terrestrial ecologist from Oak Ridge National Lab oratory, Frances Sharples, completed a study for the EPA in 1982 on the effects of the introduction of organisms with novel genotypes into the environment.

Microbial ecologist Martin Alexander of Cornell University chaired an EPA study group on biotechnology. In 1984, ecologists participated in an EPAsponsored workshop at the Banbury Center at Cold Spring Harbour, New York, on the evolutionary consequences of biotechnology. Also, in 1985, with support from the EPA, the USDA, national environmental organizations, and corporations, ecologists, geneticists, public interest representatives, and federal regulators met for four days at Cornell's biological field station in Bridgeport, New York, to develop principles of containment for genetically engineered organisms tested under field conditions.

By raising the issues of biotechnology in the national media, Rifkin also helped catalyze the concerns and perspectives of ecologists on the problems of deliberate release, particularly those who had not been brought in as consultants to the major agencies. Eventually, letters from ecologists appeared in *Science,* followed by responses from geneticists. A dialogue was opened between molecular geneticists, microbiologists, and ecologists on their approaches to risk assessment. Soon it became evident that the issues over ice minus unfolded into a larger debate about scientific culture, epistemology, and disciplinary hegemony.

PRINCIPAL DIRECTIONS OF BIOTECHNOLOGICAL DEVELOPMENT

In a recent review of agricultural technology, the Office of Technology Assessment (OTA) of the US Congress defined biotechnology to include 'any technique that uses living organisms or processes to make or modify products, to improve plants or animals or to develop micro-organisms for specific uses-it focuses upon recombinant DNA and cell fusion technologies'. Longworth concurs with this definition, with the significant addition of tissue-culture techniques, an aspect of biotechnology which has great economic, and therefore social and political, potential impact. Despite debates about whether these or any other definitions are adequate, they do touch on the main aspects of biotechnology which deserve to be considered.

The aspect of this new biotechnology which most captures the imagination and stirs the greatest controversy is gene splicing or recombinant DNA techniques, which inspire researchers to consider the possibilities of producing reproducible animals and plants markedly different from current existing species, and referred to as transgenic species. Already in higher animals and plants recombinant DNA has produced transgenic forms which are being commercially exploited. Their agricultural significance is, however, so far limited, and commercial application is concentrated in highly profitable pharmaceutical and horticultural markets. Genes have been introduced into several animal species which alter their protein synthesis to enable transgenic sheep to produce insulin in their milk, and rabbits to produce interferon.

A completely different application has originated in Denmark for salmon, where it has proved possible to introduce germplasm which enables the salmon's physiology to handle heavy metals, which are normally toxic, so opening up new locations for farm fisheries.

Much current research is directed to conferring disease immunity on animals, and holds out the prospect of widespread commercial application. In plants one achievement has been the transfer of genetic resistance to antibiotics in the petunia, and another has been the introduction of storage-protein genes from French bean plants into tobacco plants. As yet commercial progress with recombinant-DNA technology in plants appears limited, but extensive opportunities beckon, particularly because plant research is less restricted by the ethical and animal-welfare concerns which apply to research on transgenic animals.

IMPORTANT COMMERCIAL DEVELOPMENTS OF GENE

The most important commercial developments based on gene splicing have so far occurred with genetically much simpler microorganisms, and it is with these that the greatest short-to mediumrun commercial potential lies. Already genetically engineered micro-organisms are producing a variety of hormones, vaccines, enzymes and other proteins. Important examples are the production of insulin, vaccines for neo-natal diarrhoea in calves and piglets, and the bovine growth hormone BST identical to that produced in cows which can stimulate a 20 per cent increase in milk yield. BST is already a source of problems for legislators in the European Community, and has provoked strong reactions from the media and milk consumers. As far as larger animals, and cattle in particular, are concerned, it is developments in embryo transfer and in many processes for manipulating reproduction which hold out the prospect of continuing increases in yields, feed conversion efficiency and general economic efficiency. Already, apparently over 1 per cent of dairy calves in the USA are from embryo transplants, despite the high costs still associated with this procedure.

Widespread adoption of these sophisticated technologies would further distance livestock farming from its traditional rural simplicity, and from the natural mating of animals to produce offspring. It would place technical demands upon the operators which favour large-scale company farms capable of supporting a range of highly trained specialists. Longworth identifies new techniques of tissue and cell culture as having 'the potential for enormous advances in crop improvement in the next couple of decades'. Those techniques 'can both increase the genetic diversity and greatly increase selection efficiency', and they permit innumerable plants to be reproduced asexually from single cells or small pieces of tissue. The particular technique known as 'callus culture' has been used for many years to clone highly valued horticultural plants such as orchids, and cloning of cuttings is widely practised for tree crops such as tea and palm oil as well as by millions of gardeners for garden plants.

According to Longworth 'cell culture' of single cells has unexpectedly, and so far inexplicably, resulted in plants with different properties being regenerated from the same clump of parent tissue. Sugar cane, maize and potato plants regenerated in this way have been found which are resistant to important pathogens. This application of cell culture with the capacity to generate vast numbers of seedlings rapidly has the potential to simplify greatly the hitherto labourious procedures of plant breeding and selection. However, in terms of current and immediate commercial importance, it is through micro-organisms that biotechnology has its greatest impact. Microbial fermentation processes have been of great commercial significance for centuries, for example in bread, wine and cheese production, and there is the prospect of considerable development. Two recent examples of new processes indicate the sorts of impact that such developments can have.

In the late 1960s genetically engineered bacteria were developed which were able to digest corn starch to produce high-fructose corn syrup (HFCS) and which left as a residue corn gluten, which is now an important protein feed for livestock. HFCS has made considerable inroads in the United States and in Japan. It has largely replaced sugar as a sweetener in Coca Cola as well as many other food and drink products. This has helped depress sugar prices to Third World growers. In the European Community steps have been taken to prevent HFCS and other new powerful sweeteners from being produced and from undermining the market for domestically grown beet sugar, which is supported by the Common Agricultural Policy.

A second important application of microbial fermentation has been the production of ethanol from sugar cane in Brazil and from maize in the USA. The commercial viability of these processes is critically dependent upon the price of oil as the main non-renewable source of fuel, and, to date, massive subsidies have been required to maintain the Brazilian and USA ethanol programmes. Longworth states, however, that a new biotechnology, Sucrotech, is being patented, which will not only reduce the cost of producing ethanol from sugar cane, but will simultaneously produce fructose at a cost which will be competitive with HFCS and may thereby reclaim part of the sweetener market for sugar cane. That such possibilities are in prospect indicates how volatile the future might be; biotechnology has tilted the competitive balance from sugar cane to maize, causing economic pressure and even disruption to sugar-cane-dependent economies and may in future switch it back again.

In the long run the capacity to ferment fuels microbially from renewable agricultural feedstocks points to an important long-term reorientation of agriculture if non-renewable oil becomes uncompetitively expensive as a fuel for cars and feedstock for certain chemicals. It suggests that eventually there will be an increased emphasis on agricultural production of industrial feedstocks at the same time as continually increasing food output will be required to feed the expanding world population. All this will require considerable increases in agricultural productivity, to which biotechnology will increasingly contribute, but it will at the same time impose great strains on the natural environment and the structure of agriculture.

PRIVATE-SECTOR CONTROL OF BIOTECHNOLOGY DEVELOPMENT IN AGRICULTURAL TECHNOLOGY

It has been the role of the public sector to undertake research and development of agricultural technology as a public service to firms which might profit from translating that R and D into a commercial product or process, to farmers who might profit from adoption of the technology, and perhaps most importantly to consumers at home and abroad who benefited from the lower prices resulting from greater abundance. This was particularly true of the phase of change dominated by improvements in plant and animal breeding, the 'old' biotechnology. While the public sector had a role in basic research for chemical and mechanical technologies, the benefits of research expenditure in these areas were easier to capture by private companies investing in R and D, so that increasingly the public sector has taken a smaller role in these areas although maintaining a strong regulatory role in regard to agricultural chemicals in particular. Where it is impossible to prevent others from escaping payment for the research costs, either because the product is easily copied (seeds which can be regenerated by farmers or other firms) or because proposals for new methods can be readily implemented, private firms are understandably unwilling to invest. Machines, insecticides, fungicides and other manufactured inputs do lend themselves more readily to private exploitation.

Nevertheless the returns to R and D in these products do depend upon the degree of difficulty potential competitors would have in copying the product.

In some cases there are inherent technical difficulties in copying the process, or it would be prohibitively expensive, but in other cases it is the ability to obtain patent protection which creates legal barriers to potential competitors' ability to become 'free-riders' and which protects incentives to private investment in R and D. Traditionally, however, it has been impossible for plant and animal breeders to obtain patent rights for their products, which is one reason why public-sector R and D has remained so important in this area. For, as under the European Patent Convention (EPC) of 1973, it has been judged impossible for plant varieties to satisfy one of the key criteria to qualify for a patent, namely proof of 'an inventive step'.

The application of standard breeding practices to generate new varieties by crossing existing plants or animal strains has not been deemed to be invention. Thus under the EPC one set of exclusions from patentability is 'Plant or animal varieties or essentially biological processes for microbiological processes or the products thereof.'

In the absence of patent rights plant-breeding firms in particular have worked hard to obtain other means of protection and royalties for their products. The history of the development of Plant Breeders' Rights (PER) is presented by Mooney, and is of particular interest because of the concerns he expresses about the consequences of allowing the basic genetic stock, which underpins agriculture and hence our whole society, from becoming private rather than public property. While Mooney's concerns are expressed in relation to 'old' biotechnology varieties of plants, they are of particular importance with respect to the products of the 'new' biotechnology. For bio-engineered products are capable of meeting the inventive-step criterion of patentability, and both micro-organisms and transgenic animals have now been patented in the USA.

Before examining developments in the seed industry it is worth touching upon the implications of changes in research policy which stress increasing reliance upon private R and D to develop and exploit the new biotechnology and other technologies. In the UK government has decided that it should withdraw from funding what it terms 'near-market research', that is, research beyond the basic phase and which is preparation for commercial exploitation. This policy is based in part upon the arguments that industry should be investing more heavily in R and D, and that research strategies at the near-market stage should be driven by assessments of likely commercial success which can best be made by the firms involved, and will therefore lead to greater efficiency in allocating research funds.

This has led to the closure of a number of government-financed agricultural research institutes, the scaling down of others and the sale of the National Seeds Organization by auction to Unilever; Unilever won against competition with BP and other major public companies.

This deliberate attempt to switch an increasing proportion of agricultural R and D expenditure from the public to the private sector carries with it a number of risks. In the first place there is a controversy about whether there is under-investment in R and D so that the returns to extra expenditure are high, or whether the converse is the case. If there is under-investment, then withdrawal of public support for applied research will exacerbate this, as the private sector will only undertake R and D expenditure on those products and processes from which exclusive benefits can be captured by the investor.

Thus in relation to biotechnology in the USA, where again public support for applied biotechnology research is relatively small, Stallman and Schmid (1987) argue that there will be emphasis on technologies which are applied in factory conditions where secrecy and control can be maintained. For technologies which cannot be confined to factories, such as seeds, Stallman and Schmid state that 'Firms are also considering mechanisms which "scramble" the genome of a plant in the second generation', in order to prevent farmers from reproducing seed with the enhanced, engineered characteristics.

Clearly such actions are designed to frustrate the maximum spread of benefits from the technology and to maximize private profit for companies investing in research. Another facet of commercially orientated biotechnology research is that it will aim at the most important crops and livestock products, those produced by the largest farming units, and those which are most heavily subsidized. In the latter case this will worsen the budgetary problems of adjusting agricultural policy in OECD countries.

The probable neglect of minor crops, difficult habitats and small farmers means that the public sector will have a defined role in agricultural technology research, but one in which it will be relegated to the second division and where it is unlikely to prove successful in terms of the commercial yardstick of rates of return which is increasingly emphasized by public-research policy.

6

Gene Transfer Methods in Plant Biotechnology

GENE TRANSFER METHODS IN PLANTS

To achieve genetic transformation in plants, we need the construction of a vector (genetic vehicle) which transports the genes of interest, flanked by the necessary controlling sequences *i.e.* promoter and terminator, and deliver the genes into the host plant. The two kinds of gene transfer methods in plants are:

VECTOR-MEDIATED OR INDIRECT GENE TRANSFER

Among the various vectors used in plant transformation, the Ti plasmid of Agrobacterium tumefaciens has been widely used. This bacteria is known as "natural genetic engineer" of plants because these bacteria have natural ability to transfer T-DNA of their plasmids into plant genome upon infection of cells at the wound site and cause an unorganized growth of a cell mass known as crown gall. Ti plasmids are used as gene vectors for delivering useful foreign genes into target plant cells and tissues. The foreign gene is cloned in the T-DNA region of Ti-plasmid in place of unwanted sequences.

To transform plants, leaf discs (in case of dicots) or embryogenic callus (in case of monocots) are collected and infected with Agrobacterium carrying recombinant disarmed Ti-plasmid vector. The infected tissue is then cultured (co-cultivation) on shoot regeneration medium for 2-3 days during which time the transfer of T-DNA along with foreign genes takes place.

After this, the transformed tissues (leaf discs/calli) are transferred onto selection cum plant regeneration medium supplemented with usually lethal concentration of an antibiotic to selectively eliminate non-transformed tissues. After 3-5 weeks, the regenerated shoots are transferred to root-inducing medium, and after another 3-4 weeks, complete plants are transferred to soil following the hardening (acclimatization) of regenerated plants. The molecular techniques like PCR and southern hybridization are used to detect the presence of foreign genes in the transgenic plants.

VECTORLESS OR DIRECT GENE TRANSFER

In the direct gene transfer methods, the foreign gene of interest is delivered into the host plant cell without the help of a vector. The methods used for direct gene transfer in plants are:

Chemical mediated gene transfer *e.g.* chemicals like polyethylene glycol (PEG) and dextran sulphate induce DNA uptake into plant protoplasts. Calcium phosphate is also used to transfer DNA into cultured cells. Microinjection where the DNA is directly injected into plant protoplasts or cells (specifically into the nucleus or cytoplasm) using fine tipped (0.5 - 1.0 micrometer diameter) glass needle or micropipette. This method of gene transfer is used to introduce DNA into large cells such as oocytes, eggs, and the cells of early embryo.

Electroporation involves a pulse of high voltage applied to protoplasts/cells/ tissues to make transient (temporary) pores in the plasma membrane which facilitates the uptake of foreign DNA.

The cells are placed in a solution containing DNA and subjected to electrical shocks to cause holes in the membranes. The foreign DNA fragments enter through the holes into the cytoplasm and then to nucleus.

Particle gun/Particle bombardment - In this method, the foreign DNA containing the genes to be transferred is coated onto the surface of minute gold or tungsten particles (1-3 micrometers) and bombarded onto the target tissue or cells using a particle gun (also called as gene gun/shot gun/ microprojectile gun).The microprojectile bombardment method was initially named as biolistics by its inventor Sanford. Two types of plant tissue are commonly used for particle bombardment- Primary explants and the proliferating embryonic tissues.

Transformation - This method is used for introducing foreign DNA into bacterial cells *e.g.* E. Coli. The transformation frequency (the fraction of cell population that can be transferred) is very good in this method. *E.g.* the uptake of plasmid DNA by E. coli is carried out in ice cold $CaCl_2$ (0-50C) followed by heat shock treatment at 37-450C for about 90 sec. The transformation efficiency refers to the number of transformants per microgram of added DNA. The $CaCl_2$ breaks the cell wall at certain regions and binds the DNA to the cell surface.

Conjuction - It is a natural microbial recombination process and is used as a method for gene transfer. In conjuction, two live bacteria come together and the single stranded DNA is transferred via cytoplasmic bridges from the donor bacteria to the recipient bacteria.

Liposome mediated gene transfer or Lipofection - Liposomes are circular lipid molecules with an aqueous interior that can carry nucleic acids. Liposomes encapsulate the DNA fragments and then adher to the cell membranes and fuse with them to transfer DNA fragments. Thus, the DNA enters the cell and then to the nucleus. Lipofection is a very efficient technique used to transfer genes in bacterial, animal and plant cells.

SELECTION OF TRANSFORMED CELLS FROM UNTRANSFORMED CELLS

The selection of transformed plant cells from untransformed cells is an important step in the plant genetic engineering. For this, a marker gene (*e.g.* for antibiotic resistance) is introduced into the plant along with the transgene followed by the selection of an appropriate selection medium (containing the antibiotic). The segregation and stability of the transgene integration and expression in the subsequent generations can be studied by genetic and molecular analyses.

TRANSGENIC PLANTS WITH BENEFICIAL TRAITS

During the last decades, a tremendous progress has been made in the development of transgenic plants using the various techniques of genetic engineering. The plants, in which a functional foreign gene has been incorporated by any biotechnological methods that generally are not present in the plant, are called transgenic plants.

As per estimates recorded in 2002, transgenic crops are cultivated worldwide on about 148 million acres (587 million hectares) land by about 5.5 million farmers. Transgenic plants have many beneficial traits like insect resistance, herbicide tolerance, delayed fruit ripening, improved oil quality, weed control etc.

STRESS TOLERANCE

Biotechnology strategies are being developed to overcome problems caused due to biotic stresses (viral, bacterial infections, pests and weeds) and abiotic stresses (physical actors such as temperature, humidity, salinity etc).

Abiotic Stress Tolerance

The plants show their abiotic stress response reactions by the production of stress related osmolytes like sugars (*e.g.* trehalose and fructans), sugar alcohols (*e.g.* mannitol), amino acids (*e.g.* proline, glycine, betaine) and certain proteins (*e.g.* antifreeze proteins). Transgenic plants have been produced which over express the genes for one or more of the above mentioned compounds. Such plants show increased tolerance to environmental stresses. Resistance to abiotic stresses includes stress induced by herbicides, temperature (heat, chilling, freezing), drought, salinity, ozone and intense light. These environmental stresses result in the destruction, deterioration of crop plants which leads to low crop productivity. Several strategies have been used and developed to build ressitance in the plants against these stresses.

HERBICIDE TOLERANCE

Weeds are unwanted plants which decrease the crop yields and by competing with crop plants for light, water and nutrients. Several biotechnological strategies for weed control are being used *e.g.* the over-production of herbicide target enzyme (usually in the chloroplast) in the plant which makes the plant insensitive to the herbicide. This is done by the introduction of a modified gene that encodes for a resistant form of the enzyme targeted by the herbicide in weeds and crop plants. Roundup Ready crop plants tolerant to herbicide-Roundup, is already being used commercially.

The biological manipulations using genetic engineering to develop herbicide resistant plants are:

- Over-expression of the target protein by integrating multiple copies of the gene or by using a strong promoter.,
- Enhancing the plant detoxification system which helps in reducing the effect of herbicide.,
- Detoxifying the herbicide by using a foreign gene., and
- Modification of the target protein by mutation.

Some of the examples are:

Glyphosate resistance - Glyphosate is a glycine derivative and is a herbicide which is found to be effective against the 76 of the world's worst 78 weeds. It kills the plant by being the competitive inhibitor of the enzyme 5-enoyl-pyruvylshikimate 3- phosphate synthase (EPSPS) in the shikimic acid pathway. Due to it's structural similarity with the substrate phosphoenol pyruvate, glyphosate binds more tightly with EPSPS and thus blocks the shikimic acid pathway.

Certain strategies were used to provide glyphosate resistance to plants:

- It was found that EPSPS gene was overexpressed in Petunia due to gene amplification. EPSPS gene was isolated from Petunia and introduced in to the other plants. These plants could tolerate glyphosate at a dose of 2- 4 times higher than that required to kill wild type plants.
  - *By using mutant EPSPS genes*: A single base substitution from C to T resulted in the change of an amino acid from proline to serine in EPSPS. The modified enzyme cannot bind to glyphosate and thus provides resistance.
  - The detoxification of glyphosate by introducing the gene (isolated from soil organism- Ochrobactrum anthropi) encoding for glyphosate oxidase into crop plants. The enzyme glyphosate oxidase converts glyphosate to glyoxylate and aminomethylphosponic acid. The transgenic plants exhibited very good glyphosate ressitance in the field.

Another Example is of Phosphinothricin Resistance

Phosphinothricin is a broad spectrum herbicide and is effective against broad-leafed weeds. It acts as a competitive inhibitor of the enzyme glutamine synthase which results in the inhibition of the enzyme glutamine synthase and accumulation of ammonia and finally the death of the plant. The disturbace in the glutamine synthesis also inhibits the photosynthetic activity. The enzyme phosphinothricin acetyl transferase (which was first observed in Streptomyces sp in natural detoxifying mechanism against phosphinothricin) acetylates phosphinothricin, and thus inactivates the herbicide. The gene encoding for phosphinothricin acetyl transferase (bar gene) was introduced in transgenic maize and oil seed rape to provide resistance against phosphinothricin.

OTHER ABIOTIC STRESSES

The abiotic stresses due to temperature, drought, and salinity are collectively also known as water deficit stresses. The plants produce osmolytes or osmoprotectants to overcome the osmotic stress. The attempts are on to use genetic engineering strategies to increase the production of osmoprotectants in the plants. The biosynthetic pathways for the production of many osmoprotectants have been established and genes coding the key enzymes have been isolated. *E.g.* Glycine betaine is a cellular osmolyte which is produced by the participation of a number of key enzymes like choline dehydrogenase, choline monooxygenase etc. The choline oxidase gene from Arthrobacter sp. was used to produce transgenic rice with high levels of glycine betaine giving tolerance against water deficit stress.

Scientists also developed cold-tolerant genes (around 20) in Arabidopsis when this plant was gradually exposed to slowly declining temperature. By introducing the coordinating gene (it encodes a protein which acts as transcription factor for regulating the expression of cold tolerant genes), expression of cold tolerant genes was triggered giving protection to the plants against the cold temperatures.

INSECT RESISTANCE

A variety of insects, mites and nematodes significantly reduce the yield and quality of the crop plants. The conventional method is to use synthetic pesticides, which also have severe effects on human health and environment. The transgenic technology uses an innovative and eco-friendly method to improve pest control management.About 40 genes obtained from microorganisms of higher plants and animals have been used to provide insect resistance in crop plants

The first genes available for genetic engineering of crop plants for pest resistance were Cry genes (popularly known as Bt genes) from a bacterium Bacillus thuringiensis. These are specific to particular group of insect pests, and are not harmful to other useful insects like butter flies and silk worms. Transgenic crops with Bt genes (*e.g.* cotton, rice, maize, potato, tomato, brinjal, cauliflower, cabbage, etc.) have been developed. This has proved to be an effective way of controlling the insect pests and has reduced the pesticide use.

The most notable example is Bt cotton (which contains CrylAc gene) that is resistant to a notorious insect pest Bollworm (Helicoperpa armigera).. There are certain other insect resistant genes from other microorganisms which have been used for this purpose. Isopentenyl transferase gene from Agrobacterium tumefaciens has been introduced into tobacco and tomato. The transenic plants with this transgene were found to reduce the leaf consumption by tobacco hornworm and decrease the survival of peach potato aphid.

Certain genes from higher plants were also found to result in the synthesis of products possessing insecticidal activity. One of the examples is the Cowpea trypsin inhibitor gene (CpTi) which was introduced into tobacco, potato, and oilseed rape for develping transgenic plants. Earlier it was observed that the wild species of cowpea plants growing in Africa were resistant to attack by a wide range of insects. It was observed that the insecticidal protein was a trypsin inhibitor that was capable of destroying insects belonging to the orders Lepidoptera, Orthaptera etc. Cowpea trypsin inhibitor (CpTi) has no effect on mammalian trypsin, hence it is non-toxic to mammals.

VIRUS RESISTANCE

There are several strategies for engineering plants for viral resistance, and these utilizes the genes from virus itself (*e.g.* the viral coat protein gene). The virus-derived resistance has given promising results in a number of crop plants such as tobacco, tomato, potato, alfalfa, and papaya. The induction of virus resistance is done by employing virus-encoded genes-virus coat proteins, movement proteins, transmission proteins, satellite RNa, antisense RNAs, and ribozymes. The virus coat protein-mediated approach is the most successful one to provide virus resistance to plants. It was in 1986, transgenic tobacco plants expressing tobacco mosaic virus (TMV) coat protein gene were first developed. These plants exhibited high levels of resistance to TMV.

The transgenic plant providing coat protein-mediated resistance to virus are rice, potato, peanut, sugar beet, alfalfa etc. The viruses that have been used include alfalfa mosaic virus (AIMV), cucumber mosaic virus (CMV), potato virus X (PVX), potato virus Y (PVY) etc.

RESISTANCE AGAINST FUNGAL AND BACTERIAL INFECTIONS

As a defence strategy against the invading pathogens (fungi and bacteria) the plants accumulate low molecular weight proteins which are collectively known as pathogenesis-related (PR) proteins.

Several transgenic crop plants with increased resistance to fungal pathogens are being raised with genes coding for the different compounds. One of the examples is the Glucanase enzyme that degrades the cell wall of many fungi. The most widely used glucanase is beta-1,4-glucanase. The gene encoding for beta-1,4 glucanase has been isolated from barley, introduced, and expressed in transgenic tobacco plants. This gene provided good protection against soil-borne fungal pathogen Rhizoctonia solani. Lysozyme degrades chitin and peptidoglycan of cell wall, and in this way fungal infection can be reduced. Transgenic potato plants with lysozyme gene providing resistance to Eswinia carotovora have been developed.

DELAYED FRUIT RIPENING

The gas hormone, ethylene regulates the ripening of fruits, therefore, ripening can be slowed down by blocking or reducing ethylene production. This can be achieved by introducing ethylene forming gene(s) in a way that will suppress its own expression in the crop plant. Such fruits ripen very slowly (however, they can be ripen by ethylene application) and this helps in exporting the fruits to longer distances without spoilage due to longer-shelf life.

The most common example is the 'Flavr Savr' transgenic tomatoes, which were commercialized in U.S.A in 1994. The main strategy used was the antisense RNA approach. In the normal tomato plant, the PG gene (for the enzyme polygalacturonase) encodes a normal mRNA that produces the enzyme polygalacturonase which is involved in the fruit ripening.

The complimentary DNA of PG encodes for antisense mRNA, which is complimentary to normal (sense) mRNA. The hybridization between the sense and antisnse mRNAs renders the sense mRNA ineffective. Consequently, polygalacturonase is not produced causing delay in the fruit ripening. Similarly strategies have been developed to block the ethylene biosynthesis thereby reducing the fruit ripening. *E.g.* transgenic plants with antisense gene of ACC oxidase (an enzyme involved in the biosynthetic process of ethylene) have been developed. In these plants, production of ethylene was reduced by about 97 per cent with a significant delay in the fruit ripening. The bacterial gene encoding ACC deaminase (an enzyme that acts on ACC and removes amino group) has been transferred and expressed in tomato plants which showed 90 per cent inhibition in the ethylene biosynthesis.

CHARACTERISTICS OF THE GENES

When genetics was a young science most of us considered the genes as the determinants for the "unit-characters", each gene responsible for a separate quality of an organism. The facts of Mendelian inheritance showed us that there are very small hereditary differences—differences between a contrasting pair of characters.

And it seemed obvious to assume that in such a case the genes themselves occur in pairs, one member of the pair causing white flower-colour and one blue colour, one causing waxy endosperm in corn and one horny. It is very curious that this idea should have taken such a firm grip upon the imagination of the earliest geneticists. If we compare genetics with other sciences—with physics, chemistry, astronomy—we generally see that it is not self-evident to assume that where we see a difference, or where we see two contrasting qualities, two different contrasting causes occur. We speak of hot and cold, of light and dark, of wet and dry. But we do not really think of a separate cause for wetness and a separate one for dryness—a relative presence or absence of water does very well as an explanation—and the same is true in all the other examples. To assume the existence of pairs of genes in each pair of which the resultant action would be contrasting is an unnecessary complication.

The alternative is a much simpler hypothesis which was proposed by Bateson and by myself, and which went under the name of the "presence- and-absence "theory of the gene. We assumed that in a monogenetic difference—the smallest "Mendelizing" difference—we were dealing with the absence or presence of just one gene.

At first this hypothesis met with a veritable storm of protests. To some of my colleagues it seemed the height of folly that the absence of anything should ever determine an inheritable quality. On the face of it, this objection seems very strange; for it seems quite logical to speak of silence as determined by a purely negative absence of noise, of dryness determined by the absence of water. But of course the apparent absurdity of thinking of the absence of a gene as causing an inheritable quality grew out of Weismann's conception of determinants, inherited protoplasmic entities that should be each responsible for just one particular property of a living being.

The idea, now generally accepted by biologists, that the qualities of an organism develop as a result of growth and development, this development being caused by a co-operation of thousands of factors, some of these factors being environmental, while some others—the genes— influence the course of development from within. If we compare a plant having pale-pink flowers with one having dark-pink ones, we assume that the development which leads to the production of the plant with the pale flowers is due to the co-operation of thousands of causes and substances. But in the other plant one additional substance is present, whose action at the appropriate developmental stage allows a further transformation of the pink colour into a darker one. It is therefore not the absence of this gene which directly causes pink colour; it is the presence of the thousand-and-one genes which both plants have in common.

If we knock off a man's hat we may see that he is red-headed. He does not become red-headed as a result of the absence of the hat, but we can see the result of all the other things which cause his hair to be red, provided we take away that hat. If in the course of this story we substitute an inherited gene which helps to transform red hair into black, the reasoning remains essentially the same.

Very soon after the original presence-absence hypothesis was first published it was found that in its original shape this hypothesis was not adequate to explain all the facts. It was found that very often in inheritance we are not dealing with just two contrasting qualities, but with three or more alternatives. In snapdragons we have a series of varieties that gradually grade into each other, starting from yellow through different intermediate stages to red. If we call those forms A, B, C, we find that any pair of them between them show a difference in just one gene. When we mate A with B, we will get only A and B plants in following generations; if we combine B with Z, no 's or C"s, but only B's and Z>'s will be found in the second generation.

The best explanation of such series of multiple alleles is this: that in such cases we are dealing not with the presence and complete absence of one gene, but with different quantities of one and the same gene in the one spot on the chromosome where our gene is located. Instead of the presence-absence hypothesis, we now prefer to speak of a hypothesis about quantitative differences in inheritance.

It is highly probable that in certain instances we are really concerned with presence versus complete absence; in other cases, however, intermediate stages may occur in which fewer molecules of the gene are present per cell-nucleus.

This quantitative gene-theory is closely bound up with the hypothesis and which is now generally accepted by geneticists—namely, that the genes are not living protoplasma globules, but that they are non-living chemical substances, that have only this in common: that they are each the ferment for their own synthesis. With Alexander, we may speak of reproductive catalysis. We assume that each molecule of a gene, between every time the cells, in growing tissue, divide, attracts and helps to combine from the ambient building stones just one new molecule that will in every particular be its exact duplicate.

The genes are combined longitudinally in long series—the chromosomes—and it is the chromosomes whose movements and combinations give rise to the phenomena we notice as Mendelian segregation and redistribution of genes. The number of the chromosomes is highly constant for every plant species; and sometimes, as in wheat and barley, different sections of the group are distinguished by different numbers of chromosomes. In that case the numbers are always multiples

Many genes of different kinds of a common basis number, and it is very probable that in such a series the higher chromosome numbers may have originated by reduplication of entire chromosome sets during the course of evolution of the group. In the cells of the plants those chromosomes, made up of genes, occur in pairs.

The homologous chromosomes of each pair are not necessarily identical. We know that in a plant heterozygous in respect to one gene, the chromosomes of the corresponding pair are different in that particular spot ("locus "). The two chromosomes of a homologous pair have originally been derived from the two parents, one from the pollen-cell and one from the ovule.

The same is true of all the other pairs. The somatic (body) cells have a double set of chromosomes; the germ-cells have only a single set—one of each kind. In ordinary growing tissue every chromosome doubles its mass (and probably the quantity of every gene present on it) between two cell divisions. The two new chromosomes each consisting of part of the two original ones are identical, the break occurring on each at exactly the same spot. And when cell-division commences, the chromosomes are drawn apart to different poles of the cell in such a way that each daughter cell obtains a complete set of chromosomes that is identical with the original set.

The exception is that a "reduction division" occurs when germ-cells are formed. In this case there is a second cell division without a preceding chromosomal doubling. At this reduction division the two chromosomes of a homologous pair become separated one from the other. The result is that in a plant, heterozygous in respect to one gene, Ay in which all the somatic cells are Aa, the germ-cells will be either A or a or only half the germ-cells contain the full quantity of gene A.

When the break does not happen to occur at exactly the same "locus" in the two original chromosomes, one of the new chromosomes may have one gene in duplicate and the other one may be lacking it. Mutations may arise in this way from small irregularities during crossing-over. for this reason, be of four kinds: ABy Ab, aB and ab. To a certain extent this redistribution of distinct genes over different germ-cells results in a redistribution of the qualities which were shown by the parents of the heterozygous plant. If we cross a plant with AB to an ab plant, the hybrid will be AaBb and as this hybrid will be making four kinds of germ-cells, there will be all sorts of offspring, with different combinations of the qualities in which the grandparents differed.

We were just speaking of genes located in different chromosomal pairs. But sometimes plants are heterozygous in respect to two genes that happen to be located in the same chromosome. In this case those chromosomes tend to be transmitted as such, in their entirety, so that the distribution of the two genes is not mutually independent. Such genes are "coupled ".

All kinds, all degrees of coupling have been met with, from complete or almost complete coupling to very loose coupling. Morgan proposed a hypothesis which is now very well proven and established—namely, that the degree of coupling is dependent on the greater or lesser degree of contiguity of the genes. Genes that lie close together are closely coupled, those that are far apart must lie farther apart on the same chromosome.

The mechanism which is thought to account for coupling is the phenomenon of "crossing-over ". Sometimes chromosomal strands seem to touch at one or more places, and when the strands break apart, the top half of one strand may be attached to the bottom part of the other.

A new pair of chromosomal strands originates, but if on both sides of a break there are loci where the strands differ in chromosomal content, those loci are redistributed, so that, as far as the independent inheritance of separate genes is concerned, crossing-over may help to redistribute the genes, even if they are located in the same chromosome.

GENETIC VARIATION

Our main interest is to isolate genes that are responsible for the intra- and interspecific morphological variation in plants. Therefore, the analysis of natural genetic variation is the starting point for the majority of individual projects in our lab.

For the isolation of genes involved in morphological variation, exploiting natural genetic variation offers some distinct advantages as compared to classical mutagenesis experiments. Most importantly, it allows the identification of naturally occurring, differentially active gene alleles that are likely targets for the evolution of morphological variation. Because naturally occurring lines do not carry major deleterious mutations that impair their survival in the wild, the approach also counter-selects against the isolation of mutant alleles that result in strong phenotypes as commonly detected in mutagenesis experiments.

This applies in particular to wild populations with a high level of inbreeding, such as in our favourite model organism, *Arabidopsis thaliana*.

So why has this approach not been followed earlier? Mainly because the isolation of genes that modify quantitative traits has technically not been feasible. However, the rich biological and genomic resources, such as the fully sequenced genome, have turned Arabidopsis into an ideal model system to investigate the molecular basis of natural genetic variation at the gene level.

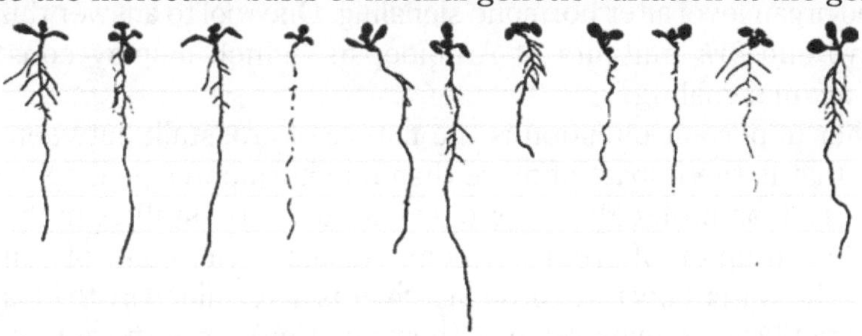

Fig. Natural morphological variation of root system architecture between wild isolates of Arabidopsis. Most of this variation results from the influence of genetic factors.

Plant organs are formed in a continuous, post-embryonic manner by ordered cell divisions and expansions. The extent of growth ultimately determines organ shape. Thus, we are primarily interested in genes that modulate growth rate. To isolate such genes, we focus on the root system.

This is because of our interest in root biology per se, but also because roots grow from a so-called meristematic region at their tip, which produces concentric tissue layers through controlled cell proliferation, elongation and differentiation. This feature enables us to reduce the problem of accurately measuring growth from three dimensions to a single one.

Nevertheless, analysis of root system growth requires care and highly reproducible growth conditions, because root system development is very plastic and responds to many macro- and micro-environmental cues. The natural variation in root system morphology between Arabidopsis accessions collected from different locations and grown in identical growth chamber conditions on tissue culture media.So far, we have isolated an important regulator of root system architecture from the accession Uk-1. This gene is a major so-called quantitative trait locus (QTL), which we gave the name *BREVIS RADIX (BRX)*, latin for "short root". Other efforts in this area include attempts to isolate two other QTLs for root growth located in other accessions.

ACTION OF PLANT HORMONES

Plant hormones are at the heart of many plant developmental processes. They determine critical growth parameters, such as the rate of cell division or elongation, as well as developmental decisions, for instance where along a primary root a lateral root should be formed.

Different plant hormones control different developmental processes to varying degrees, sometimes in the same, sometimes in opposite directions. Whether and how the overlapping influence of different hormones on traits is coordinated is largely unknown. Part of the research in our lab aims to determine how the major plant hormone auxin, arguably the most important in shaping the plant, regulates downstream events that lead to discrete morphological change. After all, those changes require structural rearrangements on the cellular and organ level after hormone signaling. One tool to answer this question are the *hy5* and *hyh* mutants of Arabidopsis, which display constitutively increased auxin signaling.

Another important question is the nature of crosstalk between hormone pathways, that is the impact of more than one hormone signaling pathway on the same trait or molecular event. Does such crosstalk simply reflect a quantitative add up of different hormone signals at the point of convergence (In the way the water flow rate through a river is determined by the contribution of its two tributaries downstream of their junction), or is there real crosstalk happening upstream (The two tributaries that will meet have built-in dams, and if so, do the dam gatekeepers communicate with each other to intentionally regulate the water flow beyond the river junction)?Much to our surprise, we isolated an important regulator of hormone pathway interaction in the root through our natural variation project: loss of function of the *BRX* gene results in reduced root growth because *BRX* is required for the biosynthesis of the hormone brassinolide, whose concentration is in turn rate-limiting for auxin action. Since *BRX* is itself under auxin control, this creates a feedback loop required for optimal root growth.

Whether an equivalent loop works in the shoot, and whether this feedback is conserved in other species are some of the questions we are currently interested in with regard to this project, generously funded by the Swiss National Science Foundation.

Variation

If you have grown Fast Plants you will have noticed variation among the individual plants within the group. Variation can range from a little to a lot.

Variation is one of the fundamental characteristics of life. All organisms exhibit some variation among individuals. Understanding the ways that variation is manifested in organisms, how it comes to be expressed through the development of the individual, and how it is transmitted from one individual to the next generation of individuals are central themes in biology.

Working with Fast Plants will enrich a student's understanding of variation. By observing the growth and development of a Fast Plants through the various stages in the life cycle, students will become aware of many visible features, or phenotypes, that make up the organism.

Only, however, upon close observation of a population of plants, will they become aware that the characteristics observed on one plant vary more or less on other plants. Such is the nature of variation.

PHENOTYPIC VARIATION

Phenotypic variation, e.g. plant height at a particular stage of development, is considered to be the expression of the genetic makeup (genotype) of the individual as it interacts with the environment. Variation in plant height among individuals in a population is therefore due to variation in the interaction between the genotype and the environment, as expressed through the development of each individual plant.

In order to be useful in an experiment the phenotype must be described using terms that are widely understood and easily communicated.

For these reasons scientists have agreed upon various standards or descriptors to describe characteristics in the natural world. The choice of how to describe what you observe is important, because it will determine the kinds of descriptors used and establish the basis for recording, analyzing and communicating results.

ENVIRONMENTAL VARIATION

Much can be learned about the role of light, temperature, and nutrients on plant development from experiments in which one or more environmental parameters are changed. Because Fast Plants are highly responsive to changes in the environment, they are ideal for examining the role of the environment on the expression of phenotypic variation.

Although Fast Plants are able to grow within a wide range of environmental conditions, for most investigations it is recommended that they be grown under uniform and ideal conditions. In this way variation arising from sub-optimal conditions of environment will be minimized. The Wisconsin Fast Plants Information Document (WFPID) Understanding the Environment describes how to provide and maintain the various physical, chemical and biotic components of the environment that are most suitable for Fast Plants.

GENOTYPIC VARIATION

How can students use Fast Plants to investigate the contribution of the genotype to the phenotype?

Fast Plants, rapid cycling Brassica rapa, are genotypically variable in that they have a genetically controlled mating system that prevents self-fertilization and favours out-crossing among individuals.

As a consequence, even seed stocks selected for uniformity of specific phenotypes and genotypes exhibit considerable variation for other traits.

Fig. Rosette, dwarf, basic, elongated internode

The Wisconsin Fast Plants® Programme has developed a number of genetic stocks of rapid cycling Brassica rapa for genetic investigations on the nature and inheritance of variation.

Some WFP stocks contain distinctive mutant phenotypes, *e.g.*, anthocyaninless plant, anl, yellow green plant, ygr1, rosette, ros, and male sterile, mst2, that are suitable for Mendelian genetics.

Other stocks exhibit phenotypes whose expression may vary continually and which may be quantified as discrete or countable units, *e.g.*, number of hairs, or as estimates of size, *e.g.* petite dwarf, dwf1, or of intensity of colour saturation, *e.g.*, purple anthocyanin.

These quantitative phenotypes may be conditioned by a few or many genes and normally require numerical descriptions in which the statistical notations of population size, (n), range (r), arithmetic mean (x), and standard deviation (s) are applied.

Yet other stocks have been developed to combine both simply inherited mutant genotypes and quantitatively expressed phenotypes. An important part of WFP is the continuing development and improvement of seed stocks for uses in genetics.

GENE INTERACTIONS

FINDING THE GENES

Between 1884 (the year Mendel died) and 1888 details of mitosis and meiosis were reported, the cell nucleus was identified as the location of the genetic material, and "qualities" were even proposed to be transmitted on chromosomes to daughter cells at mitosis. In 1903 Walter Sutton and Theodore Boveri formally proposed that chromosomes contain the genes. The Chromosome Theory of Inheritance is one of the foundations of genetics and explains the physical reality of Mendel's principles of inheritance.

Fig. Fruit Fly, *Drosophila Melanogaster*, (SEM X60)

The location of many genes (Mendel's factors) was determined by Thomas Hunt Morgan and his coworkers in the early 1900's. Morgan's experimental organism was the fruit fly (*Drosophila melanogaster*). Fruit flies are ideal organisms for genetics, having a small size, ease of care, susceptibility to mutations, and short (7-9 day) generation time. The role of chromosomes in determination of sex was deduced by Morgan from work on fruit flies. During Metaphase I, homologous chromosomes will line up. A karyotype can be made by cutting and arranging photomicrographs of the homologous chromosomes thus revealed at Metaphase I. Two types of chromosome pairs occur. Autosomes resemble each other in size and placement of the centromere, for example pairs of chromosome 21 are the same size, while pairs of chromosome 9 are of a different size from pair 21.

Sex chromosomes may differ in their size, depending on the species of the organism they are from. In humans and*Drosophila*, males have a smaller sex chromosome, termed the Y, and a larger one, termed the X. Males are thus XY, and are termed heterogametic. Females are XX, and are termed homogametic. In grasshoppers, which Sutton studied in discovering chromosomes, there is no Y, only the X chromosome in males. Females are XX, while males are denoted as XO.

Other organisms (notably birds, moths and butterflies) have males homogametic and females heterogametic. Males (if heterogametic) contribute either an X or Y to the offspring, while females contribute either X. The male thus determines the sex of the offspring. Remember that in meiosis, each chromosome is replicated and one copy sent to each gamete. Morgan discovered a mutant eye colour and attempted to use this mutant as a recessive to duplicate Mendel's results. He failed, instead of achieving a 3:1 F2 ratio the ratio was closer to 4:1 (red to white). Most mutations are usually recessive, thus the appearance of the white mutant presented Morgan a chance to test Mendel's ratios on animals.

The F1 generation also had no white eyed females. Morgan hypothesized that the gene for eye colour was only on the X chromosome, specifically in that region of the X that had no corresponding region on the Y. White eyed fruit flies were also more likely to die prior to adulthood, thus explaining the altered ratios.

Normally eyes are red, but a variant (white) eyed was detected and used in genetic study. Cross a homozygous white eyed male with a homozygous red eyed female, and all the offspring have red eyes. Red is dominant over white. However, cross a homozygous white eyed female with a red eyed male, and the unexpected results show all the males have white eyes and all the females red eyes. This can be explained if the eye colour gene is on the X chromosome.

Explanation

If the gene for eye colour is on the X chromosome, the red eyed male in the second cross will pass his red eyed X to only his daughters, who in turn received only a recessive white-carrying X from their mother. Thus all females had red eyes like their father. Since the male fruit fly passes only the Y to his sons, their eye colour is determined entirely by the single X chromosome they receive from their mother (in this case white). Thus all the males in the second cross were white eyed.

These experiments introduced the concept of sex-linkage, the occurrence of genes on that part of the X that lack a corresponding location on the Y. Sex-linked recessives (such as white eyes in fruit flies, hemophilia, baldness, and colourblindness in humans) occur more commonly in males, since there is no chance of them being heterozygous. Such a condition is termed homozygous.

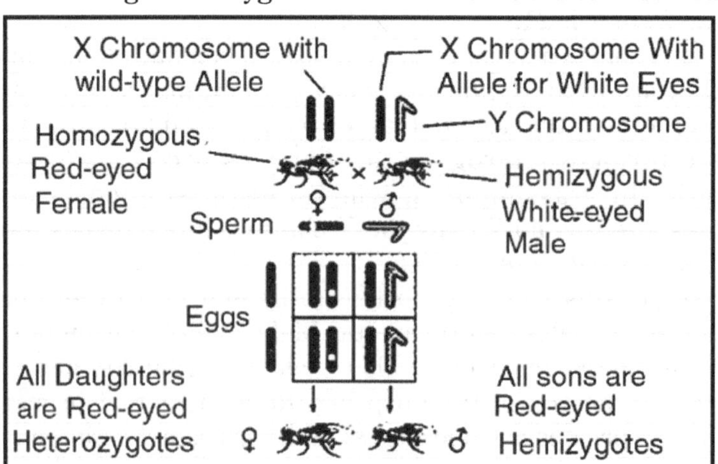

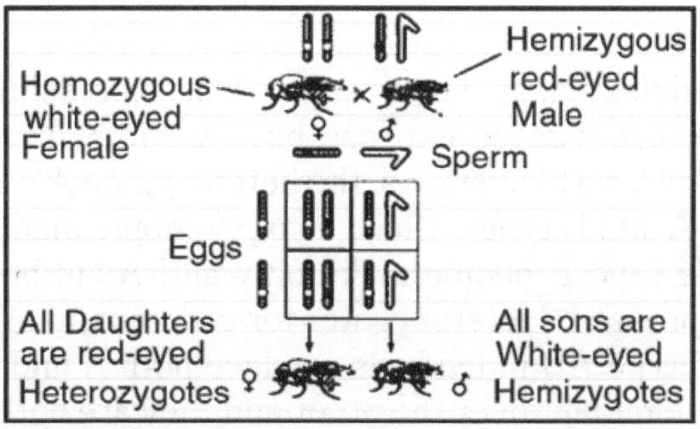

Fig. Inheritance of Eye Colour in Fruit flies.

Characteristics of X-linked Traits

- Phenotypic expression more common in males
- Sons cannot inherit the trait from their fathers, but daughters can.

Sons inherit their Y chromosome from their father.

Only a few genes have been identified on the Y chromosome, among them the testis-determining factor (TDF) that promotes development of the male phenotype.

Barr bodies are interpreted as inactivated X chromosomes in mammalian females. Since females have two X chromosomes, the Lyon hypothesis suggests that one or the other X is inactivated in each somatic (non-reproductive) cell during embryonic development. Cells mitotically produced from these embryonic cells likewise have the same inactivated X chromosome.

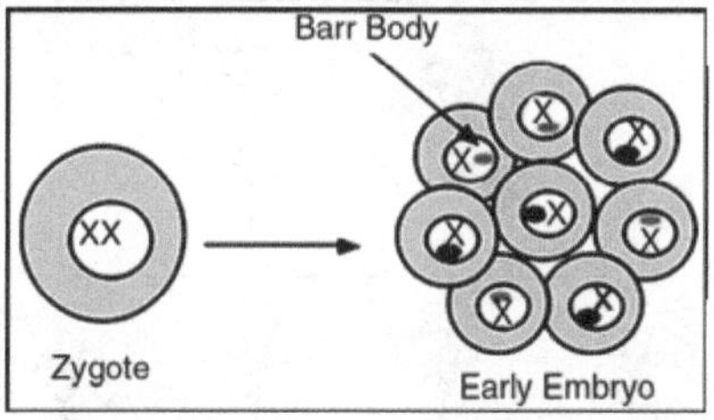

Fig. The Role of Deactivated X Chromosomes in Mammalian Female Development.).

Calico cats (sometimes called tortoiseshell) are almost always female since the calico trait is caused by some areas of the cat's fur expressing one allele and others expressing the other colour. Can there be a male tortoiseshell cat? How would such a cat get its genes? Remember that fur colour in cats is a sex-linked feature. Would the male calico be fertile or sterile?

THE MODERN VIEW OF THE GENE

While Mendel discussed traits, we now know that genes are segments of the DNA that code for specific proteins. These proteins are responsible for the expression of the phenotype. The basic principles of segregation and independent assortment as worked out by Mendel are applicable even for sex-linked traits.

Codominant alleles

Codominant alleles occur when rather than expressing an intermediate phenotype, the heterozygotes express both homozygous phenotypes. An example is in human ABO blood types, the heterozygote AB type manufactures antibodies to both A and B types. Blood Type A people manufacture only anti-B antibodies, while type B people make only anti-A antibodies. Codominant alleles are both expressed. Heterozygotes for codominant alleles fully express both alleles. Blood type AB individuals produce both A and B antigens. Since neither A nor B is dominant over the other and they are both dominant over O they are said to be codominant.

Incomplete Dominance

Incomplete dominance is a condition when neither allele is dominant over the other. The condition is recognized by the heterozygotes expressing an intermediate phenotype relative to the parental phenotypes. If a red flowered plant is crossed with a white flowered one, the progeny will all be pink. When pink is crossed with pink, the progeny are 1 red, 2 pink, and 1 white.

Flower colour in snapdragons is an example of this pattern. Cross a true-breeding red strain with a true-breeding white strain and the F1 are all pink (heterozygotes). Self-fertilize the F1 and you get an F2 ratio of 1 red: 2 pink: 1 white. This would not happen if true blending had occurred (blending cannot explain traits such as red or white skipping a generation and pink flowers crossed with pink flowers should produce ONLY pink flowers)

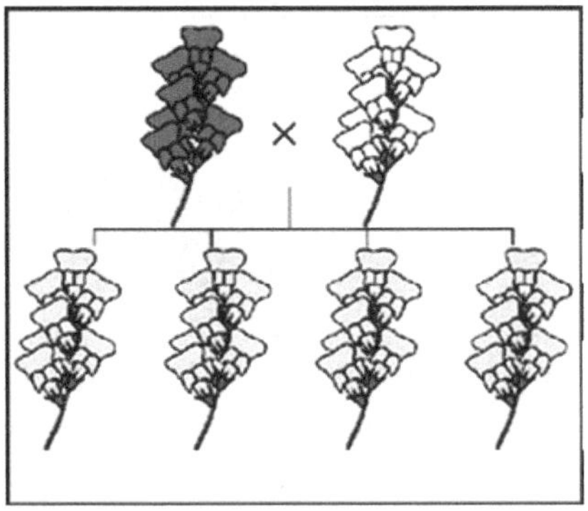

Fig. Inheritance of Flower Colour in Snapdragons.

Multiple Alleles

Many genes have more than two alleles (even though any one diploid individual can only have at most two alleles for any gene), such as the ABO blood groups in humans, which are an example of multiple alleles. Multiple alleles result from different mutations of the same gene. Coat colour in rabbits is determined by four alleles. Human ABO blood types are determined by alleles A, B, and O. A and B are codominants which are both dominant over O. The only possible genotype for a type O person is OO. Type A people have either AA or AO genotypes. Type B people have either BB or BO genotypes. Type AB have only the AB (heterozygous) genotype. The A and B alleles of gene I produce slightly different glycoproteins (antigens) that are on the surface of each cell. Homozygous A individuals have only the A antigen, homozygous B individuals have only the B antigen, homozygous O individuals produce neither antigen, while a fourth phenotype (AB) produces both A and B antigens.

INTERACTIONS AMONG GENES

While one gene may make only one protein, the effects of those proteins usually interact (for example widow's peak may be masked by expression of the baldness gene). Novel phenotypes often result from the interactions of two genes, as in the case of the comb in chickens. The single comb is produced only by the rrpp genotype. Rose comb (b) results from R_pp. (_ can be either R or r). Pea comb (c) results from rrP_. Walnut comb, a novel phenotype, is produced when the genotype has at least one dominant of each gene (R_P_).

Epistasis

Epitasis is the term applied when one gene interferes with the expression of another (as in the baldness/widow's peak mentioned earlier). Bateson reported a different phenotypic ratio in sweet pea than could be explained by simple Mendelian inheritance. This ratio is 9:7 instead of the 9:3:3:1 one would expect of a dihybrid cross between heterozygotes. Of the two genes (C and P), when either is homozygous recessive (cc or pp) that gene is epistatic to (or hides) the other. To get purple flowers one must have both C and P alleles present.

ENVIRONMENT AND GENE EXPRESSION

Phenotypes are always affected by their environment. In buttercup (*Ranunculus peltatus*), leaves below water-level are finely divided and those above water-level are broad, floating, photosynthetic leaf-like leaves. Siamese cats are darker on their extremities, due to temperature effects on phenotypic expression. Expression of phenotype is a result of interaction between genes and environment. Siamese cats and Himalayan rabbits both animals have dark coloured fur on their extremities. This is caused by an allele that controls pigment production being able only to function at the lower temperatures of those extremities. Environment determines the phenotypic pattern of expression.

POLYGENIC INHERITANCE

Polygenic inheritance is a pattern responsible for many features that seem simple on the surface. Many traits such as height, shape, weight, colour, and metabolic rate are governed by the cumulative effects of many genes. Polygenic traits are not expressed as absolute or discrete characters, as was the case with Mendel's pea plant traits. Instead, polygenic traits are recognizable by their expression as a gradation of small differences (a continuous variation). The results form a bell shaped curve, with a mean value and extremes in either direction.Height in humans is a polygenic trait, as is colour in wheat kernels. Height in humans is NOT discontinuous. If you line up the entire class a continuum of variation is evident, with an average height and extremes in variation (very short [vertically challenged?] and very tall [vertically enhanced]). Traits showing continuous variation are usually controlled by the additive effects of two or more separate gene pairs. This is an example of polygenic inheritance. The inheritance of EACH gene follows Mendelian rules.

Usually polygenic traits are distinguished by:

- Traits are usually quantified by measurement rather than counting.
- Two or more gene pairs contribute to the phenotype.
- Phenotypic expression of polygenic traits varies over a wide range.

Human polygenic traits include:

- Height
- SLE (Lupus)
- Weight
- Eye Colour
- Intelligence
- Skin Colour
- Many forms of behaviour

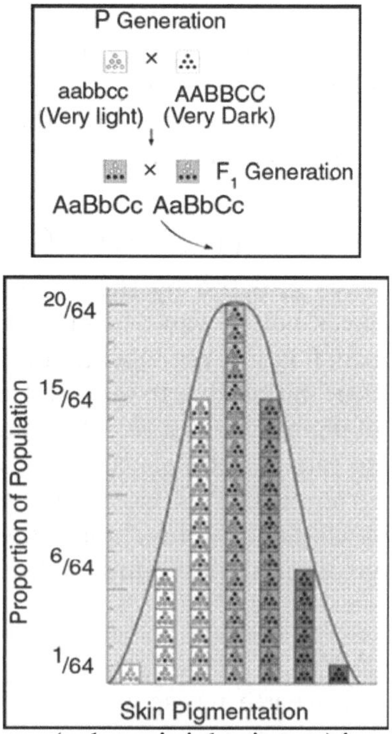

The role of many genes (polygenic inheritance) in production of a continuum of phenotypes. Images.

PLEIOTROPY

Pleiotropy is the effect of a single gene on more than one characteristic. An example is the "frizzle-trait" in chickens. The primary result of this gene is the production of defective feathers. Secondary results are both good and bad; good include increased adaptation to warm temperatures, bad include increased metabolic rate, decreased egg-laying, changes in heart, kidney and spleen.

Cats that are white with blue eyes are often deaf, white cats with a blue and an yellow-orange eye are deaf on the side with the blue eye. Sickle-cell anemia is a human disease originating in warm lowland tropical areas where malaria is common. Sickle-celled individuals suffer from a number of problems, all of which are pleiotropic effects of the sickle-cell allele.

GENES AND CHROMOSOMES

Linkage occurs when genes are on the same chromosome. Remember that sex-linked genes are on the X chromosome, one of the sex chromosomes. Linkage groups are invariably the same number as the pairs of homologous chromosomes an organism possesses. Recombination occurs when crossing-over has broken linkage groups, as in the case of the genes for wing size and body colour that Morgan studied. Chromosome mapping was originally based on the frequencies of recombination between alleles. Since mutations can be induced (by radiation or chemicals), Morgan and his coworkers were able to cause new alleles to form by subjecting fruit flies to mutagens (agents of mutation, or mutation generators). Genes are located on specific regions of a certain chromosome, termed the gene locus (plural: loci). A gene therefore is a specific segment of the DNA molecule. Alfred Sturtevant, while an undergraduate student in Morgan's lab, postulated that crossing-over would be less common between genes adjacent to each other on the same chromosome and that it should be possible to plot the sequence of genes along a fruit fly chromosome by using crossing-over frequencies. Distances on gene maps are expressed in map units (one map unit = 1 recombinant per 100 fertilized eggs; or a 1 per cent chance of recombination). The map for *Drosophila melanogaster* chromosomes is well known. Note that eye colour and aristae length are far apart, as indicated by the occurrence of more recombinants (crossing-overs) between them, while wing length is closer to eye shape (as indicated by the low frequency of recombination between these two features).

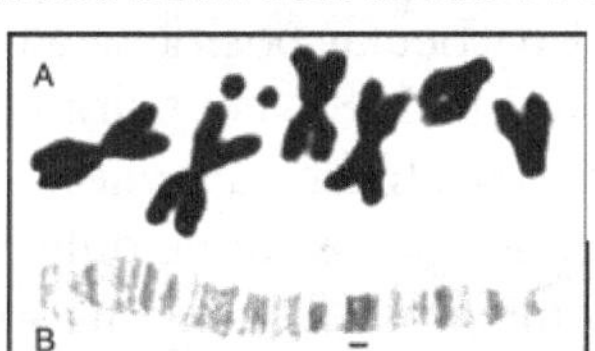

Fig. A Illustrates the Fruit fly Chromosomes at Metaphase; figure B shows a Polytene Chromosome.

CHROMOSOME ABNORMALITIES

Chromosome abnormalities include inversion, insertion, duplication, and deletion. These are types of mutations. Since DNA is information, and information typically has a beginning point, an inversion would produce an inactive or altered protein. Likewise deletion or duplication will alter the gene product.

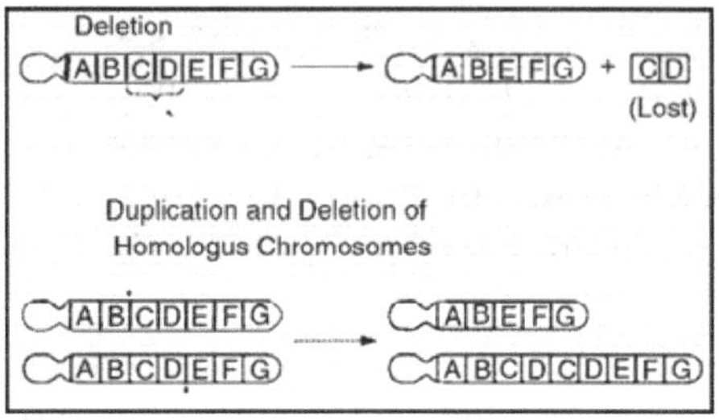

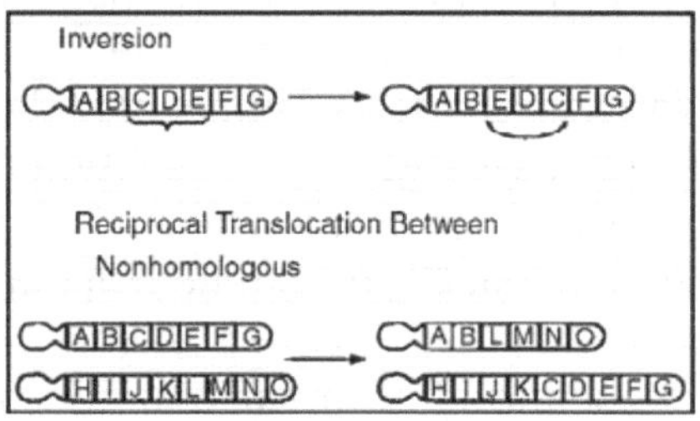

GENE MAPPING

An important, though by no means an essential, step in genetic analysis is to produce genetic maps of the marker loci. Such maps are often referred to as 'framework maps' because they provide a framework within which important genes can be located, as well as providing a means of comparing chromosome organization in other closely or distantly related species.

There are two stages to mapping. Firstly, to arrange the markers in a linear sequence separated by an appropriate map distance, *i.e.* to construct a linkage map. The second is to relate the linkage maps to particular recognizable chromosomes. The latter is often the most difficult and generally not essential for breeding or conservation work. We will, therefore, concentrate on the former.

Chromosomes contain a single linear molecule of DNA and hence the markers on that chromosome occur at particular positions along that molecule. Typically, each chromosome contains 107 to 108 base pairs (bp), *i.e.* 104 to 105 kilo-base pairs (kbp) of DNA, while a typical structural gene, coding for a polypeptide chain, would be between 1 and 2 kbp long. Assuming only 10 per cent of the genome is coding DNA, then a chromosome probably contains something of the order of 1000 to 10 000 genes. Fortunately, the fact that the chromosome is a linear molecule means that the genes need to be mapped in only one dimension.

Currently, the only useful method of gene mapping, at least as far as breeders and conservationists are concerned, relies on recombination between homologous chromosomes resulting from genetic exchange, which can be seen as chiasmata at diplotene through first metaphase of meiosis. A single chiasma on a particular chromosome results in half the gametes from such a meiosis being recombinant for that chromosome: half the gametes will contain a copy of that chromosome that is part maternal and part paternal in origin, while the remaining gametes will be entirely parental. Very few plant species have a sufficiently low number of clearly recognizable chromosomes that the number of chiasmata on a particular chromosome can be compared in different nuclei.

However, in many species the total number of chiasmata in each diplotene nucleus can be counted, at least in pollen meiosis, and the average number of chiasmata per nucleus calculated. The number and position of chiasmata on a particular pair of homologous chromosomes will vary from nucleus to nucleus; it is as though there are a large number of potential sites of exchange, but that in any given meiosis only a few sites are actually involved.

The longer the chromosome the more potential sites there might be. There will invariably be one chiasma and there may be two, three, four or more; a typical set of results is shown in Table. A chromosome that has one chiasma on average is said to be 50 centimorgans (cM) long – a map length unit named after the American geneticist Thomas Hunt Morgan.

This relates to the fact mentioned above that one chiasma results in 50 per cent recombinant chromosomes. By extension, a chromosome with an average of 2.5 chiasmata is said to be 125 cM (*i.e.* 2.5 350 cM) long. It appears that, as a rough rule of thumb, each chromosome has on average two chiasmata and so is 100 cM long. It follows, therefore, that providing the haploid chromosome number is known (*n*), the total map length will be approximately 2 3 *n* 3 50 cM. The average number of chiasmata per nucleus has been calculated in many species and this rule generally holds.

Table. Chiasma Fraquency and Mapping.Chiasma Frequency in Chromosomes of Secale Cereale.

Number of chiasmata **per chromosome**	**Frequency**	Percentage unrecombined **Chromosomes**
1	0.267	50.0
2	0.716	25.0
^{3}3	0.017	12.5
Mean	–	22.0

This rough rule is important because it gives the geneticist an idea of the total genetic map length that should be expected as more genes or markers are mapped. It provides a guide to the extent to which the currently mapped markers cover the full genetic map. Because the map is based in chiasma units, it is not the same as a physical map. The distribution of chiasmata is not uniform over a given chromosome in a species and will vary with the chromosome and the species.

Although a knowledge of the number of chromosomes and, ideally, the mean chiasma frequency per nucleus for the species, indicates the total map length, the actual map has to be constructed from examining the frequencies of progeny in crosses; *i.e.* maps are constructed from information based on the consequences of chiasmata, namely recombination of genetic markers. Recombination of genetic markers is identified by the presence of gametes that contain recombined markers, and such gametes can be recognized from the phenotypes of progeny of a cross.

The frequency of such gametes is an estimate of the recombination frequency between the two markers concerned. The closer two markers are to each other on the chromosome, the less likely a chiasma will occur between them and the less likely they are to recombine. Providing the two markers are sufficiently close that either none or one chiasma occurs between them but never more, then the recombination frequency as a percentage is equal to the map distance in centimorgans as described above.

The problem arises when the markers are sufficiently far apart for two or more chiasmata to occur between them in some meioses. It can be shown that not only does a single chiasma between a pair of markers result in 50 per cent recombination between them, but so also do two, three or more chiasmata, on average.It is necessary to say 'on average', because there are a variety of possible consequences with two, three or more chiasmata in any given meiosis but, in practice, when one is looking at the progeny of a cross or self, each progeny will be the result of gametes from different meioses.

This fact implies that although map distance increases linearly with the number of chiasmata, the same is not true with the frequency of recombination, which can never be more than 50 per cent for a pair of markers even though they might be 100 cM apart at opposite ends of a chromosome.These relationships are illustrated in Fig.. In order to overcome this problem, mapping functions have been devised to correct for multiple chiasmata.

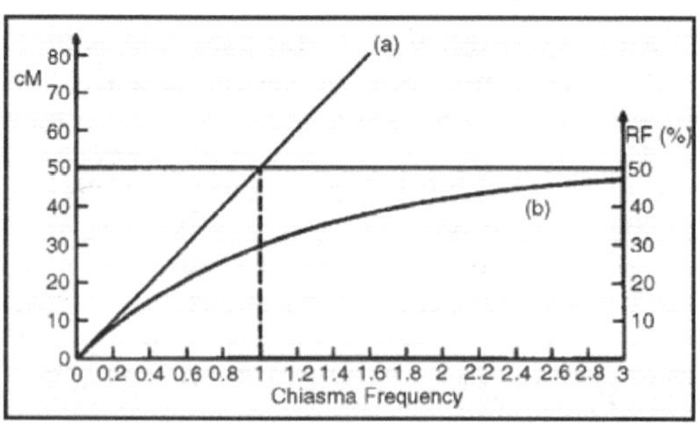

Fig. The Relationship of Chiasma Frequency With (a) Genetic Map Length and(b)Recombination Frequency.

The most common of these are the Haldane (1919) and Kosambi (1943) mapping functions. Haldane's assumes that the probability of none, one, two or more chiasmata in a given interval follows a Poisson distribution, *i.e.* that the chiasmata are independent, random events. Kosambi's method allows for a certain degree of Non-independence in chiasma occurrences. It has long been known by cytogeneticists that chiasma interference occurs over quite large regions of a chromosome.

Interference means that if a chiasma occurs at a particular point on the chromosome, the next one will never occur closer than some fixed distance, the interference distance, from it. Beyond this distance, there is a short distance in which the interference disappears and the next chiasma can then occur randomly outside this range.

Observational data in various species of plants are confirming this and suggest that the interference distance is generally between 15 and 20 cM, *i.e.* 15 to 20 per cent of a typical chromosome on either side of the chiasma. This has important and useful consequences for genetic and breeding work as will be shown later.

Recombination frequencies between pairs of markers can be scored in a wide variety of different crosses. The simplest to use are generations derived from an F1 because only two alleles are segregating, and their distribution in the chromosomes of the parents of the F1 can be determined. The generations that can be used are F2, backcrosses (Bc), recombinant inbred lines (RILs) or doubled haploid (DH) lines, with F2s being the most informative.

General formulae for calculating recombination frequencies are given by Allard (1956). In outbreeding species where F1s are not available, a given individual can be considered as an F1 and its selfed progeny as being an F2, even though the inbred parents do not exist. If it cannot be selfed but controlled crossing to another individual is possible, then again those genes segregating in the cross can be mapped although the situation is more complex.

There could be from two to four alleles segregating at each locus and, with respect to any one locus, the cross could represent an F2, Bc1 or Bc2. Moreover, the distribution of alleles in the two parents cannot be known and has to be inferred from the progeny. This complexity is compounded by the large number of marker loci that may be segregating in any given cross.

Fortunately, a range of software packages are available to estimate these recombination frequencies, identify linkage groups, assign the markers to the most likely order and space them in map units (cM) on these linkage groups; examples are MAPMAKER and JOINMAP. Ideally the number of linkage groups should be equal to the chromosome number in the gametes but, unless the markers available provide a good coverage of the genome, it is likely that those markers on a given chromosome may appear as two or more separate linkage groups simply because the subsets are not sufficiently close for them to be recognized as being together on one chromosome.

A typical marker framework map is shown in Fig. for the chromosomes of *Brassica oleracea*. Clearly these maps provide a very detailed coverage of the chromosome. Many of the markers are very close, *i.e.* less than 10 cM, and over this distance it matters little which of the two mapping functions is used. With the more widely spaced markers, *i.e.* recombination frequency >15 per cent, Haldane's function will exaggerate their distance apart because it will allow for more double crossovers than actually occur.

As more markers are mapped, the total length should converge on the value predicted by the chiasma frequency, *i.e.* approximately 2 3 *n* 3 50 cM. It is important to remember that genetic map distances have standard errors which are dependent on the size and type of population used to construct the map, the mapping population.

The map obtained is that which best fits, given the data and the assumptions underlying it, but there may well be other maps which also fit the data well, though are slightly less likely. Any particular map always develops a greater aura of respectability once it is published, frequently without stating its reliability, and subsequent yet different maps are treated with suspicion. A recombination frequency of true value p has a standard error of ?[p(1 2 p)/N], where N is the size of the gamete population. For example, if p is 0.1 (*i.e.* RF = 10 per cent) and$N$ = 100, the standard error is 0.03 or 3 per cent. In other words the 95 per cent confidence interval of the estimate lies approximately between 4 per cent and 16 per cent.

There are other reasons, apart from statistical sampling, why maps could be wrong. The most obvious is the accuracy with which the data are scored and recorded. Autoradiographs and banding patterns on gels can easily be misread and research workers are loath to discard data even though its interpretation is ambiguous. All data should be checked independently by two or more people and any ambiguities removed. Wrong scores will bias the data and often exaggerate map lengths as they suggest double recombination; wherever double recombination appears to have occurred in two short, adjacent intervals, the data should be checked. Changes in methylation rather than scoring errors could cause a restriction site to appear or disappear so giving the false impression of double crossing- over, or it could simply be an error.

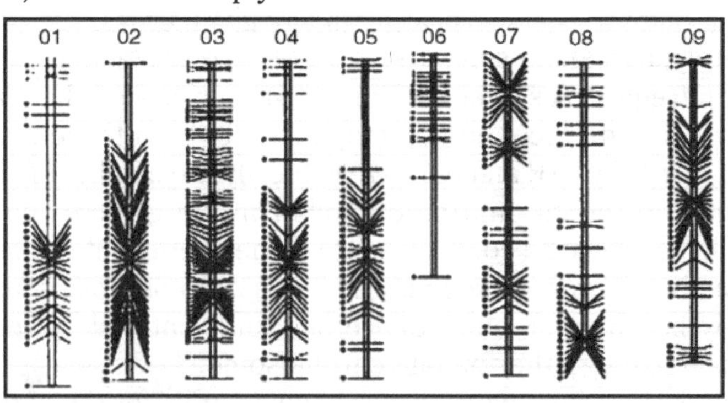

Fig. Typical Molecular Marker Framework Map.The Nine Groups of Brassica Oleracea Based on RFLP, Isozyme and Morphological Markers

Where maps appear to have major inconsistencies in different crosses, different chromosomal structural arrangements may be responsible, such as translocations or inversions. There is considerable evidence that chiasma frequencies, and hence recombination frequencies, may be quite different in male and female meioses even on the same plant although in other studies this may not be so.

Thus although the order of markers on a linkage group may remain the same, the relative distances between them may be quite different if the map derives from male versus female meioses. For example, in *Brassica* it appears that the genetic map obtained from a backcross where the F1 is the female parent is 60 per cent longer than when the F1 is the male parent. It is also clear that environmental factors during meiosis, particularly temperature, can affect the number and distribution of chiasmata, and this could be very critical if the crosses are set up at different times of the year.

Two or more different populations will almost certainly be segregating for different combinations of polymorphic markers, although some at least should be in common. These common ones can be used to overlay the two or more maps and a consensus map produced from an amalgamation of these. Again software is available to produce these consensus maps. Despite the various error-causing factors discussed above a considerable degree of consensus can be produced from such populations and, considerable similarity of chromosome sequence is conserved between even distantly related species, allowing the potential for cross-species genetic transfer.

It was stated above that the total map length should be dictated by the chiasma frequency, in so far as this can be accurately measured. When the first few genes were initially put onto genetic maps in the first 40 years of this century, they only covered a small part of the genome – except in wellstudied species such as *Drosophila* and maize. As more and more genes were placed on the map, so the total map lengths increased towards the asymptote dictated by the chiasma frequency. However, during the first decade following the use of molecular markers, the map lengths of some species appeared to exceed that expected, and it was even argued by some that the chiasma theory of recombination might be wrong.

No one would wish to argue that chiasma frequency data are free from error but there would have had to have been quite excessive underscoring to result in the disparity which was occasionally found. It would appear, however, that the excessive lengths were due in part to errors in scoring and to the use of small mapping populations, because subsequent map sizes have shown a progressive decrease towards the predicted asymptote. Scoring errors bias estimates of RF generally upwards, while RFs have to be large to be detectable in small populations. Our own experience with mapping in cereals and brassicas has also shown a progressive decrease in estimated map length as more markers can be found to fill some of the large gaps in the map and as errors are removed from the mapping data.

LOCATING GENES OF MAJOR EFFECT

The use of extensive molecular and other markers as described above provides a general framework map. Although this has value in its own right for comparing linkage groups in different species as well as possible structural variation within a species, the main value lies in providing a set of markers to locate genes of economic or special scientific interest.

Locating individual major genes is a straightforward development of the procedures above. Let us assume that a cultivar is identified which contains an allele of interest which changes the phenotype in a clearly recognizable fashion and which appears to segregate as a single gene in crosses to other cultivars. Using a knowledge of the positions of existing marker loci, a small subset of markers can be chosen which provide a reasonably even coverage of the genome, say every 20 to 30 cM, *i.e.* approximately five per chromosome.

These then have to be shown to differ between the two cultivars and any which are monomorphic replaced by polymorphic markers.A rough position of the gene of interest can then be obtained by bulked segregant analysis. This involves taking the mapping population, *e.g.* a Bc, and dividing the plants into two groups depending on whether or not they show the phenotype of interest. Bulk samples of DNA are taken from each group, and the two samples assayed for the marker systems chosen as in the paragraph above.

Any marker which is unlinked to the gene to be located will produce similar banding patterns in both samples because it will be segregating independently of the target gene.

Conversely, should one of the markers be very closely linked to the target gene, it will co-segregate and the two samples will show quite different patterns for that marker.In practice, complete co-segregation with any one marker is not very likely but, given a reasonable spread of markers over the genome, one or two are likely to show a strong association with the two groups.

Thus although bands associated with both marker alleles will be found in each sample, their relative intensities will be quite different because of the association with the target gene.

Having located the chromosome and the region on the chromosome, the actual position can be identified in a segregating population using the principal marker associated in the bulked segregant analysis together with other polymorphic markers situated about 10 cM on either side of the principal marker.

The target gene is then mapped with respect to the three markers. This approach can be used for a wide variety of populations and traits. Even if the trait has poor penetrance, the affected group will clearly show the marker banding pattern even if the non-affected group consists of a mixture due to misclassification.

MAPPING GENES CONTROLLING QUANTITATIVE TRAITS

Quantitative traits such as yield, quality, height and flowering time, which are controlled by several genes and are greatly influenced by the environment, create particular difficulties for gene mapping. The problems arise because the genotype for the trait concerned can never be clearly identified from the phenotype; many different genotypes could produce the same or very similar phenotypes whilst the same genotype can result in different phenotypes depending on what may be elusive factors in the environment.

Whereas with a single gene difference controlling a major effect, two or more Mendelian phenotypic classes might be recognized in ratios of 3:1, etc., quantitative traits resulting from the joint segregation of many genes show a continuous, often normally distributed range of phenotypes.

The total phenotypic variation in an F2, for example, VP, is made up of genetic and environmental components, VG and VE. The genes underlying such traits are variously called polygenes, effective factors or, more recently, quantitative trait loci or QTL.

However, it is still possible to map and measure the effects of the QTL by techniques analogous to those used for single major genes, *i.e.* by looking for the effects of co-segregation of particular marker loci with differences in the quantitative trait.

Various mapping populations can be used as before, F_2, Bc, RIL, DH or open-pollinated full sib (FS) or half sib (HS) families, and the individuals or families are scored both for their phenotype for the quantitative trait(s) and for their genotype at the marker loci.

The type of population that is used will depend on the traits studied and the breeding system of the plant species but, other things being equal, an F_2 population will normally provide the most information for a given size.

Unfortunately, most traits of economic importance such as yield, quality and disease resistance are not usefully scored in an F_2 population because one is interested in how the material performs in something similar to the high-density monoculture that it will encounter in agricultural practice.

Measuring yield in an F_2 population with spaced plants is straightforward and statistically very powerful but it may well not provide any insight into how yield is controlled in agricultural practice.

It is therefore necessary to use a plot structure for most traits in an attempt to simulate commercial conditions. For this reason most plant breeders will prefer to work with genotypes that can be extensively and easily replicated such as RILs, DHs or F_3s. It is also easier to explain the principles underlying QTL mapping using a population of DH lines, and so for both reasons the methodology will be illustrated with DH lines, although the same principles apply to all populations. Because several trait loci are concerned, probably on different chromosomes, bulk segregant analysis is of limited applicability, at least initially.

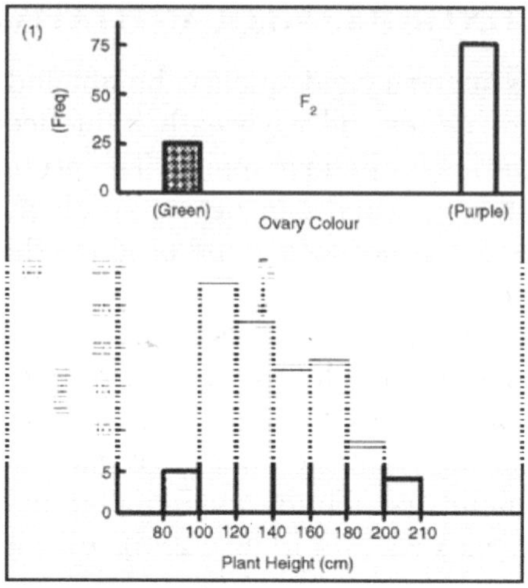

Fig. Contrast Between Distribution of a Qualitive and Quantitative trait in an F_2. Data for NIcotiana Rustica,Ovary Colour and Height.

THEORY OF QTL MAPPING IN DOUBLED HAPLOID LINES

Doubled haploid (DH) lines can be produced parthenogenetically from an F1 by microspore or ovule culture and, in some species, such as wheat and barley, by wide outcrossing. Each original DH plant is derived from a single gamete, which subsequently becomes homozygous and disomic either naturally or by treatment with the drug colchicine, which inhibits spindle formation, and hence chromosome disjunction, at metaphase of mitosis.

Selfed seeds from such a plant produce a DH line of identical individuals so the line becomes effectively immortal. If we consider an F1 that is heterozygous for a marker locus M1/M2 and for a linked QTL locus Q1/Q2, where the 1/2 indicates the allele with an increasing or decreasing effect on the trait, this will produce the gametes and hence the DH lines shown in Table.

If the DH lines are scored for some trait for which the Q+/Q+ homozygotes increase the trait above the overall mean, *m*, by *a* units while the Q2/Q2 homozygotes decrease the mean by *a*, then the means of the four gametic types are as shown in Table, where *R* is the recombination frequency between the QTL and the marker. From this it follows that the mean trait score of all those DH lines which are M1/M1 (M1M1) is *m* 1 *a*(1 2 2*R*), while the mean of all those DH lines which are M2/M2 (M2M2) is *m* 2 *a*(1 2 2*R*). In other words, half the difference between these two means (M1M1 2 M2M2) is *a*(1 2 2*R*). Clearly, if the marker is so close to the QTL that it never recombines, *R* will be 0 and the marker difference will represent the QTL effect, *a*. Conversely if the two loci are unlinked, *i.e.* *R* = 0.5, then the difference will be zero. This effect is, of course, the basis of bulked segregant analysis. In general, the magnitude of the marker effect will decline linearly with the distance of the marker from the QTL in terms of recombination frequency, and so with several linked markers, their individual trait effects will be as shown in Fig.. Similar Arguments Apply to Other Populations Such as F_2s and RILs.

Table. QTL model for market analysis in doubles haploid (DH)lines from and F_1.(a)The F_1 constitution; M=Market and Q=QLT genotype.(b)DH genotypes,frequencies and genetic values.(c) Expertations of marker means and differences.R=recombination frequency between M and Q; m and a are the mean and additive genetic effects of the QLT;g is the additive genetic deviation at the market locus

(a)

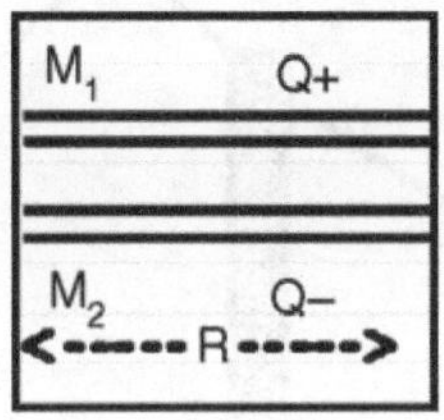

(b)

	F_1 gametes			
	$M_1 Q^{\cdot}$	$M_1 Q^{-}$	$M_2 Q^{\cdot}$	$M_2 Q^{-}$
Frequency	a(1–R)	a(R)	a(R)	a(1–R)
DH genotype	$M_1M_1 Q^{\cdot}Q^{\cdot}$	$M_1M_1 Q^{-}Q^{-}$	$M_2M_2 Q^{\cdot}Q^{\cdot}$	$M_2M_2 Q^{-}Q^{-}$
Genetic value	m . a	m – a	m . a	m – a

(c)

$$M_1M_1 \ @\{a(1-R)(m\ .\ a)\ .\ a(R)(m-a)\}a$$
$$@m\ .\ (1-R)a$$
$$M_2M_2 \ @\{a(R)(m\ .\ a)\ .\ a(R)(1-R)(m-a)\}a$$
$$@m-(1-R)a$$
$$g \quad @(\overline{M_1M_1}-\overline{M_2M_2})/2$$
$$@(1-2R)a$$

If the trait means of the markers and their map are known very accurately, and only one QTL existed on the chromosome, then as Fig. shows, it would be relatively easy to locate the QTL. Figure shows the distribution of flowering time among a number of doubled haploid lines of *Brassica oleracea* containing alternative alleles at a marker located very close to a QTL. In practice neither the map positions of the markers nor their means are known with very great accuracy because the number of DH lines and replicated plots are generally few and the heritability of quantitative traits is generally low. Moreover, linked QTLs create difficulties because their individual effects can combine either to create the impression of an intermediate, ghost QTL, or to conceal their effects altogether.

Various analytical procedures have been developed to maximize the efficiency and accuracy of locating individual QTLs and to separate linked QTLs. They all rely on the relationships described in the previous paragraph and use either maximum likelihood or weighted least squares regression procedures to estimate the QTL locations and effects that best fit the observed trait scores for each genotype.

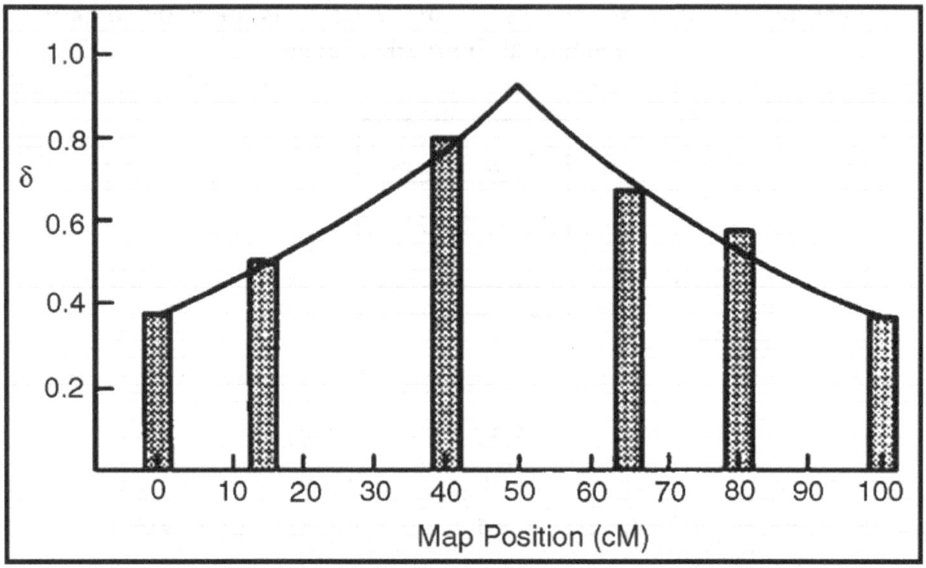

Fig. Quantitative Trait Effects (d) Associated with Marker Loci.The Bars Indicate the Observed Effects,While the Curve Illustrates the Expected Relationship if a QTL of d = 1.0 Existed at 50 cM

The main difficulty is that there are two unknown parameters with respect to each marker score; the QTL effect, *a*, and map position. This means that it is necessary, in all methods, to use an iterative approach which involves trying all possible QTL locations along each chromosome and identifying the position or positions of the QTL which best fit the observed data as indicated by the size of the likelihood or the residual regression variance. As a result, a very large number of statistical tests are performed with the concomitant risk of false positive results.

All methods provide estimates of QTL locations and effects with similar precision in terms of confidence intervals, and this precision drops rapidly as the heritability of individual QTLs declines, *i.e.* the estimates are accurate only when there are just two or three unlinked QTLs with large effects.

For example, Hyne *et al.* (1995) showed that the 95 per cent confidence interval for the location of a QTL contributing 10 per cent to the phenotypic variance of an F2 could be as large as 35 cM. However, conservationists and breeders who are concerned with introducing QTL into commercial cultivars from other cultivars or more distantly related species, would normally only be interested in those cases where a few QTLs of major effect are to be manipulated. There is no evidence that dense maps are required for initial

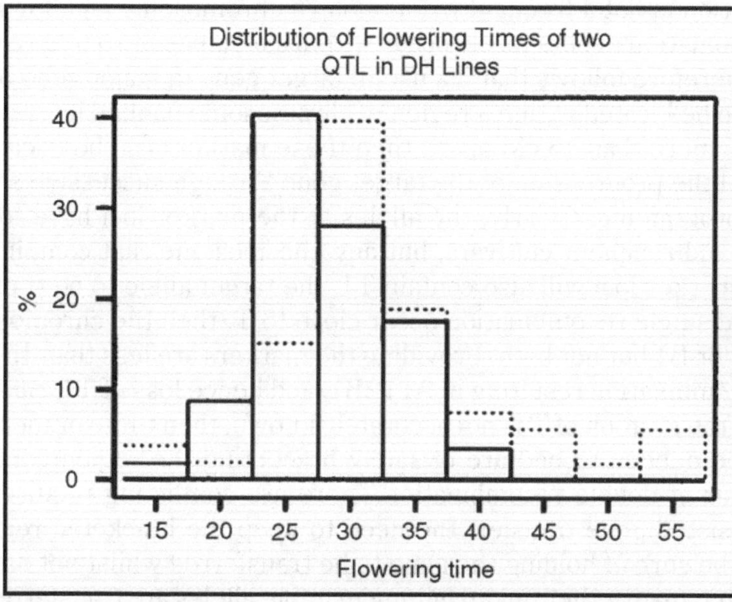

Fig. Distribution of Flowering Times Among Doubled Haploid (DH)Lines in Brassica Oleracea Containing alrernative Alleles of a Marker Linked to a QLT.The two sets Significantly Different Means Indicating a QLT with an Effect of Apporoximately ± 2 Days

QTL location. It is better to have large populations and a few (5–10) markers per chromosome.There is evidence that many of the larger QTLs located so far map closely to previously known major genes. This is often used as evidence that there are, in fact, very few QTLs, but this may well be misleading for several reasons. The QTL map locations are sufficiently imprecise that they could well appear spuriously close to candidate loci. Only QTLs of large effect will stand much chance of being located and, by definition, they are likely to be major genes. It is likely that many cases where QTLs have very large effects could be due to the chance association of alleles of like effect at several QTLs along a chromosome, which cannot be separated as individual effects.

Much effort has been devoted to increasing the precision of QTL location, in particular in improving the power of the test to reduce the failure to detect genuine QTL. A long-established problem with breeding for quantitative traits in plants is the genotype 3 environment interaction. Using QTL location techniques, it is now possible to explore these effects at the level of individual QTLs although it is often difficult to know whether genotype 3 environment or poor repeatability is responsible.

ALIEN GENE TRANSFER

Once useful genes have been located on a framework map, it is then possible to use the molecular markers to facilitate the transfer of the useful genes between species and strains in an efficient manner. As stated earlier, recombination is not a frequent event along a chromosome and, furthermore, two recombination events rarely occur within a distance of 15 to 20 cM of each other. It therefore follows that if a useful target gene (a major gene or a QTL) is known to be located within a region of chromosome flanked by two markers that are no more than 15 cM apart, then these markers can be used to follow and control the progression of the target gene through successive stages of a breeding programme. Clearly, the alleles at the marker loci have to differ in the donor and recipient cultivars, but any chromosome that contains A1 and B1 at the marker loci will also contain T1, the target allele to be transferred.

Should single recombination occur close to T, then the chromosome will contain A1 or B1 but not both. Providing the markers are less than 15 cM apart double recombination resulting in A1T2B1, and hence loss of T1, can be safely ignored. If the location of T is not accurately known, then three or more marker loci may be needed to be sure of safely bracketing the region containing T without fear of double recombination.There are conflicting requirements in marker-assisted gene transfer: the need to keep the bracketed region large enough to be sure of holding the gene to be transferred whilst not having it so large that too many other linked but undesirable alleles are transferred with it. Let us now consider the basic procedure of gene transfer.

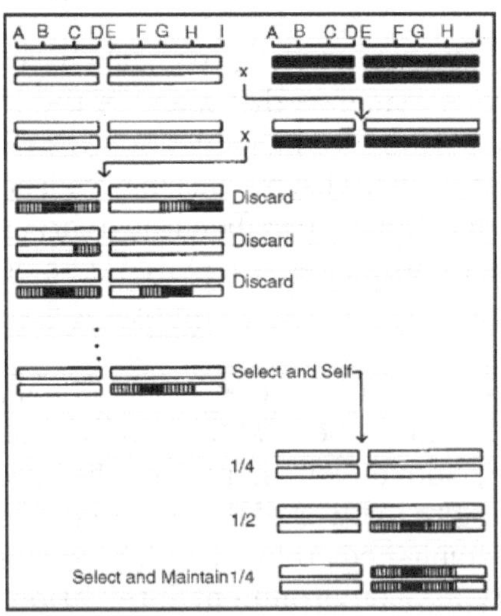

Fig. The use of Molecular Markers to Facilitate Gene Introgression.

Traditionally, breeders have introduced a useful allele into their cultivars by backcrossing the original F1 to the commercial cultivar whilst selecting at each generation for the desired allele. As was shown earlier, this can result in a very large region of chromosome around the target gene surviving even after ten generations of backcrossing – a phenomenon known as linkage drag. With markers, however, it is possible to reduce linkage drag considerably whilst simultaneously reducing the number of backcrossing generations to two or three instead of the normal six to eight. Moreover, the amount of plant material raised at each generation can be reduced because much of the molecular genotyping can be achieved at the juvenile stage.

The aim is to hold the target allele, T1, heterozygous throughout the backcrossing process, by selecting for the flanking markers, whilst simultaneously selecting for the genotype of the recurrent, commercial cultivar at all other loci. The latter can readily be achieved by having as few as four or five well-spaced markers on all chromosomes and selecting for recurrent parent alleles at each generation.

Depending on the chromosome number, some individuals in the first backcross generation will be homozygous for several whole chromosomes whilst still being heterozygous for the target gene. Those which have most of the recurrent parent chromosomes are selected and backcrossed again, and if necessary the process is repeated for a third generation. For each generation, fewer markers need to be screened as those found to be homozygous in the previous backcross no longer have to be checked.

Unless the position of the target gene is very accurately defined, it is prudent to have several markers covering its likely position, again with the proviso that they need to be separated by no more than 15 cM. At the final stage of the introgression process, a backcross individual which is homozygous for most of the recurrent parent alleles is selfed or, if this is not possible, intercrossed with a similar genotype.

From among the progeny, individuals are chosen which have different combinations of markers bracketing the target region as shown in figure.

These can be selfed again and homozygotes for the different sequences identified and multiplied for reassessment for the target trait. Some will fail to show the target trait because of recombination, but of those that do, the lines with the shortest sequence of donor markers are selected for further trials. The approach can obviously be adapted to the simultaneous introgression of several genes: no new principles are involved but the screening process becomes more elaborate.

It is always possible at a later stage to reduce the length of the introgressed region around T by looking at other, more closely linked, markers. Similarly, it is wise to check the origins of the other chromosomes to avoid having introgressed unwanted regions through missing double recombinants in the central regions of individual chromosomes or single recombinants towards their ends.

If such errors are found they can easily be corrected by another round of backcrossing to the recurrent parent. When attempting to reduce the length of donor chromosome surrounding the target gene, it is important to realise that this should be done in two successive generations. If the sites of the required recombination events are less than 15 cM apart, then the simultaneous double event will not occur. Even if they are far enough apart for the double event to occur, the probability of such an event may be too low to be practicable. If two recombination events each have a probability of 0.05 (*i.e.* 5 per cent), their combined probability is at best 0.0025. Therefore, it would be necessary to genotype approximately 1200 individuals to be 95 per cent sure of having at least one double recombinant. If it was tackled in two stages then it would require only the genotyping of 86 (= 28 in the first round plus 58 in the second; less being needed in the first round because the recombination could occur on either side of T), a very considerable saving in effort of 93 per cent, or 1114 plants!

Marker-aided selection has been successfully tried for quantitative traits in several crops while the efficiency of the procedure has been examined by Lande and Thompson (1990).

MAP-BASED GENE CLONING

Providing a target gene can be located to within a pair of marker loci, it should be possible to use standard cloning techniques to walk along the chromosome between the markers and to find the target gene. The feasibility of this depends on the distance between the markers and the ease with which the target gene can be recognized from among the clones.

Depending on the species, a genetic length of 1 cM can, on average, consist of as few as 280 kbp of DNA in rice, or 2220 kbp in maize. It is not always easy to locate a target gene with sufficient accuracy to be able to say that it is between two markers as close as 1 cM, and hence the actual lengths of DNA between them can often be much greater. With QTLs one can seldom achieve anything approaching this accuracy by existing methods of conventional mapping. Furthermore, these are average distances per centimorgan; some chromosomal regions could be much longer or, indeed, less.

Increases in mapping reliability can be obtained by using what are called near isogenic lines (NILs), which are lines obtained by selfing or backcrossing and are known to differ from some standard genotype by just a short defined section of chromosome. Providing this section is delineated by accurately mapped flanking markers, and the NILs which do and do not contain this region can clearly be shown to differ for the target gene, then the target gene has to be in that region. The genetic and physical proximity of the markers can be established by conventional mapping and by gene cloning.

Map-based gene cloning involves starting from one of the markers and walking to the next by means of a series of partially overlapping cosmid clones. Cosmid clones are necessary because they can be up to 50 kb in length and hence require fewer cloning steps in order to cover the DNA between the markers (280 to 2200 kb per cM in the examples cited above). Identifying the target gene is more difficult. Clearly, if the gene is between the two flanking markers it has to be on one of the cloned regions and could possibly be on two overlapping clones. It could be identified by transforming plants with each clone separately and identifying which transformant expresses the effect. If the gene concerned exhibits classical dominance it would be necessary to transform the homozygous recessive with the dominant allele or the transformant would not be recognized. Alternatively, the gene has to be introduced into a species that does not normally express the gene. Once the effective cloned fragment is identified, the actual location of the gene within the fragment could be identified by successively transforming sub-fractions of the clone or, following sequencing, by looking for potential open reading frames. In those cases where the target gene produces a known product, cDNA clones derived from tissue likely to be rich in the appropriate mRNA of the target gene can be used to identify likely sequences in the previous overlapping clones.

Once the gene has been identified by one of these methods, its structure and activity can be studied in detail. Disease resistance genes are obvious candidates for such studies as their action is very specific and their location on the genome of many species is well documented. However, unless suitable candidate loci.

DEVELOPMENTS IN THE SEED INDUSTRY

Many major food crops upon which we depend originated and were first cultivated in Third World countries rather than in the industrialized countries where they are most productively exploited today and in which they have been improved by traditional plant breeding.

The potato originated in South America, wheat in Ethiopia and the Near East, important maize varieties in South America, and rice in Asia. Moreover the so-called Vavilov centres, which contain the greatest variety of living species and which have been the source of much important gene material, are mainly in Third World countries. The wealth of the industrial countries owes much to their past ability to exploit agriculturally crops and animals originating in other countries. In the colonial era the success of British scientists in smuggling rubber plants from Brazil to Sri Lanka, Singapore and Malaysia and the comparable more complex route by which coffee was introduced to Latin America from Ethiopia are instances of colonial powers obtaining significant wealth by exploiting plant material from the Third World. In one way or another the process by which institutions and firms in developed countries have continued to collect new plant and animal varieties from less-developed countries has continued.

But what is still a major source of friction is the procedures by which legislative protection is developed to give companies in the richer countries commercial rights to exclusive exploitation of varieties developed from plants freely collected from other countries, and to extract high rates of profit from the sale of those varieties. Most extreme is the situation whereby under the

USA Plant Patent Act of 1930, which covered asexually propagated plants such as certain fruits, flowers, ornamental shrubs and trees, it has been possible for plants discovered and smuggled out of other countries to be given patent protection in the USA.

The key to this is that the Act accepted the unfeasibility of requiring proof of an 'inventive-step' for the granting of a patent, and employed the criterion that as far as can be determined the plant variety is new. Since some novel variety found in the rain forests of South America may satisfy this criterion, and the source of the plant can be concealed from the authorities, the theft and smuggling of plants can be said to have been encouraged.

Seed companies in developed countries have successfully pressed their case for protection, and have progressively extended Plant Breeders' Rights (PBR) through both national legislation and the Union for the Protection of New Varieties of Plant (UPOV), which is an international agreement among signatory countries to honour a system of varietal rights. In the case of food crops, in the absence of patentability, PBR operates by registering distinct varieties. Pressure has been exerted upon less-developed countries to become signatories of UPOV and thus acknowledge the rights of breeders, most of whom are from developed countries. However, the majority have stalled or rejected membership, and have been fundamentally unhappy about accepting the principle of creating monopoly rights for others over plant material which in some cases originated in their own country.

The whole issue of intellectual property rights has been raised to a prominent political level by the USA's insistence that it be included as one of the thirteen areas for negotiation in the Uruguay Round of multilateral bargaining of the General Agreement on Trade and Tariffs.

While the issue is a much broader one than that of plant and animal breeders' rights, it reflects a basic political conflict between the richer countries, determined to obtain secure returns on R and D expenditures by companies, and LDCs struggling to catch up and anxious to avoid having to pay royalties on products protected by PBR and patents. This conflict has broader moral and ethical dimensions when it relates to the issue of charging royalties for the seeds of food crops to countries where malnutrition is widespread.

Against the previous argument has to be balanced the need of those incurring R&D expenditure to capture enough of the returns to provide an incentive for R and D. At least that is certainly the case where R and D is to be undertaken by the private sector, which is politically the increasingly preferred solution in the UK, the USA and elsewhere.

This, however, leads to another area of concern, which is that a small number of large multinational chemical and pharmaceutical firms have increasingly come to control the seed industry. Companies such as Shell, Sandoz, Dekallb-Pfizer, Ciba-Geigy and BP have become owners of most of what were small independent seed companies.

This linking between chemicals and seeds within firms has a strong commercial logic. It facilitates the marketing of packages consisting of seed varieties plus the appropri-ate fertilizers, insecticides, etc. It also allows the companies concerned to integrate their plant breeding, biotechnology and chemical research.

One particular synergy for these companies will be to bioengineer varieties of major food crops to resist their own herbicides and weedkillers. This would permit chemical weed control to be extended to large-scale crops where this is not now possible, and is just one way in which the direction of biotechnological research may be influenced (biased) by the combination of forces giving a small number of companies extensive control over the basic means of agricultural production.

There are many other issues which could be touched on regarding corporate influence over the development of agricultural biotechnology. It is, however, understandable that the spokesmen for LDC countries, which are profoundly concerned about their dependence upon western technology and exercised by what is seen as colonial and post-colonial exploitation by the West, should be alarmed by the prospect of new rights being created over seeds and other biotechnological products.

Certainly there are widespread fears that they will only gain access to the fruits of biotechnology on unfavourable terms and that their relative dependence on and subservience to western industry will be increased so that benefits to them will be small. There are many western critics who would agree with this, and that the system by which agricultural biotechnology is delivered will favour the 'haves' rather than the 'have-nots'.

METHODS FOR CLONING PLANT TRAITS

CLONING BASED ON KNOWLEDGE OF PROTEIN STRUCTURE/ SEQUENCE

Many of the first plant traits to be cloned were those in which some degree of biochemical understanding preceded any work with DNA. The understanding may have come from plant systems or from outside the plant kingdom. Most of the cloned genes in these cases were enzymes which were part of a well-characterized biochemical pathway or had a protein structure which was otherwise well known, such as seed storage proteins. Cloning genes based on a knowledge of the protein sequence can be achieved by working backwards from the gene product (amino acid sequence) to the coding sequence or the DNA template. Such an approach has been used for cloning many seed storage proteins, such as the 2S sulphur- rich proteins from Brazil nut.

Genes of plant origin have been used in transgenic plants to improve resistance to insect pests. One of the first to be characterized was the cowpea trypsin inhibitor, a protease inhibitor which acts as an antifeedant to a range of insect pests. The protein is naturally produced in seeds of some genotypes of cowpea (*Vigna unguiculata*), which was shown to confer lepidopteran insect resistance when expressed constitutively in tobacco. Another such report from the same group at Durham

University described the use of a snowdrop lectin gene to confer aphid resistance in transgenic tobacco, the first report of genetically engineered resistance to a sucking insect in plants. The lectin from snowdrop (*Galanthus nivalis*), known as GNA, has been demonstrated to be effective against two major rice pests, the brown planthopper and the green leafhopper, as well as the peach potato aphid. When expressed at 0.1per cent of total leaf protein in tobacco, GNA is antimetabolic to homopteran pests. These levels of protein were accumulated in leaf tissues when placed under the control of the CaMV 35S promoter. However, Hilder *et al.* (1995) postulate that better control would be achieved with the use of a phloem-specific promoter, such as the sucrose synthase 1 promoter from rice.

Another significant group of genes which have been cloned based on prior knowledge of the end product are the seed storage genes, including both seed storage proteins and lipids. In both cases there was considerable biochemical knowledge of the pathways involved in biosynthesis and storage. For seed storage proteins, the genes can be cloned based on the knowledge of the amino acid sequence of the mature protein and targeting signal or transit peptide.

Molecular genetic manipulation of vegetable oils is a major area of research in the developed world. While approximately 90per cent of vegetable oils are used for human consumption, there are emerging markets for industrial applications including lubricants, paints, cosmetics, plasticizers and soaps.

With increasing emphasis on quality and dietary considerations, there is a desire in developed countries to substantially alter the types and balances of fatty acids found in the major oil crops. Six crop species (soybean, oil palm, rapeseed/canola, sunflower, cottonseed and peanut) account for 84per cent of the world's vegetable oil production, with the major fatty acids produced being palmitic, stearic, linoleic, linolenic and oleic acids.

However, almost 200 different fatty acids are produced by plants, mostly occurring in Non-domesticated plant species. Hence, genetic manipulation of the major crop species, either via conventional breeding or genetic engineering, has the potential to modify the fatty acid types produced and stored by seeds such as soybean and canola.

The biochemistry of fatty acid biosynthesis is well established for many of the major pathways, with enzymes identified which control the chain length of fatty acids, and the degree of saturation, or number of double bonds in the fatty acid chain. The oils stored in seeds have been altered by downregulation of particular enzymes using antisense RNA techniques.

The oleic acid content of rapeseed oil has been increased to 83per cent (from 62per cent in conventionally bred cultivars) by downregulating the enzyme

D12-oleate desaturase using antisense RNA. A similar result has been achieved using co-suppression, which has produced rapeseed oil with 87per cent oleic acid content.

Genetic engineering approaches have also been used to introduce totally novel fatty acids to major oilseed crops. Petroselinic acid, an isomer of oleic acid, is found in abundance in the spice plant, coriander (*Coriandrum sativum*). This oil has uses in cosmetics and pharmaceuticals, and its oxidation leads to the formation of lauric acid, used in detergents, and adipic acid, which is used in nylon production.

Hence this is a potentially valuable oil which may enable soybean and canola crops – *i.e.* renewable resources – to replace some of the Non-renewable carbon fuels which are currently used for many industrial polymer and lubrication applications. A cDNA clone from coriander encoding an acyl-ACP desaturase has been genetically engineered into tobacco, resulting in the accumulation of petroselinic acid. Oil crops such as rapeseed and soybean could be genetically modified to produce this fatty acid on a commercial scale.

Hence, here is an example whereby a minor crop plant has become a donor of genetic diversity which will enable the commercial production of a very useful array of products for food and industrial purposes.

Cloning Genes Known Only by Phenotype

However, a more challenging task has been to move beyond proteins and pathways about which sufficient biochemistry is known into cloning genes known by phenotype only. By far the majority of genes fit into this category, which includes many of the most important plant traits such as disease resistance, adaptation to abiotic stresses, phenology and flowering behaviour and fertility. Only in the past few years have genes for traits known only by phenotype been cloned. One of the most interesting areas of plant biology is the understanding of plant–pathogen interactions at the molecular level.

The major methods used for cloning genes known only by phenotype are:

- Transposon tagging.
- Map-based or positional cloning.
- T-DNA tagging.

There have been a number of plant genes cloned using all these methods or modifications of them. There are also examples where a combination of methods has been used to expedite the rate of identification of the unknown sequence.

Transposon Tagging

The proposal that specific pieces of DNA are able to autonomously excise from a particular locus and insert, theoretically at random, into another linked or Non-linked locus was first proposed by Barbara McClintock. Transposable elements are now accepted as ubiquitous, and are particularly well characterized in organisms such as *Escherichia coli*, *Drosophila* and maize. In two plant species, maize and the snapdragon (*Antirrhinum majus*), transposable elements are sufficiently well characterized to be used for mutagenesis and gene tagging. The best understood system is the maize *Ac/Ds* (*Activator/Dissociation*) family of elements. The behaviour of the *Ac* element is such that it can excise autonomously from its locus, under the control of its own transposase, and insert into another chromosomal position.

Where the *Ac* element is in or near a gene, that gene is inactivated (*i.e.* not expressed), and will appear phenotypically as a recessive mutation. When *Ac* transposes (excises and reinserts at another locus), the gene is commonly returned to normal function, and hence is termed a revertant.

Should this *Ac* element then insert into another gene (transcribed or regulatory sequence), then a new recessive mutation will manifest. Transposition can occur either germinally, which results in an altered phenotype for the whole plant, or somatically, which results in a mosaic mutation, and can often be detected as a sectoring of plant or seed colour. The *Ds* element is actually a group of elements derived from defective *Ac* elements, with a common feature being their lack of effective *Ac* transposase.

Hence these elements are Non-autonomous, and will only transpose in the presence of an *Ac* element capable of producing transposase. Equipped with sequence and map information of an *Ac* element in maize, geneticists have been able to clone important maize genes using the transposon-tagging approach.

Table. Plant Transposable Elements which have been used for Transposon-Taggine

Element	Species	Type of Element	Size
Ac	Maize	Autonomous	4.6
Ds	Maize	Non-autonomous	0.5–30
Tam1	Snapdragon	Autonomous	15
Tam2	Snapdragon	Non-autonomous	5
En/Spm	Maize	Autonomous	8.3

When a gene is inactivated by an *Ac* element, proof of which can be demonstrated by a low frequency of trait reversion, the *Ac* element can be used as a probe for 'fishing-out' the tagged gene. A tagged gene can be cloned by performing Southern blot analysis with the *Ac* element used as probe, and any polymorphism represents an excision/insertion event. The technique is described more fully in the review by Walbot. The *Ac* system has proven extremely useful for cloning maize genes for traits such as plant pigment genes (*bronze*; the anthocyanin pathway), starch biosynthesis (*opaque2*), and abscisic acid (AbA) insensitivity. Other maize transposable elements such as *Mutator* (*Mu*) and the *En/Spm* system have been used to clone maize genes, as has the *Tam* system in snapdragon. As a result, the search began for transposable element systems in other species so that a similar approach could be taken. However, it was soon demonstrated that the well-characterized *Ac/Ds* system could be genetically engineered into other plant species including tobacco, *Arabidopsis*, tomato and petunia. It was subsequently shown that the *Ac* element acted in a similar manner in these heterologous species, undergoing germinal and somatic excision and insertion, and maintaining the propensity to reinsert at a linked site (usually within 4 cM).

As a result, the frequency of mutation in the surrounding area approaches 1024. It was not until 1993 that a plant gene was cloned using the heterologous system, that of the *Ph6* gene in petunia, which is involved in floral pigmentation. Using the transposon-tagging approach, a number of plant disease resistance genes (R genes) have now been cloned. The first of these was the tomato *Cf-9* gene, which confers resistance to race 9 of the fungal pathogen *Cladosporium fulvum*, causal agent of leaf mould. The gene was tagged using a modified *Ac*/*Ds* system.The system relied on the use of a stabilized *Ac* element (s*Ac*), which was effectively a disabled *Ac* element that nevertheless produced *Ac* transposase and hence enables *Ds* transposition. The two elements are maintained in separate lines, and hence are both stable until controlled crossing is performed.

This gives a significant advantage in that *Ds* insertion mutants can be maintained indefinitely in the homozygous or hemizygous form, and revertants can be isolated by crossing lines with s*Ac* tomatoes. The transposon-tagging approach used to isolate the *Cf-9* gene, which was known by phenotype only, did have considerable advantages in that much was known of the pathogen. The hypersensitive response elicited by the virulent strain fitted into the gene-for-gene hypersensitive response proposed by Flor, with a dominant avirulence gene (*Avr9*) interacting with the *Cf-9* gene to form an incompatible or hypersensitive response. *Avr*9 had previously been cloned and fully characterized, and was known to specify a 28 amino acid peptide which alone could elicit a necrotic response in resistant tomatoes.

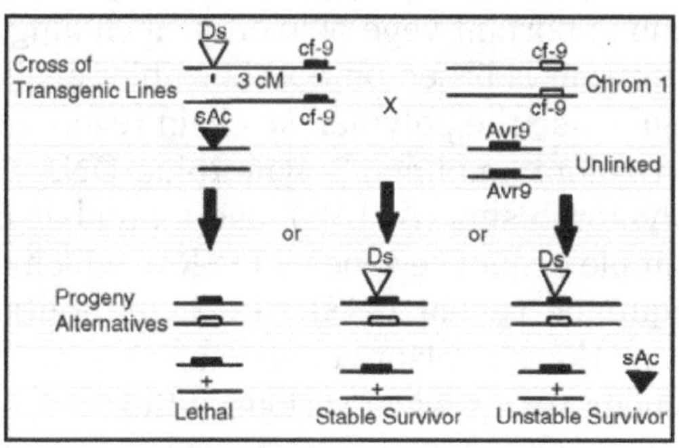

Fig. Graphical Representation of the Crosing Strategy of Transgenic Ac/Ds Tomato Lines.

Crosses were made to develop a transgenic tomato line homozygous for *Cf-9*, but heterozygous for *Ds*. Crossing was performed as outlined in Figure. As can be seen, most seedlings died from systemic hypersensitive response, with *Avr9* and *Cf-9* present in all cells. The only survivors would be individuals where *Ds* had inserted into the *Cf-9* gene. Where this happened in plants carrying s*Ac*, the seedlings were variegated for the hypersensitive response (somatic excision), and in the absence of s*Ac*, the survivors were stable (germinal excision).

This work was a major undertaking, involving the germination of approximately 160 000 progeny, yielding 118 survivors, representing 63 independent mutations. After considerable effort to characterize these, the *Cf-9* gene was cloned and characterized, with an open reading frame encoding an 863 amino acid protein.

Map-based Cloning

The previously described transposon-tagging scheme utilized to clone *Cf-9* had an element of the map-based approach by utilizing the phenomenon of *Ac/Ds* transposition to linked sites, with a *Ds* element within 3 cM of the locus of interest, which greatly enhanced the likelihood of insertional mutagenesis. However, map-based cloning requires a much tighter genetic linkage (in centimorgans or recombination fraction), which in turn needs to be converted into a physical distance (kbp of DNA sequence).

The map-based, or positional cloning concept relies on building a well-saturated genetic map of DNA sequences of the organism in question. The genetic map is built up in a polymorphic mapping population, commonly an F2, backcross (BC) or recombinant inbred, with markers such as RFLPs (restriction fragment length polymorphisms). The first RFLP map to be made was of human. Based on such maps, important human genes responsible for disorders such as cystic fibrosis and muscular dystrophy have been cloned in this manner.

Genetic maps of plant species have been made with DNA markers in virtually all important plant species, particularly the major cereals, grain legumes, oilseeds and important vegetable crops including tomato and potato. The RFLP marker system is based on Southern blot analysis, while many of the newer technologies use the polymerase chain reaction (PCR).

These include random amplified polymorphic DNA (RAPDs), amplified fragment length polymorphisms (AFLPs), and a number of different marker systems based on simple sequence repeats (SSRs), which may use PCR, such as inter simple sequence repeat (ISSR)-PCR, or Southern analysis. The applications of DNA markers to plant genetic analysis and breeding are many, and the subject has been reviewed by various authors.

The population for saturation mapping should be segregating in a Mendelian manner for the trait of interest. This requires a lot of tedious but demanding work, and it may be that particular populations such as near isogenic lines, or DNA pooling strategies, such as bulked segregant analysis, will allow the mapping effort to be more targeted to the region of interest, hence improving the rate of map saturation. It is desirable to obtain extremely close genetic linkage to the locus of interest (<1 cM), with flanking markers if possible. It can be of considerable advantage to identify a marker which co-segregates (say within 0.1 cM) with the trait. The next step is to convert the genetic map into a physical map of the chromosome region, translating distances from centimorgans to nucleotides.

The Gene of Interest (in this illustration,R) is Flanked by Markers such as RFLPs or SSRs. A Series of Overlapping YAC Contigs is used to cover th Chromosomal Region Containing the Gene.The YAC Which Contains the Complete Gene is Marked in Bold.This is then Subcloned or Matched to a cDNA Library to Idenfify the Gene(s) Contained the Region, and Ultimately Identify the R Gene.

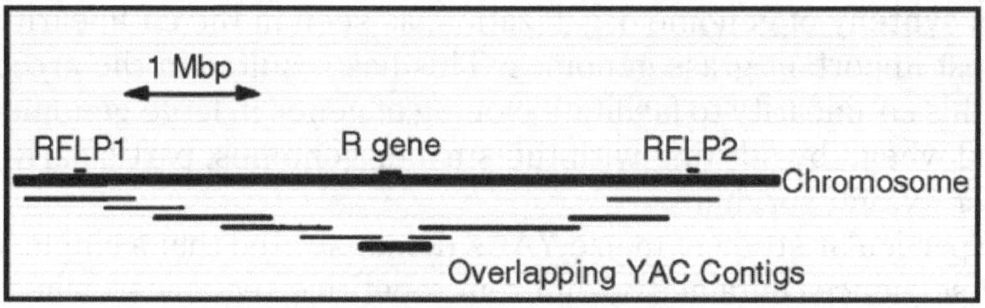

Fig. Generalized Strategy of Chromosome Walking Used in Map-based gene Cloning.

Physical mapping then involves the use of YAC or BAC clones which span the region of interest. These YAC/BAC contigs will span the region between two flanking DNA marker sequences as shown in Figure, a process known as chromosome walking. In this example, two YAC/BAC clones contain the gene of interest. Methodology is being developed to enable BAC clones to be directly transformed into plants, to find which one contains the gene of interest (in the example, the *R* gene). Alternatively, the YAC/BAC must be subcloned (for example as a cDNA library), and these subclones can be put into the susceptible plant and tested for the R phenotype.

The first R gene to be cloned in plants via map-based cloning was the *Pto* gene, which confers resistance to *Pseudomonas syringae* pv. *lycopersici*, the causal agent of bacterial speck in tomato.

The starting points for the work were:

- A high-density RFLP map of tomato.
- A tomato YAC library.

The *Pto* locus had been previously mapped to tomato chromosome 5 and one RFLP marker co-segregated with *Pto*. Hence this was a useful starting point to screen the YAC library. A 400 kb YAC was identified, and via inverse PCR, some end-specific probes were generated.

The left arm of the YAC was 1.8 cM from *Pto*, with the right arm perfectly co-segregating. This YAC was then used to screen a 920 000 plaque leaf cDNA library, from which 30 were selected. One of these was shown to co-segregate with *Pto*. This cDNA was then transformed into a susceptible tomato cultivar under the control of the CaMV 35S promoter, where it conferred resistance.

Hence the gene was identified, and could then be sequenced and characterized. The success of this approach was partly due to the small genome size of tomato (950 Mbp), and a relatively low level of repeat sequence. Such an approach is much more difficult in species with larger genomes such as maize (2300–2700 Mbp) and wheat (16 000 Mbp).

One such example of this problem is the 79 000 clone maize YAC library produced by Zeneca Seeds. As reported by K. Edwards, only 15per cent of the YACs have unique ends, with the remainder terminating in repeat sequences, which makes chromosome walking very difficult. The synteny of genomes is a major advantage in this instance. It is now well established that there is a high degree of synteny of genome organization, as seen in the co-linearity of maps of the most important grass genomes. This has resulted in the great interest in using this co-linearity to facilitate cloning of genes in large genomes such as maize and wheat, by starting with the smaller genomes, particularly rice, and to a lesser extent, sorghum.

One particular study of maize YACs demonstrated that a single YAC with the *Adh1* sequence contained 36 different repetitive sequences, whereas these repetitive sequences were largely absent in corresponding areas of the rice and sorghum genomes, a pattern seen also for the *a1–sh2* region in all three genomes. In a similar manner, many genes of interest for *Brassica* improvement are being identified in the smaller *Arabidopsis thaliana* genome.

Paterson *et al.* have recently demonstrated the potential of this approach, by identifying quantitative trait loci (QTLs) for seed mass and phenology which were co-linear on molecular marker maps of maize, sorghum and rice. They proposed that, as the maize genome is four and six times larger than the sorghum and rice genomes, respectively, the genes of interest in maize may be cloned in sorghum/rice more expediently. It must be remembered that QTLs are subject to genotype 3 environment interaction, which will be a major complicating factor in attempts to clone and characterize a QTL.

T-DNA Tagging

When T-DNA is inserted into the plant genome, it does so via illegitimate recombination. In the same way as a transposon, the insertion of T-DNA into a gene will inactivate it, which phenotypically will appear as a mutation. The comparative advantage of such a method is that stable mutations will be conferred. This can be a disadvantage, however, as reversions, such as those made possible by the *Ac/Ds* system are not possible. It is also the case that each independent transformation event will result in just one mutation event (or more if multiple independent insertions are made), whereas the heterologous transposon approach allows multiple mutation events to take place from a single initial transformant. Hence, mutated genes with T-DNA insertions can be cloned, in much the same manner as those with transposon insertions, by using part of the T-DNA as a probe in Southern blots.

The procedure for selecting out the tagged gene is quite straightforward, particularly when the plant contains a single T-DNA copy. However, this means that a large number of independent transformants must be generated. This can be achieved with species such as *Arabidopsis*, where Feldmann demonstrated that a seed-based transformation system using *Agrobacterium* can be quite efficient. Feldmann demonstrated that by inoculating approximately 1000 seeds, approximately 300 000 selfed T2 seed can be collected. They estimated that approximately 8000 transformants could be recovered from an experiment of this scale and, using this approach, cloned a gene involved in trichome development and a dwarf gene from *Arabidopsis*.

A regulatory gene in the ethylene biosynthesis pathway of *Arabidopsis* has also been cloned using the T-DNA tagging strategy. With modification, T-DNA tagging methodology can be used to clone plant promoter sequences. Koncz *et al*. designed a vector which contained a promoterless antibiotic selection gene (aminoglycoside phosphotransferase II) and transformed this into *Nicotiana* and *Arabidopsis*. In this way, promoters could be 'trapped' as plants that were regenerated on selective antibiotic were antibiotic resistant because the T-DNA was inserted downstream of a promoter such that the gene was expressed.

The most serious limitation to the T-DNA tagging approach is that large numbers of independent transformation events must be generated, which is not currently possible with most species. This is really only feasible with a system such as the seed transformation technique developed for *Arabidopsis*, or possibly the 'whole-leaf' transformation system developed for tobacco.

7

Cell Biotechnology

CELL

CYTOLOGY

A science studying a structure, the basic functions and a cell role. Aspects of studying of cytology-cells of multicellular animals, plants, unicells which number bacteria concern, the elementary and unicellular seaweed. And as the cytology studies a structure and a chemical compound of cells, intracellular structures and their functions, functions of cells in an organism of animals and plants, reproduction and development of cells, adaptation of cells to environment conditions.

Today the cytology directly depends on other sciences, as well as they from it.

The cytology has the close connections with biological sciences, for example with botany, zoology, physiology, the doctrine about evolution of the organic world, and also with molecular biology, chemistry, physics, mathematics. Cytology-young enough science, its age of all about 100 years though the term "cell" is known already more than 300 years. In the middle of XVII century Robert Hooke, surveying a thin section of a stopper by means of a microscope has seen, that the stopper consists of small cells-cells. Then it for all history of mankind has applied for the first time the term "cell".

THE CELLULAR THEORY

To the middle of XIX century enough extensive knowledge of a cell, and their basis Theodor Schwann in 1838 have collected has formulated the cellular theory. Having generalised available knowledge of a cell, it has shown, that the cell represents basic unit of a structure of all live organisms, and has declared, that cells of animals and plants are similar on the structure. These positions became the major proofs of unity of a parentage of all live organisms, unities of all organic world. Schwann has brought in a science correct understanding of a cell as independent vital unit, the least unit live-out of a cell a life cannot exist. Studying of the chemical organisation and cell structure has led to a conclusion what exactly chemical processes underlie her life and support processes occurring in it. That cells of all organisms are similar on a chemical compound, at them the basic processes of a metabolism the same proceed.

Modern cellular-the theory includes following positions:

- *The Cell*: is basic unit of a structure and development of all live organisms;
- Cells of all organisms are similar (are homologous) on the structure, a chemical compound, the basic implications of vital activity and a metabolism both unicellular, and multicellular;
- Reproduction of cells occurs by division of a maternal cell into some identical daughter cells;
- In difficult metaphytes of a cell are grouped on function carried out by them and form tissues. All organs of an organism which are closely bound among themselves and subordinated to humoural and nervous systems of regulation consist of tissues.

Cell researches are of great importance for a solution of diseases. As in cells arise and the pathological changes leading to occurrence of malignant diseases start to develop. To understand a role of cells in development of diseases, we will result some examples. One of serious diseases of the person-a diabetes. The reason of its occurrence is insufficient activity of cells of the pancreas developing a hormone insulin which participates in regulation of a sugar exchange of an organism. Malignant changes which lead to development of cancer tumours, also arise in cells.

At studying of a structure and functions of cells various unicellular and metaphytes has shown, that on the structure they are parted on two groups. The first group includes bacteria and blue-green seaweed. These organisms have more simple structure of cells. Them name procaryotes. The second group includes all other organisms: from unicellular green seaweed and the elementary to the higher floral plants, mammal, including person. These kinds of organisms have difficult arranged cells which name nuclear or eukaryotes. These cells have a nucleus and the organoids which are carrying out specific functions.

There is one more group of organisms which does not concern neither to pro-nor eukaryotes. It represents the Non-cellular form of a life in it the viruses in which studying the virology is engaged enter.

STRUCTURE AND FUNCTIONS OF A COVER OF A CELL

The cell of each of organisms, represents integrated live system which consists of three parts inseparably linked among themselves:

CELL COVER

The cell cover directly co-operates with an environment and with the next cells (in metaphytes). It consists of an external layer and a plasmatic membrane. Cells of animals and plants differ on a structure of their external layer. At plants, and also at bacteria, blue-green seaweed and mushrooms the cell is covered with a dense cover, or a cellular wall. At the majority of plants it consists of a fat. The cellular wall plays exclusively important role: it represents an external skeleton, a protective cover, provides safety of vegetative cells. Though the cellular wall dense enough, but through its wall passes water, salts, molecules of many organic substances.

A blanket of cells of animals unlike cellular walls of plants very thin, elastic. He cannot be seen in a light microscope and it consists of various polysaccharides and fibres. The blanket of animal cells has received the name a glycocalix. The glycocalix carries out first of all function of a direct connection of cells of animals with an environment, with all substances surrounding it. Having an insignificant thickness (it is less 1 microns), the external layer of a cell of animals does not carry out a basic role what is inherent to cellular walls of plants. Glycocalix formation as well as cellular walls of plants, occurs thanks to vital activity of cells.

Plasmatic Membrane

Under a glycocalix and a cellular wall of plants the plasmatic membrane adjoining immediately with a cytoplasma is located. The thickness of a plasmatic membrane is insignificant, studying of its structure and functions probably only by means of a supermicroscope.

The structure of a plasmatic membrane includes lipoproteins (zhiro-fibres). They are located ordered and bridged with each other by chemical bonds. On modern representations of a molecule of lipids in a plasmatic membrane are located in two ranks and form, so-called, bilipid a layer. Molecules of fibres settle down in a layer of lipids, plunging in it on different depth.

Fibre molecules are located not permanently, that provides dynamism of a plasmatic membrane. The plasmatic membrane carries out many the important functions from which will catch sight of vital activity of cells: barrier function, a skeleton of a cell which gives it certain structure. One of the main functions is transport of substances through a membrane. So, in a cell arrive ions amino acid hormones (not fibres), carbohydrates, Adepses, and are deduced products of synthesis and products of vital activity of a cell. Transport occurs with the assistance of fibres, because of a gradient of concentration, etc.

The cells forming at multicellular animals various tissues (epithelial, muscular, etc.), are bridged with each other by a plasmatic membrane. In junctions of two cells the membrane of each of them can form cords or excrescences which give to bonds special durability though there are bonds with the help of a cleft (synapses at animals).

Bond of cells of plants is provided by formation of thin channels which are filled by a cytoplasma and are limited by a plasmatic membrane. On such channels which are passing through cellular covers, from one cell in another nutrients, ions, carbohydrates and other bonds arrive. On a surface of many cells of animals, for example various epithelia, there are very fine thin excrescences of a cytoplasma covered with a plasmatic membrane,-microvillis. The greatest quantity of microvillis is on a surface of cells of an intestine where there is an intensive digestion and a nutrition absorption.

Phagocytosis

Large molecules of organic substances, for example fibres and polysaccharides, a particle of nutrition, a bacterium arrive in a cell by a phagocyte. In a phagocyte immediate participation is accepted by a plasmatic membrane. In that place where the cell surface adjoins to a particle of any dense substance, the membrane caves in, forms an excavation and surrounds a particle which in "membranous packing" plunges in a cell. The digestive vacuole is formed and in it the organic substances which have arrived in a cell are digested.

CYTOPLASMA

Delimited from an environment a plasmatic membrane, the cytoplasma represents internal semifluid medium of cells. In a cytoplasma of eukaryotic cells the nucleus and various organoids settle down. The nucleus, in norm, settles down in the central part of a cytoplasma. In it various includings-products of cellular activity, a vacuole are concentrated also. As a part of the basic substance of a cytoplasma fibres prevail. In a cytoplasma the basic processes of a metabolism proceed, it unites to a nucleus and all organoids in a single whole, provides their interaction, cell activity as uniform integrated live system.

Cytoplasmic Reticulum

All internal zone of a cytoplasma is filled by numerous fine channels and the cavities which walls represent the membranes similar on the structure with a plasmatic membrane. These channels branch, bridged with each other and form the network which has received the name of a cytoplasmic reticulum.

The cytoplasmic reticulum is non-uniform on the structure. Two its types-granulous and smooth are known. On membranes of channels and cavities of a granulous network the set of fine roundish little bodies-ribosomes which give to membranes a rough kind settles down. Membranes of a smooth cytoplasmic reticulum do not carry ribosomes on the surface.

The cytoplasmic reticulum carries out many various functions. The basic function of a granulous cytoplasmic reticulum-participation in synthesis of fibre which is carried out in ribosomes. On membranes of a smooth cytoplasmic reticulum there is a synthesis of lipids and carbohydrates. All these products of synthesis collect in channels and cavities, and then are transported to various organoids of a cell where are consumed or collect in a cytoplasma as cellular includings. The cytoplasmic reticulum binds among themselves the basic organoids of a cell.

Ribosomes

Ribosomes are found out in cells of all organisms. These are microscopical little bodies of the roundish form in diameter of 15-20 nanometers. Each ribosome consists of two unequal particles in the sizes, small and big.

In one cell many thousand ribosomes contain, they settle down or on membranes of a granulous cytoplasmic reticulum, or freely lay in a cytoplasma. The structure of ribosomes includes fibres and RNA. Function of ribosomes is a synthesis of fibre. Fibre synthesis-difficult process which is carried out not by one ribosome, and the whole group including to several tens of incorporated ribosomes. Such group of ribosomes name a polysoma. The synthesised fibres at first collect in channels and cytoplasmic reticulum cavities, and then are transported to organoids and cell sites where they are consumed. The cytoplasmic reticulum and the ribosomes located on its membranes, represent the uniform apparatus of biosynthesis and transportation of fibres.

Mitochondrions

In a cytoplasma of the majority of cells of animals and plants fine little bodies (0,2-7 microns)-mitochondrions contain. Mitochondrions are well visible in a light microscope with which help it is possible to survey their form, a locating, to count quantity. The internal structure of mitochondrions is studied by means of a supermicroscope. The mitochondrion cover consists of two membranes-external and internal.

The external membrane smooth, it does not form any cords and excrescences. The internal membrane, on the contrary, forms numerous cords which are referred to a mitochondrion cavity. Cords of an internal membrane name cristas (an armour. "Crista"-a crest, an excrescence) Number of cristas unequally in mitochondrions of different cells. Them can be from several tens to several hundreds, and especially it is a lot of cristas in mitochondrions of activly functioning cells, for example muscular.Mitochondrions name "power stations of"cells"as their basic function-synthesis Adenosine-triphosphate (ATP). This acid is synthesised in mitochondrions of cells of all organisms and represents the universal energy source necessary for realisation of processes of vital activity of a cell and the whole organism. New mitochondrions are formed by division of mitochondrions already existing in a cell.

Plastids

In a cytoplasma of cells of all plants there are plastids. In cells of animals of a plastid are absent. Distinguish three basic types of plastids: green-chloroplasts; red, orange and yellow-chromoplasts; colourless-leucoplasts.

Chloroplast

These organoids contain in cells of leaves and other green organs of plants, and also at various seaweed. The sizes of chloroplasts 4-6 microns, most often they have the oval form. The higher plants in one cell usually have some tens chloroplasts. Green colour of chloroplasts depends on the maintenance in them of a chlorophyll pigment. A chloroplast-the basic organoid of cells of plants in which there is a photosynthesis, *i.e.* Formation of organic substances (carbohydrates) from inorganic (CO_2 and H_2O) at use of energy of a sunlight.

On a structure chloroplasts are similar to mitochondrions. From a cytoplasma the chloroplast is delimited by two membranes-external and internal. The external membrane smooth, without cords and excrescences, and internal forms many the plaited excrescences referred in a chloroplast. Therefore in a chloroplast the considerable quantity of the membranes forming special structures-grans is concentrated. They are combined like a rouleau.

In membranes a gran chlorophyll molecules therefore here there is a photosynthesis settle down. In chloroplasts ATP is synthesised also. Between internal membranes of a chloroplast DNA, RNA and ribosomes contain. Hence, in chloroplasts as well as in mitochondrions, there is a synthesis of the fibre necessary for activity of these organoids. Chloroplasts breed division.

Chromoplasts are in a cytoplasma of cells of different parts of plants: in colours, foetuses, stalks, leaves. Presence of chromoplasts yellow, orange and red colouring of crowns of colours, foetuses, autumn leaves speaks. Leucoplasts. Are in a cytoplasma of cells of unpainted parts of plants, for example in stalks, roots, tubers. The form of leucoplasts is various.

Chloroplasts, chromoplasts and leucoplasts are capable a cell to mutual transition. So at maturing of foetuses or change of colouring of leaves in the autumn chloroplasts turn to chromoplasts, and leucoplasts can turn to chloroplasts, for example, at gardening of tubers of a potato.

Golgi Apparatus

In many cells of animals, for example in nervous, it has the form of the difficult network located round a nucleus. In cells of plants and the elementary Golgi apparatus is presented by separate little bodies of the crescent or rhabdoid form. The structure of this organoid is similar in cells of vegetative and animal organisms, despite a variety of its form.

Into Golgi apparatus structure enter: the cavities limited to membranes and located groups (till 5-10); the large and fine blisters located on the extremities of cavities. All these elements make a uniform complex. Golgi apparatus carries out many the important functions. On cytoplasmic reticulum channels to it products of synthetic activity of a cell-fibres, carbohydrates and Adepses are transported. All these substances at first collect, and then in the form of large and fine blisters arrive in a cytoplasma and or are used in the cell in the course of its vital activity, or deduced from it and used in an organism. For example, in cells of a pancreas of mammals digestive enzymes which collect in organoid cavities are synthesised.

Then the blisters filled with enzymes are formed. They are deduced from cells in a pancreat duct, whence flow in an intestine cavity. One more important function of this organoid consists that on its membranes there is a synthesis of Adepses and carbohydrates (polysaccharides) which are used in a cell and which are a part of membranes. Thanks to Golgi apparatus activity there is an updating and growth of a plasmatic membrane.

Lysosomes

Represent small roundish little bodies. From the Cytoplasma each lysosome is delimited by a membrane. In a lysosome there are enzymes, proteolytic, Adepses, carbohydrates, nucleic acids. To the alimentary particle which has arrived in a cytoplasma, lysosomes approach, merge with it, and one digestive vacuole in which there is the alimentary particle surrounded with enzymes of lysosomes is formed. The substances formed as a result of digestion of an alimentary particle, arrive in a cytoplasma and are used by a cell.

Possessing ability to active digestion of alimentary substances, lysosomes participate at a distance parts of cells dying off in the course of vital activity, the whole cells and organs. Formation of new lysosomes occurs in a cell constantly. The enzymes containing in lysosomes, as well as any other fibres are synthesised on cytoplasma ribosomes. Then these enzymes arrive on cytoplasmic reticulum channels to Golgi apparatus in which cavities lysosomes are formed. In such kind of a lysosome arrive in a cytoplasma.

The cellular centre. In cells of animals near to a nucleus there is an organoid which name the cellular centre. The basic part of the cellular centre is made by two small little bodies-the centrioles located in a small site of the condensed cytoplasma. Each centriole has the form of the cylinder in length to 1 micron. Centrioles play the important role at cell division; they participate in formation of a spindle of division.

Cellular includings. Carbohydrates, Adepses concern cellular includings and fibres. All these substances collect in a cytoplasma of a cell in the form of drops and grains of various size and the form. They are periodically synthesised in a cell and used in the course of a metabolism.

Cell Nucleus

Cell nucleus. Each cell of unicellular and multicellular animals, and also plants contains a nucleus. The form and the sizes of a nucleus depend on the form and the size of cells. In the majority of cells there is one nucleus, and such cells name uninuclear. There are also cells with two, three, with several tens and even hundreds nucleuss. These are multinuclear cells.

CELL BIOLOGY

Far away in the distant past, when the primeval seas bubbled and steamed, when ultraviolet light of great intensity streamed onto the surface of the earth, molecules of polypeptides, of long-chain fatty acids, steroids of various sorts and purines and pyrimidines ultimately to form part of the nucleic acids, went through the seething pangs of synthesis. Both purines and pyrimidines have recently been found in the interior of meteorites —thus they are not unique to this world and indicate that nucleic acid synthesis may be occurring elsewhere in space. These compounds, floating free in a warm soup, performed no organized activity. They reacted indiscrim-inately with each other, blindly, with no form or purpose. Then, groups of molecules responding to electrical and surfaceforming forces became orientated in a way which produced a membrane.

Presumably these membranes which were forming on the surface of the water were at first simply flat structures, but "One day" through inter-reaction of the various molecules a membrane formed a little vesicle. Many other vesicles became formed and here we had for the first time the potentialities for the development of living structures, for the membrane cut off the contents of the bag from the environment. What could have been the composition of such membranes? Perhaps the amino acids were formed into proteins at the time that this happened. Proteins, in general, have a remarkable property of spreading in a very thin film upon the surface of water—even when they are themselves in solution they are capable of forming these thin films. Compounds such as steroids and fats, *i.e.*, compounds which have a preponderance of hydrocarbon groupings which makes them only slightly soluble in water, also have the ability to spread on surfaces and to have a particular orientation so that their polar groups will, in general, be in contact with water or with other polar groups and those parts which have hydrocarbon groups will be out of contact with the water. Furthermore, lipoprotein films which combine both these two types of compounds, can even be formed artificially, for instance cholesterol and the protein gliadin have been used to produce such a film. These films have considerable elasticity and a good deal of strength.

Within the little vesicles we have mentioned, the reorganization of the molecules went on and eventually a vesicle formed within the parent vesicle which formed a structure similar to the nucleus. Combinations between proteins and nucleic acids formed a basis for inheritance of some of the characters of the cell in a way which we do not yet understand fully. By a slow and gradual process of evolution a complicated structure we now know to be a cell was built up and from it developed the multicellular organisms, animal and plant. At the stage when the multicellular organisms were evolving, it is not even certain that the cell was completely developed in the way we know it at the moment, it probably had a much simpler character than present day cells.

CHARACTERISTIC OF THE LIVING CELL

The bags or vesicles contained proteins, fats, phospholipids, carbohydrates, minerals, enzymes, vitamins and water. Now if one normally shook such components up in a little bag not much would happen, although no doubt there would be some linkup of some of the molecules to form structures. For example, if collagen is disaggregated with citrate and then is left to stand after the citrate had been removed, many of these groups of molecules reform into collagen units or even collagen Fibres. Something of this sort could occur in the type of vesicle we have described if the right molecules got together. On the other hand, in the beginning very little activity characteristic of the living cell would develop. In the presence of their substrates, for instance, many of the enzymes present would have some action, but for some of the enzymes the pH would be wrong and in such a mixture there would probably be only one uniform pH and the temperature would not be optimal. There would be no mechanism for removal of reaction products from the site of their formation and the reactions would be rapidly saturated with them and would grind to a stop.

The secret of the complex activity of the living cell lies in the isolation or partial isolation or even the timed isolation of its various activities. This is achieved by the presence of various membranes. The nuclear membrane guards the activity of the nucleus and the Golgi apparatus and the mitochondria have membranes which permit them to carry out activities in partial but not complete seclusion.

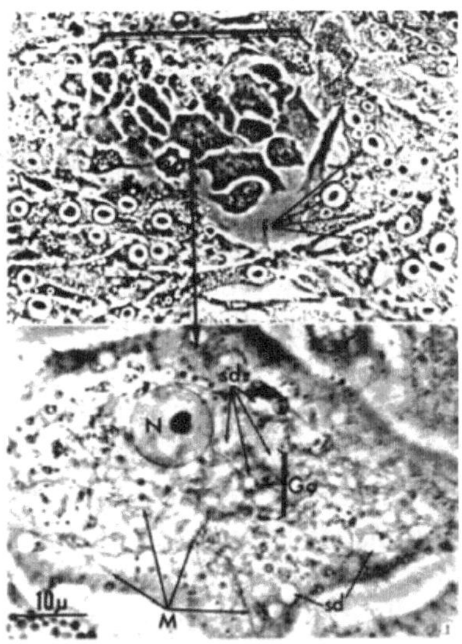

Fig. The Living Cells Under the Optical Microscope.

It is of interest that natural, fresh water frequently contains films and these are probably derived, according to Goldacre, from decomposed biological material. They appear to contain protein or lipoprotein and in many ways possibly resemble the type of film just described from which cells were probably originally formed. It is of further interest that films such is these, upon collapsing, can spontaneously form vesicles or cylinders, *e.g.*, the wind may cause a collapse of the film. In the case of water passing under a floating barrier, for example, when a stream disappears underground, the surface film is carried

Towards the point of entry and according to Goldacre, may be concentrated. Compression of this film causes a wrinkling and if the wrinkles become sufficiently pinched the walls finally touch underneath the wrinkles and complete cylinders may be detached which separate from the rest of the film.

These cylinders themselves can form vesicles which would resemble the primeval vesicles previously mentioned; some may contain air and some the underlying water. Sometimes instead of cylinders, Fibres may be produced. It is of interest that the properties of these collapsed lipoprotein films (which can actually be produced in the laboratory) have, according to Goldacre, many similarities to those of the membranes of living cells. Proteins and other macromolecules, according to Brownell, seem to be able to form in nature without any special difficulty.

Presumably small organic molecules could form quite easily under the circumstances in which the earth existed at one time. Carbon monoxide, carbon dioxide, hydrogen, nitrogen and other gases including some inorganic catalysts such as various metals and ions can produce small organic chemicals which, once formed may go through a process of polymerization to build up larger and more complex molecules. This, in fact, has been done in the laboratory. It is very likely that amino acids and proteins existed in the soupy seas of the earth for a long time, perhaps before membranes formed; on the other hand, it is possible that as the former substances were produced many of them became organized into surface films which became membranes and subsequently divided up into vesicles.

As mentioned earlier, artificial membranes can easily be formed and mixtures of fatty substances, alkalis, proteins and inorganic salts are very suitable for this purpose. Hereira has even formed in this way, what he describes as artificial amoebae which were persuaded to produce types of amoeboid movement. But, as Goldacre points out, such movements are uncontrolled and what is so important in cells is the fact that there is a very considerable degree of control by the cell over itself. Nevertheless the studies of Hereira help us perhaps to understand how objects comparable to cells, in other words, isolated vesicles containing macromolecules could have been produced in the first place.

Goldacre himself points out that the essential step in the evolution of the cell was the formation around a chemical system of what he describes as a relatively impermeable envelope in the beginning.

But, as he also points out, what was presumably a relatively simple structure at first must have undergone a considerable amount of further development and membranes of present day cells are relatively complicated structures which take quite an active part in regulating the compounds and ions that go in and out of the cell. For this purpose they seem to have absorbed, adsorbed, or incorporated enzymes in some way onto or into their surfaces and these enzymes play a part in the process of active transport.

THE ARCHITECTURE OF CELLS

Although generalizations in biology usually lack the theoretical underpinnings found in physics, there are very clear commonalities among living systems that give biology a unity. One is the style of cellular construction. The biological universe consists of two types of cells — *prokaryotic cells,* which lack a defined nucleus and have a simplified internal organization, and *eukaryotic cells,* which have a more complicated internal structure including a defined, membrane-limited nucleus.

Detailed analysis of the DNA from a variety of prokaryotic organisms in recent years has revealed two distinct types: bacteria (often called "true" bacteria or eubacteria) and archaea (also called *archaebacteria* or *archaeans*). As we discuss the archaea are in some respects more similar to eukaryotic organisms than to the true bacteria.

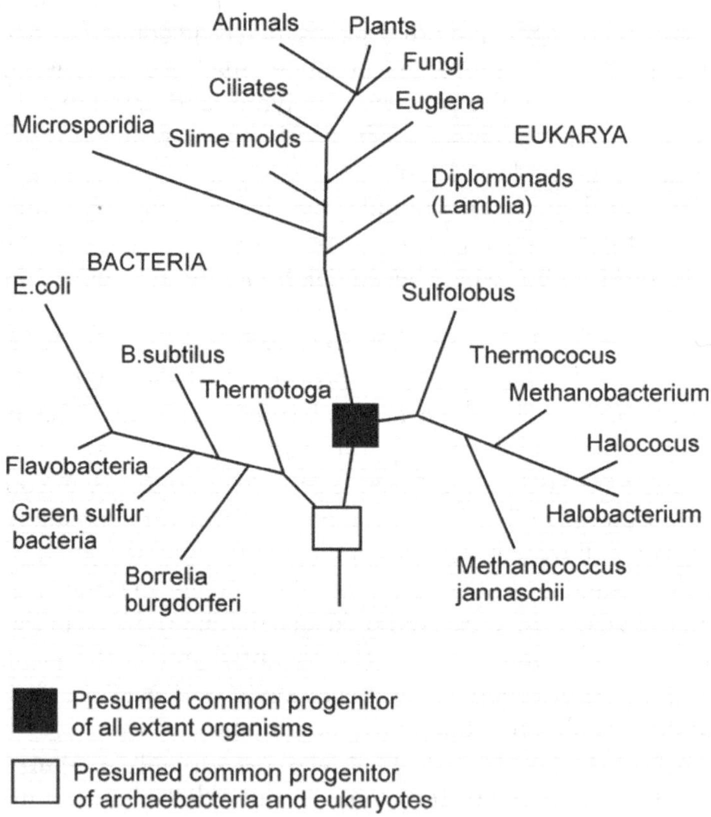

Fig. The three kingdoms of organisms are related through common sequences of their ribosomal RNAs.

Based on the assumption that organisms with more similar genes evolved from a common progenitor more recently than those with more dissimilar genes, researchers have developed the lineage tree. According to this tree, the archaea and eukarya (eukaryotes) are thought to have diverged from the bacteria before they diverged from each other. Despite the differences in the organization of prokaryotic and eukaryotic cells, all cells share certain structural features and carry out many complicated processes in basically the same way.

CELLS ARE SURROUNDED BY WATER-IMPERMEABLE MEMBRANES

A cell, because it is a limited space, must have an outer border. The construction of that border represents one of the most fundamental considerations in biological organization. The outer shell of cells, like any shell, is built to keep the interior contents from leaking out into the surrounding environment. The chemical processes of cellular life generally take place in a watery solution, and the intracellular constituents of cells are largely molecules that are easily dissolved in water. Similarly, the environment around cells is a watery one, the blood and other bodily fluids being solutions in water. Cells then, in order to maintain their integrity, need to be surrounded by an environment through which water cannot flow. A membrane composed of fatty molecules serves this purpose.

We all know from common experience that "oil and water don't mix." That maxim is all one needs to appreciate how a cell is constructed. When oil is poured on water, the oil spreads into a thin film; that film is analogous to the film of fat that surrounds cells, called the plasma membrane. Biological membranes differ from a pure oil film in that the molecules that make the membrane have both oily and watery portions; they have long fatty chains, but they also have a head group that is water-soluble by virtue of being electrically charged.

Thus membranes are formed because these bipartite molecules, called phospholipids, spontaneously orient themselves to form a double layer, or bilayer, having a fatty interior with external surfaces bonded to the surrounding water by the charged head groups. The membrane is given rigidity by interspersion of cholesterol, a molecule we have come to hate because of its association with heart disease, but one that is required to build the outer membrane of all our cells. Hence from an understanding of the contrasting properties of watery solutions and oily layers, an understanding of cellular construction emerges.

In spite of the rigidity provided by cholesterol, membranes composed of fat are not very strong, so numerous mechanisms for strengthening the borders of cells have evolved. In plants the plasma membrane is surrounded by a rigid cell wall. Although most animal cells lack a cell wall, proteins attached to their exterior surfaces provide some stability; the linking of cells together through these proteins helps maintain the integrity of tissues. Tissues and organs are often covered by strong networks of proteins and other molecules that strengthen and protect them, and also wall off the various compartments of the body. Single-celled organisms, like bacteria, have special outer coats to protect them.

MEMBRANES SERVE FUNCTIONS OTHER THAN SEGREGATION

Although membranes are valuable as a way to segregate the watery interior of the cell from its environment, or to segregate intracellular events from one another, they have other important functions, including energy storage. Because membranes separate watery compartments from one another, if an ion or a molecule dissolved in water is moved through a membrane into a new cellular compartment, it will not be able to diffuse freely out of the compartment into which it was moved. It takes energy to move the molecule, but once moved, the molecule stores that energy by virtue of its entrapment. Formally, this storage of energy is just like the storage of energy in a battery. Therefore, membranes not only delineate compartments, but also serve as active participants in the cell's dynamism.

The functions of many proteins depend on their mode of association with membranes. For instance, the passage of water-soluble molecules through membranes is carried out by protein transporters that are embedded in the membrane. Also, cells send information to one another by releasing signaling molecules. The outer membranes of cells have proteins, known appropriately as *receptors,* that bind the circulating signaling molecules. These signaling molecules allow the individual activities of the many cells in the body to be coordinated. The receipt of a signaling molecule by a receptor causes the transient organization of particular types of intracellular proteins, called *signal-transduction proteins,* into an activated complex at the interior face of the cell's outer membrane, from which it directs alterations of events in the cell's cytoplasm or nucleus.

PROKARYOTES COMPRISE A SINGLE MEMBRANE LIMITED COMPARTMENT

All prokaryotes are single-celled organisms, or protists. The bacterial lineage includes *Escherichia coli,* found in animal intestines and a favourite experimental organism, and the photosynthetic organisms formerly known as *blue-green algae* but better known today as *cyanobacteria.* (Because most prokaryotes studied in laboratories are bacteria, discussions of prokaryotic structure or metabolism throughout this book refer to these organisms, not archaeans, unless noted otherwise.) Many members of the archaeal lineage grow in unusual, often extreme, environments. For instance, the halophiles require high concentrations of salt to survive, and the thermoacidophiles grow in hot (80°C) sulfur springs, where a pH of less than 2 is common. Other archaeans, called *methanogens,* live in oxygen-free milieus and generate methane (CH_4) by the reduction of carbon dioxide.

Archaeal cells have a similar structure. In general, prokaryotes consist of a single closed compartment containing the cytosol and bounded by the plasma membrane. Although bacterial cells do not have a defined nucleus, the genetic material, DNA, is condensed into the central region of the cell. In addition, most ribosomes — the cell's protein-synthesizing particles — are found in the DNA-free region of the cell. Some bacteria also have an invagination of the cell membrane, called a *mesosome,* which is associated with synthesis of DNA and secretion of proteins. Thus bacterial cells are not completely devoid of internal organization.

Bacterial cells possess a cell wall, which lies adjacent to the external side of the plasma membrane. The cell wall is composed of layers of peptidoglycan, a complex of proteins and oligosaccharides; it helps protect the cell and maintain its shape. Some bacteria (*e.g.*, *E. coli*) have a thin cell wall and an unusual outer membrane separated from the cell wall by the periplasmic space. Such bacteria are not stained by the Gram technique and thus are classified as gram-negative. Other bacteria (*e.g.*, *Bacillus polymyxa*) that have a thicker cell wall and no outer membrane take the Gram stain and thus are classified as gram-positive.

EUKARYOTIC CELLS CONTAIN MANY ORGANELLES AND A COMPLEX CYTOSKELETON

Eukaryotes comprise all members of the plant and animal kingdoms, including the unicellular fungi (*e.g.*, yeasts, mushrooms, molds) and protozoans. Eukaryotic cells, like prokaryotic cells, are surrounded by a plasma membrane. However, unlike prokaryotic cells, most eukaryotic cells also contain extensive internal membranes that enclose specific compartments, the organelles, and separate them from the rest of the cytoplasm, the region of the cell lying outside the nucleus.

Most organelles are surrounded by a single phospholipid membrane, but several, including the nucleus, are enclosed by two membranes. Each type of organelle plays a unique role in the growth and metabolism of the cell, and each contains a collection of specific enzymes that catalyze requisite chemical reactions. The membranes defining these subcellular compartments control their internal ionic composition so that it commonly differs from that of the cytosol (the portion of the cytoplasm outside the organelles) and among the various organelles.

The largest organelle in a eukaryotic cell is generally the nucleus, which houses most of the cellular DNA. In addition to the nucleus, several other organelles are present in nearly all eukaryotic cells: the mitochondria, in which much of the cell's energy metabolism is carried out; the rough and smooth endoplasmic reticula, a network of membranes in which glycoproteins and lipids are synthesized; Golgi vesicles, which direct membrane constituents to appropriate places in the cell; and peroxisomes, in which fatty acids and amino acids are degraded. Animal cells, but not plant cells, contain lysosomes, which degrade worn-out cell constituents and foreign materials taken in by the cell. Chloroplasts, where photosynthesis occurs, are found only in certain leaf cells of plants and some single-celled organisms. Both plant cells and some single-celled eukaryotes contain one or more vacuoles, large, fluid-filled organelles in which nutrients and waste compounds are stored and some degradative reactions occur.

The cytosol of eukaryotic cells contains an array of fibrous proteins collectively called the cytoskeleton. Three classes of Fibres compose the cytoskeleton: microtubules (20 nm in diameter), built of polymers of the protein tubulin; microfilaments (7 nm in diameter), built of the protein actin; and intermediate filaments (10 nm in diameter), built of one or more rod-shaped protein subunits. The cytoskeleton gives the cell strength and rigidity, thereby helping to maintain cell shape. Cytoskeletal Fibres also control movement of structures within the cell; for example, some cytoskeletal Fibres connect to organelles or provide tracks along which organelles move.

The rigid cell wall, composed of cellulose and other polymers, that surrounds plant cells contributes to their strength and rigidity. Fungi are also surrounded by a cell wall, but its composition differs from that of bacterial or plant cell walls.

CELLULAR DNA IS PACKAGED WITHIN CHROMOSOMES

The DNA in the nuclei of eukaryotic cells is distributed among 1 to more than 50 long linear structures called chromosomes. The number and size of the chromosomes are the same in all cells of an organism, but vary among different types of organisms. Each chromosome comprises a single DNA molecule associated with numerous proteins, and the total DNA in the chromosomes of an organism is referred to as its genome. Chromosomes, which stain intensely with basic dyes, are visible in the light microscope only during cell division when the DNA becomes tightly compacted.

In all prokaryotic cells, most of or all the genetic information resides in a single circular DNA molecule, about a millimeter in length; this molecule lies, folded back on itself many times, in the central region of the cell. Although the large genomic DNA molecule in prokaryotes is associated with proteins and often is referred to as a chromosome, the arrangement of DNA within a bacterial chromosome differs greatly from that within the chromosomes of eukaryotic cells.

The concept that genes are like “beads” strung on a long “string,” the chromosome, was proposed early in the 1900s based on genetic work with the fruit fly *Drosophila*. The early *Drosophila* workers could position, or map, the genes responsible for various mutant traits on a chromosome, even though they did not yet know that genes were segments of DNA or that the function of a gene was due to a protein whose sequence was encoded by that gene!

THE LIFE CYCLE OF CELLS

A cell in an adult organism can be viewed as a steady-state system. The

DNA is constantly read out into a particular set of mRNAs, which specify a particular set of proteins. As these proteins function, they are also being degraded and replaced by new ones, and the system is so balanced that the cell neither grows, shrinks, nor changes its function. This static view of the cell, however, misses the all-important dynamic aspects of cellular life.

The dynamics of a cell can best be understood by examining the course of its life. A new cell arises when one cell divides or when two cells, like a sperm and an egg cell, fuse. Either event sets off a cell-replication programme that is encoded in the DNA and executed by proteins. This programme usually involves a period of cell growth, during which proteins are made and DNA is replicated, followed by cell division, when a cell divides into two daughter cells. Whether a given cell will grow and divide is a highly regulated decision of the body, assuring that an adult organism replaces worn out cells or makes more cells in response to a new need. Examples of the latter are the growth of muscle in response to exercise or damage, and the proliferation of red blood cells when a person ascends to a higher altitude and needs more capacity to capture oxygen. However, in one major and devastating disease — cancer — cells multiply even though they are not needed by the body. To understand how cells become cancerous, biologists have intensely studied the mechanisms that control the growth and division of cells.

THE CELL CYCLE FOLLOWS A REGULAR TIMING MECHANISM

Most eukaryotic cells live according to an internal clock; that is, they proceed through a sequence of phases, called the cell cycle, during which DNA is duplicated during the synthesis (S) phase and the copies are distributed to opposite ends of the cell during mitotic (M) phase. Progress along the cycle is controlled at key checkpoints, which monitor the status of a cell, for instance, the internal amount of DNA or the presence of extracellular nutrients. When certain conditions are met, the cell proceeds to the next checkpoint. The cycle begins after the cell divides into two daughter cells, each containing an identical copy of the parental cell's genetic material.

The cell cycle of prokaryotes is simple and fast. Replication of the single chromosome begins at a particular DNA sequence, the replication origin, which is anchored to the cell membrane. Once DNA replication is complete, assembly of new membrane and cell wall forms a septum, which eventually divides the cell in two. Because the origins of the two newly formed chromosomes are anchored to different membrane sites, each daughter cell receives one chromosome. In ideal growth conditions, the bacterial cell cycle is repeated every 30 minutes.

Only a few types of eukaryotic cells can grow and divide as quickly as bacteria. Most growing plant and animal cells take 10 – 20 hours to double in number, and some duplicate at a much slower rate. Many cells in adult animals, such as nerve cells and striated muscle cells, do not divide at all. They have temporarily exited from the cell cycle after mitosis and entered a "paused or quiescent" state called G_0. Because eukaryotic cells are larger and more complex than prokaryotic cells, a specialized mechanism coordinates their replication of genomic DNA, distribution of chromosomes, and cell division.

MITOSIS APPORTIONS

Mitosis is the mechanism in eukaryotes for partitioning the genome equally at cell division. To accomplish this complex task, plant and animal cells build a specialized machine, called the mitotic apparatus, which captures the chromosomes and then pushes and pulls them to opposite sides of the dividing cell. Remarkably, the mitotic apparatus is a temporary structure that exists only during mitosis to distribute the genetic material.

Although the events of mitosis unfold continuously, they are conventionally divided into four substages representing phases of chromosome movement. During the first substage, prophase, the replicated chromosomes, each comprising two identical chromatids, are condensed into compact packets and then released to the cytoplasm when the nuclear membrane breaks down.

During metaphase and anaphase, the chromosomes are sorted, and each chromatid of a pair moves to opposite sides of the cell. The end of mitosis is marked by re-formation of a membrane around each set of chromosomes (telophase). Division of the cytoplasm, called cytokinesis, then yields two daughter cells, each with a $2n$ complement of genetic material. Cell division in plant and animal cells differs mainly at cytokinesis. Animal cells divide in two by pinching of the cytoplasm. However, because a plant cell is surrounded by a rigid cell wall, daughter cells are formed by building a new cell membrane and cell wall between the two daughter nuclei, thereby cutting the cytoplasm into two portions.

CELL DIFFERENTIATION

The most complicated example of cellular dynamics occurs when a cell changes, or *differentiates,* to carry out a specialized function. This process often is marked by a change in the microscopic appearance, or *morphology,* of the cell. For example, the different structures of a nerve cell and a muscle cell reflect their respective functions in long-distance communication and contraction, highlighting the biological principle that "form follows function."

Cell differentiation creates the diversity of cell types that arise during the development of an organism from a fertilized egg. This is a process of extensive cell multiplication and differentiation. A mammal that starts as one cell becomes an organism with hundreds of diverse cell types such as muscle, nerve, and skin. Here we see at its most dramatic the power of DNA to control cellular behaviour: development is a DNA-orchestrated set of cellular changes (easily tens of thousands of them) that occur virtually without fail. The almost perfect resemblance of "identical" twins is a testament to the programme encoded by DNA to reproducibly direct the development of a human being.

Nowhere is the variety of cellular activities and responses better illustrated than in the body's immune system. It is there that many cell types come together in organized tissues specifically designed to allow the body to distinguish its own cells from those of foreign invaders. Within the immune system, we see both development of specialized cells that can recognize invading cells and formation of tissues from cells that originate in various parts of the body. The immune-system cells not only actively survey their environment with surface receptor proteins like antibodies, but also change their properties when they encounter a foreign substance, allowing the body to rid itself of invaders.

CELLS DIE BY SUICIDE

Unchecked cell growth and multiplication produce a mass of cells, a tumor. *Programmemed cell death* plays the very important role of population control by balancing cell growth and multiplication. In addition, cell death also eliminates unnecessary cells. For example, during embryogenesis, the digits of our fingers and toes are sculpted by the death of cells in the intervening spaces. If these cells remained alive, our hands and feet would become webbed. Thus the timing and location of cell death, as well as cell growth and division, must be precisely controlled.

Cell death follows an internal programme of events called apoptosis, in which all traces of a cell vanish. The first visible sign of apoptosis is condensation of the nucleus and fragmentation of the DNA. The cell soon shrivels and is consumed by macrophages. A cell is directed to commit suicide when an essential factor is removed from the extracellular environment or when an internal signal is activated. Thus, the default state of the cell is to remain alive. The discovery of genes that suppress the growth of tumors by activating cell death stimulated an exciting new line of cancer research that may lead to more effective treatment strategies

INTEGRATING CELLS INTO TISSUES

The evolution of multicellular organisms permitted specialized cells and tissues to form; a flowering plant has at least 15 cell types, and a vertebrate hundreds. In both plants and animals, cells that are specialized to carry out a particular task are found together in the tissues in which the task is performed: a xylem or meristem; a liver, a muscle, or a nerve ganglion. Different types of cells in a tissue are often arranged in precise patterns of staggering complexity. For instance, the hundreds of different types of neurons in the human brain are interconnected to one another through a network of some 10^{15} synaptic connections! The coordinated functioning of many types of cells within tissues, and of multiple specialized tissues, permits the organism as a whole to move, metabolize, reproduce, and carry out other essential activities. A key step in the evolution of multicellularity must have been the ability of cells to contact tightly and interact specifically with other cells.

Various integral membrane proteins, collectively termed cell-adhesion molecules (CAMs), enable many animal cells to adhere tightly and specifically with cells of the same, or similar, type; these interactions allow populations of cells to segregate into distinct tissues. Following aggregation, cells elaborate specialized cell junctions that stabilize these interactions and promote local communication between adjacent cells.

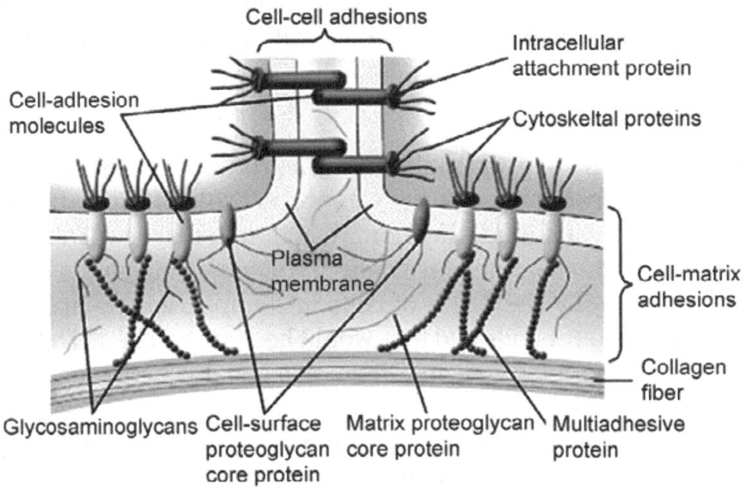

Fig. Integrating Cells into Tissues.

Animal cells also secrete a complex network of proteins and carbohydrates, the extracellular matrix (ECM), that creates a special environment in the spaces between cells. The matrix helps bind the cells in tissues together and is a reservoir for many hormones controlling cell growth and differentiation. The matrix also provides a lattice through which cells can move, particularly during the early stages of differentiation. Defects in these connections lead to cancer and developmental malformations.

The extracellular matrix has three major protein components: highly viscous proteoglycans, which cushion cells; insoluble collagen Fibres, which provide strength and resilience; and soluble multiadhesive matrix proteins, which bind these components to receptors on the cell surface.

Different combinations of these components tailor the strength of the extracellular matrix for different purposes. For example, animals contain many types of extracellular matrices, each specialized for a particular function such as strength (in a tendon), cushioning (in cartilage), or adhesion. In the case of smooth muscle cells that surround an artery, the extracellular matrix must provide strong but flexible connections. The extracellular matrix is not just an inert framework or cage that supports or surrounds cells. The matrix also communicates directly and indirectly with the intracellular signaling pathways that direct a cell to carry out specific functions.

For example, the ability of hepatocytes—the principal cells in the liver—to express liver-specific proteins depends on their association with a matrix of appropriate composition. Specific ECM components can directly activate cytosolic signal-transduction pathways by binding to cell-adhesion protein receptors in the plasma membrane.

Alternatively, by binding growth factors and other hormones, the matrix can either sequester these signals from cells or, conversely, present them to cells, thereby indirectly inducing or inhibiting signaling pathways. Morphogenesis—the later stage of embryonic development during which form is achieved by cell movements and rearrangements—also is critically dependent on ECM components, which are constantly being remodeled, degraded, and resynthesized locally. Even in adults—in areas of wounding, for example degradation and resynthesis of ECM components occurs.

In this chapter, we focus on the structure of and interactions between ECM components, cell-adhesion molecules, and cell-adhesion junctions—the main structures that permit animal cells to form organized tissues. Plant cells are surrounded by a cell wall that is thicker and more rigid than the extracellular matrix. Although the plant cell wall and the extracellular matrix serve many of the same functions, they are structurally very different.

SUGARS ARE COMMON IN THE CELL

Sugars serve three basic purposes in the cell: as carbon and energy sources, as reservoirs of carbon and energy, and as parts of cellular structures. Large amounts of energy can be extracted from sugars by processes referred to as catabolism. This may explain why many microorganisms show a preference for sugars if given a choice of energy sources. The term carbohydrate is often used to refer to them because their chemical formula can be broken down into $[C(H2O)]x$ where x is any number greater than three.

SUGAR MONOMERS

Single sugar molecules are typically 3 to 7 carbons long and are termed Figure shows that each carbon on the sugar molecule is decorated with a hydroxyl group (OH) except for one carbon that forms a carbonyl group. If the carbonyl group is at the end of the molecule, it forms an aldehyde; if the carbonyl group is in the middle of the molecule, it forms a ketone as indicated in Figure.

All sugars can exist as linear molecules in solution and those greater than 5 carbons long can also circularize with the carbonyl group attacking a hydroxyl on one of the other carbons. The circular sugar contains 5 to 7 members in the ring with one of the members being oxygen. Glucose, fructose and ribose are some of the more common sugars found in the cell.

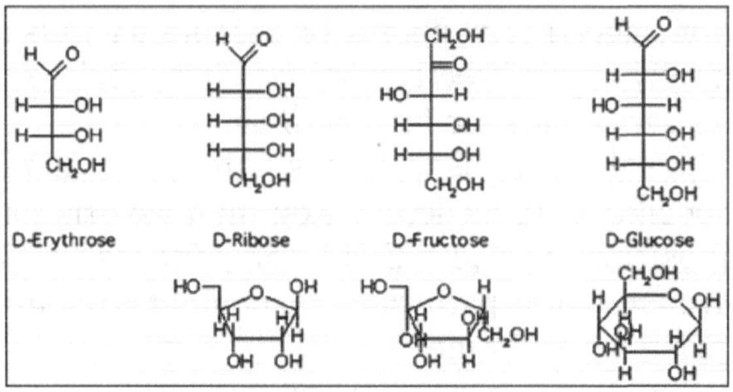

Fig. Common Monosaccharides

Sugar Polymers

Sugars readily polymerize: the combination of two sugars is called a disaccharide, while the combination of three is called a tri saccharide, and polymers of greater than three sugars are referred to as polysaccharides. Sugars are connected by d or e linkages. If the hydrogen is pointing up, it is a e linkage, if the hydrogen is pointing down, it is a d linkage as shown in Figure. This distinction may seem trivial, but it greatly influences the properties of the molecule. (The name of the linkage further depends upon the orientation of the hydrogen on the lowest numbered carbon that forms the bond. The lowest numbered carbon is decided by the rules of organic chemistry, but this is really not important for our discussion.

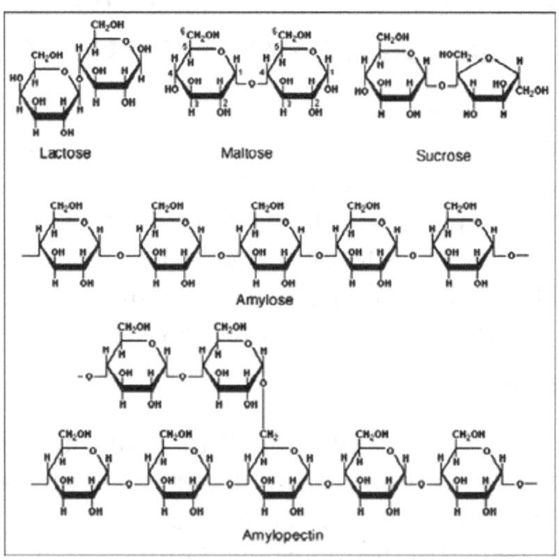

Fig. Common Disaccharides, Trisaccharides and Polysaccharides

Just realise there is a set pattern of numbering for each organic molecule and this helps determines the d or e linkage.) For example, starch is made up of glucose linked by d-1,4 bonds. Starch is also water-soluble and can serve as a food source for many organisms. In contrast, cellulose, containing glucose linked by e-1,4 bonds, is insoluble in water and is not as readily degraded by most microbes. Polysaccharides might contain only one type of sugar monomer or a variety of different ones, sometimes in repeating units. Polymers of sugars can serve as storage products for the cell. The breakdown of sugar yields a huge amount of energy, which means that they are terrific molecules for effectively storing energy for later utilization. Starch and glycogen are two examples of sugar polymers that are used in this way. Polysaccharides also serve as structural components of many different molecules in the cell including nucleic acids and the cell wall.

NUCLEIC ACIDS STORE INFORMATION AND PROCESS IT

- DNA and RNA are composed of two parts, the sugar phosphate backbone and nucleotide bases.
- The bases in DNA are purines (adenine and guanine) and pyrimidines (cytosine and thymine). RNA contains the same bases, except uracil is substituted for thymine and they contain an extra 5'-hydroxyl.
- DNA polymers can be thousands of base pairs long. RNA polymers are hundreds to thousands of bases long.
- RNA is single-stranded. DNA is double-stranded and forms a double helix in the cell.
- RNA molecules have significant secondary and tertiary structure that is important in their functions.

Deoxyribonucleic acid (DNA) and ribonucleic acid (RNA) are involved in information storage and processing. DNA serves as the cell's hereditary information, while RNA is involved in converting that information into functional products, such as proteins. RNA and DNA are long polymers of only 4 nucleotides: adenine, guanine, cytosine and thymine (or uracil for RNA). Figure lists the structure of the five nucleotides found in nucleic acids.

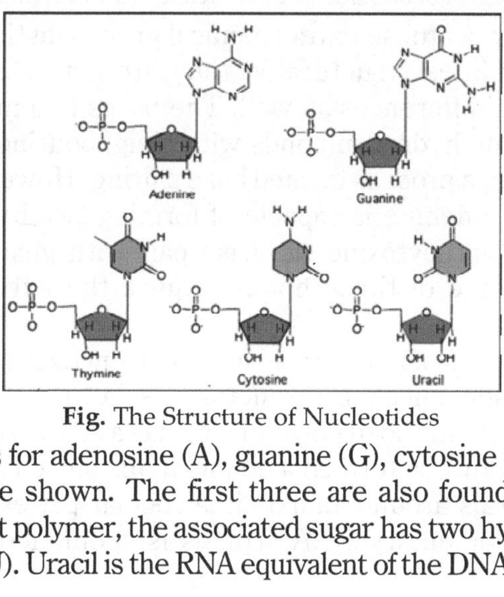

Fig. The Structure of Nucleotides

The nucleotides for adenosine (A), guanine (G), cytosine (C) and thymine (T) as found in DNA are shown. The first three are also found in RNA, but when incorporated into that polymer, the associated sugar has two hydroxyls, as shown in the model of uracil (U). Uracil is the RNA equivalent of the DNA nucleotide thymine.

The Figure Many thousands of DNA nucleotides are strung together to form genes and chromosomes. In this picture the bases of the two anti-parallel (running in opposite direction) strands of a DNA double helix are shown. Note the sugar-phosphate backbone from which the bases extend and pair with matching bases from the other strand. The dashed lines show hydrogen bonds between the base pairs.

The bases of the four nucleotides are different, but there is also a pattern. Adenine (A) and guanine (G) are purines, and therefore have a distinctive two-ring structure; they differ in the chemical groups attached to the rings. Likewise, cytosine (C), thymine (T) and uracil (U) are all pyrimidines and share a single-ringed structure, but also differ in their attached groups.

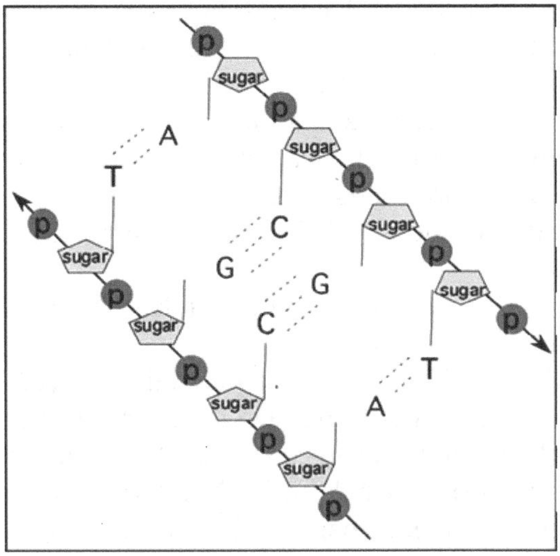

Fig. A Schematic of the Nucleic Acid Polymer

Not surprisingly, as these extra chemical groups distinguish the different purines and pyrimidines structurally, they are also responsible for their important functional differences as well. The bases in a nucleic acid polymer are capable of forming hydrogen bonds with Neighbouring bases on a second strand of nucleic acid, a process termed base pairing. However, there are rules to this association. Adenine is capable of forming two hydrogen bonds with thymine (or uracil) and cytosine can base pair with guanine, forming three hydrogen bonds. Some of these bonds require the extra chemical groups mentioned above.

If two single strands of nucleic acid have sequences that can base pair along the polymer (such sequences are sometimes said to complement), they will generate a long double-stranded polymer that has a staircase topology as shown in Figure. This structure is termed a double helix, since two strands form the molecule and it spirals around an axis in a regular pattern. Such a reaction between two complementary DNA strands is spontaneous: if you mix two complementary single strands of nucleic acid together in a test tube at a reasonable temperature, pH and salt concentration, they will find each other and anneal to form a double-stranded polymer.

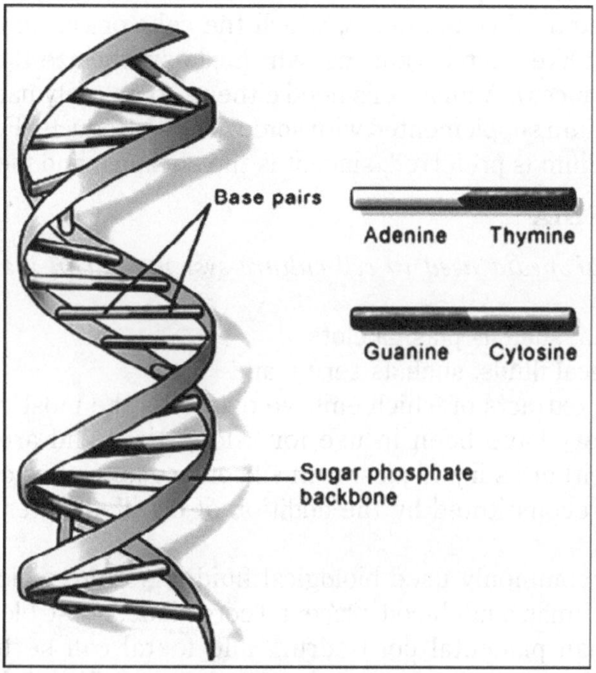

Fig. The Double Helix

Note how the two nucleic acid strands spiral around each other in a regular repeating pattern. There are 10 bases per turn of the double helix. Bases pair with one another in the centre of the helix cylinder and form what some have likened to a spiral staircase.

CELL CULTURE AND FERMENTATION TECHNOLOGY

One of the earliest workers in cell and tissue culture was Harrison who in 1907 developed a reproducible technique for tissue culture using frog. Later in 1912, Carrel used tissue embryo extracts as culture media.

They successfully demonstrated that animal cells can be grown indefinitely in vitro just like other microorganisms. Initially cold-blooded animals like frog were used, so that no incubation was required, but later warm-blooded animals like chick and rodents were used and eventually mammals became a favourite material.

Mammalian cells are grown in vitro for a number of reasons including the production of viral vaccines. Under standard tissue culture techniques, cells are immersed in a pool of medium containing essential nutrients and metabolic products. The concentration of both constituents changes as the cell population grows, thus influencing cell survival and function.

CULTURE MEDIA

Culture medium is the single most important factor for culturing cells and tissues. It provides the optimum conditions of factors like pH, osmotic pressure, etc., and the chemical constituents, which the cells or tissue are incapable of synthesizing (unlike microorganisms, which can synthesize them from simple inorganic substances). Animal cells need either a completely natural medium or an artificial medium supplemented with some natural products. For some tissues, the natural medium is preferred, since it is the cheapest and most convenient.

NATURAL MEDIA

The natural media used in cell culture systems fall in the following three categories:

1. Coagula, such as plasma clots
2. Biological fluids, such as serum and
3. Tissue extracts of which embryo extract is the most common.

Plasma clots have been in use for a long time and are now available commercially either as liquid plasma in siliconized ampoules or as lyophilized plasma, to be reconstituted by the addition of distilled water saturated with carbon dioxide.

The most commonly used biological fluid is serum, which is commonly obtained from human adult blood, placental cord blood, horse blood or calf blood. Of these, human placental cord serum and foetal calf serum seem to be particularly satisfactory. The serum is tested for sterility and toxicity before use.

Embryo extract is the most commonly used tissue extract, although effective substitutes have now been developed to replace the embryo extract as a natural medium. The most important component of the substitutes is a mixture of amino acids. Embryo extract is often prepared from chick or bovine embryos of different ages (up to 10 days).

DEFINED MEDIA

Media with serum Although natural media are very useful and convenient for a wide range of uses, they suffer from the disadvantage of poor reproducibility due to lack of knowledge of the exact composition.

For this reason, synthetic media have been designed for a variety of uses and purposes including the following:

- Media for immediate survival,
- Media for prolonged survival,
- Media for indefinite growth and
- Media for specialized functions.

For immediate survival, a "balanced salt solution" with defined osmotic pressure and pH is added. This requirement is met by the combination of certain inorganic ions, salts and glucose. For longer survival, serum may be used or the balanced salt solution may be supplemented with amino acids, oxygen, vitamins and serum proteins.

One such medium was developed and modified by Eagle and is described as 'minimum essential medium' (MEM). MEM was developed for mammalian cells grown on monolayers with more fastidious requirements. This has been used for primary mammalian cells and established cell lines. This contains essential amino acids, vitamins and salts.

Other more complex synthetic media include the following:

- 199,
- CMRL 1066,
- RPMI 1640 and
- F12.

All these media are supplemented with 5-20per cent serum. RPMI 1640 (Roswell Park Memorial Institute 1640 medium) is commonly used for human, murine normal and neoplastic white blood cells; these grow as suspension cultures and have a reduced calcium requirement.

ROLE OF SERUM IN CELL CULTURE

Serum is an extremely complex mixture of many small and large biomolecules with different physiologically balanced growpromotinth g and growth-inhibiting activities.

Some of the Major Functions of Serum are to Provide the Following:

- Basic nutrients, in solution and bound to proteins,
- Hormones and growth factors stimulating cell growth and specialized functions.
- Attachment and spreading factors.
- Binding proteins (albumin, transferrin) carrying hormones, vitamins, minerals, lipids, etc.
- Non-specific protection factors against mechanical damage and viscosity (shear forces during agitation of cell suspension)

DISADVANTAGES OF USING SERUM IN CELL CULTURE

- Serum may contain inadequate levels of cell-specific growth factors which have been supplemented, and an overabundance of others which may be cytotoxic.
- Risk of contamination with virus, fungi or mycoplasma.

SERUM-FREE MEDIA

The use of serum in culture media has several disadvantages, which led to the development of many serum-free media. Some of the disadvantages as listed in the earlier section are.

- Serum varies from batch to batch and deteriorates within one year

- Changing serum in a batch requires fresh testing, so that the replacement is as close to the previous batch as possible
- If more than one cell types are used, which may require different serum concentrations, so that a number of batches may need to be maintained and co-culturing may be different
- The demand for serum usually exceeds the supply for a variety of reasons
- When cell culture is used for downstream processing to recover cell products, the presence of serum is an obstacle to purification.
- Serum increases the cost of the medium, since its cost is ten times the cost of its constituents if used as substitute and
- For most cells, serum is not a physiological fluid which they contact in the original tissue except during wound healing and blood clotting processes, where Serum promotes fibroblast growth, but suppresses epidermal keratinocyte growth.
- Selective inhibitors, bacterial toxins and lipids found in FCS and human pregnant serum may react with polyamines to form cytotoxic polyaminoaldehydes. Hence, preference is given for horse serum. This may explain reports on the suppressor activity of these sera, and the preference of some workers for horse serum.

The major advantage of serum-free media is the ability to make medium selective for a particular cell type, since each cell type appears to require a different recipe. Examples of serum-free media include MCDB 110, MCDB 402, MCDB 153, Iscov's and LHC media.Primary cell culture can be obtained either by allowing cells to migrate out from the tissue, which is adhering to a substrate or by disaggregating the tissue mechanically or enzymatically to produce a suspension of cells. Most of the normal untransferred cells survive and proliferate to produce a primary culture and when attached to a substrate, these cells need to be obtained by disintegration. Primary cell cultures retain most of the characteristics of the cells from which they originated.

Primary cell cultures exhibit a phenomenon called contact inhibition which results in the cells lining up in strongly oriented parallel strands.With regard to chromosomal number, primary cultures usually retain their diploid karyotype. The most commonly employed tissues for preparing primary cultures are those from kidney, lung, liver, thyroid and testis. Primary cells are often used in preference to established cell lines and are regarded as a better representation of cells in vivo. Primary cells are more likely to reflect the true activity and functions that they display in their natural environment.

Disaggregation is achieved by anyone of the following methods:

- Physical disruption,
- Enzymatic digestion and
- Treatment with chelating agents. Physical disruption may often be combined with other methods.

Of the many enzymes that are used for disaggregation, trypsin and pronase are the most commonly used. Similarly, tissues like epithelium (which needs Ca^{++}, Mg^{++} ions for its integrity) can be treated with chelating agents, such as citrate and ethylene diamine tetra acetic acid (EDTA).

ENZYMATIC DISAGGREGATION

Trypsin is the most common enzyme used for disaggregation for the following reasons:

- It is tolerated by a variety of cells.
- It is effective for many tissues.
- Its residual activity is neutralized by serum of the medium or by a trypsin inhibitor in the case of a serum-free medium.

Disaggregation by trypsin can be damaging to epithelial cells or ineffective to fibrous tissues. Since intracellular matrix contains collagen, collagenase has proved effective for disaggregation of several normal and malignant tissues, which may be rather insensitive to trypsin. Crude collagenase is often used with a finely chopped tissue in complete medium.

MECHANICAL DISAGGREGATION

Enzymatic disaggregation is labour-intensive and involves damage of cells. Therefore, mechanical disaggregation of cells is sometimes preferred. In this method, tissue is carefully sliced and the cells that spill out are collected. Alternatively, the forced through a syringe and needle, or even repeatedly pipetted. Although the method may cause mechanical damage, the cell suspension is more quickly obtained than in the enzymatic disaggregation.

SEPARATION OF VIABLE AND NON-VIABLE CELLS

The dissociated cells grow well when seeded on culture in plates at high density. In the adherent culture, non-viable cells will be removed at the first change of medium. In suspension culture, on the other hand, non-viable cells are gradually diluted out, when cell proliferation stars. However, non-viable cells can also be removed from primary disaggregate by centrifuging the cells in a mixture of Ficoll and sodium metrizoate, when viable cells are collected from interface after centrifugation.

THE MOLECULAR COMPOSITION OF CELLS

Cells are composed of water, inorganic ions, and carbon-containing (organic) molecules. Water is the most abundant molecule in cells, accounting for 70per cent or more of total cell mass. Consequently, the interactions between water and the other constituents of cells are of central importance in biological chemistry. The critical property of water in this respect is that it is a polar molecule, in which the hydrogen atoms have a slight positive charge and the oxygen has a slight negative charge. Because of their polar nature, water molecules can form hydrogen bonds with each other or with other polar molecules, as well as interacting with positively or negatively charged ions. As a result of these interactions, ions and polar molecules are readily soluble in water (hydrophilic).

In contrast, nonpolar molecules, which cannot interact with water, are poorly soluble in an aqueous environment (hydrophobic). Consequently, nonpolar molecules tend to minimize their contact with water by associating closely with each other instead. Such interactions of polar and nonpolar molecules with water and with each other play crucial roles in the formation of biological structures, such as cell membranes.

The inorganic ions of the cell, including sodium (Na^+), potassium (K^+), magnesium (Mg^{2+}), calcium (Ca^{2+}), phosphate (HPO_4^{2-}), chloride (Cl^-), and bicarbonate (HCO_3^-), constitute 1per cent or less of the cell mass. These ions are involved in a number of aspects of cell metabolism, and thus play critical roles in cell function.

It is, however, the organic molecules that are the unique constituents of cells. Most of these organic compounds belong to one of four classes of molecules: carbohydrates, lipids, proteins, and nucleic acids. Proteins, nucleic acids, and most carbohydrates (the polysaccharides) are macromolecules formed by the joining (polymerization) of hundreds or thousands of low-molecular-weight precursors: amino acids, nucleotides, and simple sugars, respectively. Such macromolecules constitute 80 to 90per cent of the dry weight of most cells. Lipids are the other major constituent of cells. The remainder of the cell mass is composed of a variety of small organic molecules, including macromolecular precursors. The basic chemistry of cells can thus be understood in terms of the structures and functions of four major classes of organic molecules.

Lipids

Lipids have three major roles in cells. First, they provide an important form of energy storage. Second, and of great importance in cell biology, lipids are the major components of cell membranes. Third, lipids play important roles in cell signalling, both as steroid hormones (*e.g.*, estrogen and testosterone) and as messenger molecules that convey signals from cell surface receptors to targets within the cell.

The simplest lipids are fatty acids, which consist of long hydrocarbon chains, most frequently containing 16 or 18 carbon atoms, with a carboxyl group (COO^-) at one end. Unsaturated fatty acids contain one or more double bonds between carbon atoms; in saturated fatty acids all of the carbon atoms are bonded to the maximum number of hydrogen atoms. The long hydrocarbon chains of fatty acids contain only Non-polar C—H bonds, which are unable to interact with water. The hydrophobic nature of these fatty acid chains is responsible for much of the behaviour of complex lipids, particularly in the formation of biological membranes.

Fatty acids are stored in the form of triacylglycerols, or fats, which consist of three fatty acids linked to a glycerol molecule. Triacylglycerols are insoluble in water and therefore accumulate as fat droplets in the cytoplasm. It is noteworthy that fats are a more efficient form of energy storage than carbohydrates, yielding more than twice as much energy per weight of material broken down.

Phospholipids, the principal components of cell membranes, consist of two fatty acids joined to a polar head group. In the glycerol phospholipids, the two fatty acids are bound to carbon atoms in glycerol, as in triacylglycerols. Phospholipids are amphipathic molecules, part water-soluble and part water-insoluble. This property of phospholipids is the basis for the formation of biological membranes.

In addition to phospholipids, many cell membranes contain glycolipids and cholesterol. Glycolipids consist of two hydrocarbon chains linked to polar head groups that contain carbohydrates. They are thus similar to the phospholipids in their general organization as amphipathic molecules. Cholesterol, in contrast, consists of four hydrocarbon rings rather than linear hydrocarbon chains. The hydrocarbon rings are strongly hydrophobic, but the hydroxyl (OH) group attached to one end of cholesterol is weakly hydrophilic, so cholesterol is also amphipathic. In addition to their roles as components of cell membranes, lipids function as signalling molecules, both within and between cells. The steroid hormones (such as estrogens and testosterone) are derivatives of cholesterol. Derivatives of phospholipids also serve as messenger molecules within cells, acting to convey signals from cell surface receptors to intracellular targets.

Carbohydrates

The carbohydrates include simple sugars as well as polysaccharides. These simple sugars, such as glucose, are the major nutrients of cells. Their breakdown provides both a source of cellular energy and the starting material for the synthesis of other cell constituents. Polysaccharides are storage forms of sugars and form structural components of the cell. In addition, polysaccharides and shorter polymers of sugars act as markers for a variety of cell recognition processes, including the adhesion of cells to their Neighbours and the transport of proteins to appropriate intracellular destinations.

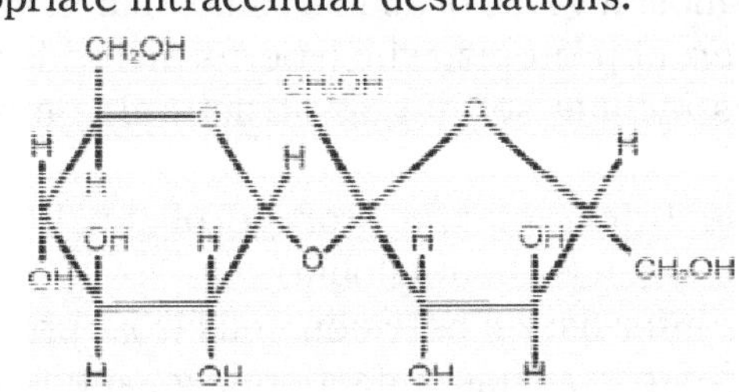

Fig. Carbohydrates

The basic formula for simple sugars (monosaccharides) is $(CH_2O)_n$, from which the name carbohydrate is derived (C= "carbo" and H_2O= "hydrate"). The six-carbon (n= 6) sugar glucose ($C_6H_{12}O_6$) is especially important in cells, since it provides the principal source of cellular energy. Other simple sugars have between three and seven carbons, with three- and five-carbon sugars being the most common. Sugars containing five or more carbons can cyclize to form ring structures, which are the predominant forms of these molecules within cells.

Monosaccharides can be joined together by dehydration reactions, in which H_2O is removed and the sugars are linked by a glycosidic bond between two of their carbons. If only a few sugars are joined together, the resulting polymer is called an oligosaccharide. If a large number (hundreds or thousands) of sugars are involved, the resulting polymers are macromolecules called polysaccharides.

Two common polysaccharides—glycogen and starch—are the storage forms of carbohydrates in animal and plant cells, respectively.

The structures of glycogen and starch are thus basically similar, as is their function: to store glucose. Cellulose, in contrast, has a quite distinct function as the principal structural component of the plant cell wall. Perhaps surprisingly, then, cellulose is also composed entirely of glucose molecules. The glucose residues in cellulose, however, are in the e rather than the d configuration, and cellulose is an unbranched polysaccharide. The linkage of glucose residues by e (1 † 4) rather than d (1 † 4) bonds causes cellulose to form long extended chains that pack side by side to form fibres of great mechanical strength.

In addition to their roles in energy storage and cell structure, oligosaccharides and polysaccharides are important in a variety of cell signalling processes. For example, oligosaccharides are frequently linked to proteins, where they serve as markers to target proteins for transport to the cell surface or incorporation into different subcellular organelles. Oligosaccharides and polysaccharides also serve as markers on the surface of cells, playing important roles in cell recognition and the interactions between cells in tissues of multicellular organisms.

Proteins

Proteins are the most diverse of all macromolecules, and each cell contains several thousand different proteins, which perform a wide variety of functions. The roles of proteins include serving as structural components of cells and tissues,

- Acting in the transport and storage of small molecules (*e.g.*, the transport of oxygen by hemoglobin),
- Transmitting information between cells (*e.g.*, protein hormones), and
- Providing a defence against infection (*e.g.*, antibodies).
- The most fundamental property of proteins, however, is their ability to act as enzymes, catalyse nearly all the chemical reactions in biological systems.

Thus, proteins direct virtually all activities of the cell. The central importance of proteins in biological chemistry is indicated by their name, which is derived from the Greek word *proteios*, meaning "of the first rank."

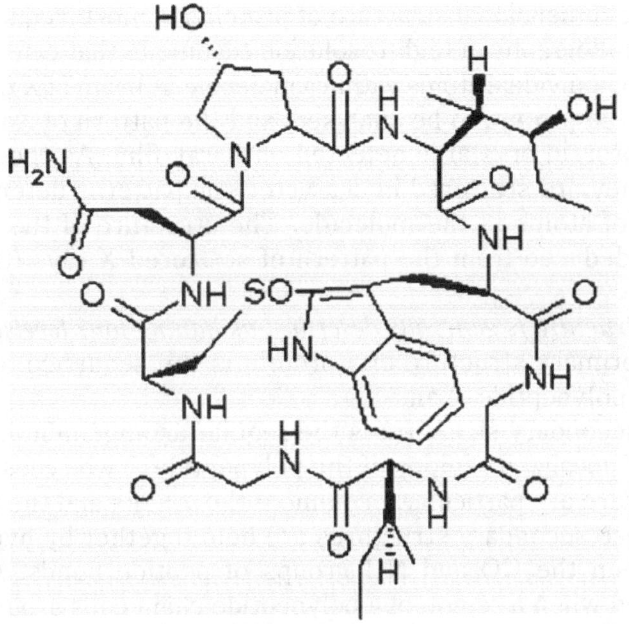

Fig. Proteins

Proteins are polymers of 20 different amino acids. Each amino acid consists of a carbon atom (called the d carbon) bonded to a carboxyl group (COO^-), an amino group (NH_3^+), a hydrogen atom, and a distinctive side chain. The specific chemical properties of the different amino acid side chains determine the roles of each amino acid in protein structure and function.

Amino acids are joined together by peptide bonds between the d amino group of one amino acid and the d carboxyl group of a second. Polypeptides are linear chains of amino acids, usually hundreds or thousands of amino acids in length. Each polypeptide chain has two distinct ends, one terminating in an d amino group (the amino, or N, terminus) and the other in an d carboxyl group (the carboxy, or C, terminus). Polypeptides are synthesized from the amino to the carboxy terminus, and the sequence of amino acids in a polypeptide is written (by convention) in the same order.

The defining characteristic of proteins is that they are polypeptides with specific amino acid sequences. Proteins are currently sequenced using automated methods, and the complete amino acid sequences of over 100,000 proteins are now known. Each consists of a unique sequence of amino acids, determined by the order of nucleotides in a gene. The amino acid sequence of a protein is only the first element of its structure. Rather than being extended chains of amino acids, proteins adopt distinct three-dimensional conformations that are critical to their function. These three-dimensional conformations of proteins are the result of interactions between their constituent amino acids, so the shapes of proteins are determined by their amino acid sequences.

The three-dimensional structure of proteins is most frequently analysed by X-ray crystallography, a high-resolution technique that can determine the arrangement of individual atoms within a molecule. A beam of X rays is directed at crystals of the protein to be analysed, and the pattern of X rays that pass through the protein crystal is detected on X-ray film. As the X rays strike the crystal, they are scattered in characteristic patterns determined by the arrangement of atoms in the molecule. The structure of the molecule can therefore be deduced from the pattern of scattered X rays (the diffraction pattern).

Protein structure is generally described as having four levels.

- The primary structure of a protein is the sequence of amino acids in its polypeptide chain.
- The secondary structure is the regular arrangement of amino acids within localized regions of the polypeptide. Two types of secondary structure are particularly common: the *a* helix and the *b* sheet. Both of these secondary structures are held together by hydrogen bonds between the CO and NH groups of peptide bonds. An *a* helix is formed when a region of a polypeptide chain coils around itself, with the CO group of one peptide bond forming a hydrogen bond with the NH group of a peptide bond located four residues downstream in the linear polypeptide chain. In contrast, a *b* sheet is formed when two parts of a polypeptide chain lie side by side with hydrogen bonds between them. Such *b* sheets can be formed between several polypeptide strands, which can be oriented either parallel or antiparallel to each other.
- Tertiary structure is the folding of the polypeptide chain as a result of interactions between the side chains of amino acids that lie in different regions of the primary sequence. In most proteins, combinations of *a* helices and *b* sheets, connected by loop regions of the polypeptide chain, fold into compact globular structures called domains, which are the basic units of tertiary structure. Small proteins, such as ribonuclease or myoglobin, contain only a single domain; larger proteins may contain a number of different domains, which are frequently associated with distinct functions.
- The fourth level of protein structure, quaternary structure, consists of the interactions between different polypeptide chains in proteins composed of more than one polypeptide. Hemoglobin, for example, is composed of four polypeptide chains held together by the same types of interactions that maintain tertiary structure.

The distinct chemical characteristics of the 20 different amino acids thus lead to considerable variation in the three-dimensional conformations of folded proteins. Consequently, proteins constitute an extremely complex and diverse group of macromolecules, suited to the wide variety of tasks they perform in cell biology.

Nucleic Acids

The nucleic acids—DNA and RNA—are the principal informational molecules of the cell. Deoxyribonucleic acid (DNA) has a unique role as the genetic material, which in eukaryotic cells is located in the nucleus. Different types of ribonucleic acid (RNA) participate in a number of cellular activities. Messenger RNA (mRNA) carries information from DNA to the ribosomes, where it serves as a template for protein synthesis. Two other types of RNA (ribosomal RNA and transfer RNA) are involved in protein synthesis. Still other kinds of RNAs are involved in the processing and transport of both RNAs and proteins. In addition to acting as an informational molecule, RNA is also capable of catalyzing a number of chemical reactions. In present-day cells, these include reactions involved in both protein synthesis and RNA processing.

DNA and RNA are polymers of nucleotides, which consist of purine and pyrimidine bases linked to phosphorylated sugars. DNA contains two purines (adenine and guanine) and two pyrimidines (cytosine and thymine). Adenine, guanine, and cytosine are also present in RNA, but RNA contains uracil in place of thymine. The bases are linked to sugars (2¢-deoxyribose in DNA, or ribose in RNA) to form nucleosides. Nucleotides additionally contain one or more phosphate groups linked to the 5¢ carbon of nucleoside sugars.

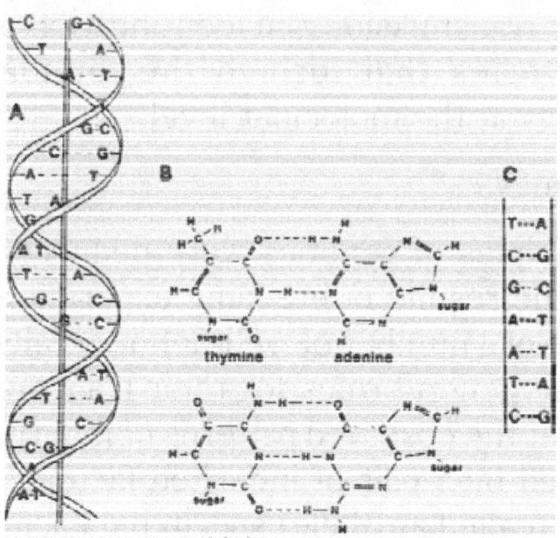

Fig. Nucleic Acids

The information in DNA and RNA is conveyed by the order of the bases in the polynucleotide chain. DNA is a double-stranded molecule consisting of two polynucleotide chains running in opposite directions. The bases are on the inside of the molecule, and the two chains are joined by hydrogen bonds between complementary base pairs—adenine pairing with thymine and guanine with cytosine. The important consequence of such complementary base pairing is that one strand of DNA (or RNA) can act as a template to direct the synthesis of a complementary strand. Nucleic acids are thus uniquely capable of directing their own self-replication, allowing them to function as the fundamental informational molecules of the cell. The information carried by DNA and RNA directs the synthesis of specific proteins, which control most cellular activities.

TECHNIQUES OF MOLECULAR BIOLOGY

Since the late 1950s and early 1960s, molecular biologists have learned to characterize, isolate, and manipulate the molecular components of cells and organisms. These components include DNA, the repository of genetic information; RNA, a close relative of DNA whose functions range from serving as a temporary working copy of DNA to actual structural and enzymatic functions as well as a functional and structural part of the translational apparatus; and proteins, the major structural and enzymatic type of molecule in cells.

HEREDITY AND ITS PROPAGATION

Perhaps the most fundamental property of all living things is the ability to reproduce. All organisms inherit the genetic information specifying their structure and function from their parents. Likewise, all cells arise from Pre-existing cells, so the genetic material must be replicated and passed from parent to progeny cell at each cell division. How genetic information is replicated and transmitted from cell to cell and organism to organism thus represents a question that is central to all of biology. Consequently, elucidation of the mechanisms of genetic transmission and identification of the genetic material as DNA were discoveries that formed the foundation of our current understanding of biology at the molecular level.

Genes and Chromosomes

The classical principles of genetics were deduced by Gregor Mendel in 1865, on the basis of the results of breeding experiments with peas. Mendel studied the inheritance of a number of well-defined traits, such as seed colour, and was able to deduce general rules for their transmission. In all cases, he could correctly interpret the observed patterns of inheritance by assuming that each trait is determined by a pair of inherited factors, which are now called genes. One gene copy (called an allele) specifying each trait is inherited from each parent. For example, breeding two strains of peas—one having yellow seeds, and the other green seeds—yields the following results. The parental strains each have two identical copies of the gene specifying yellow (*Y*) or green (*y*) seeds, respectively. The progeny plants are therefore hybrids, having inherited one gene for yellow seeds (*Y*) and one for green seeds (*y*). All these progeny plants (the first filial, or F_1, generation) have yellow seeds, so yellow

(*Y*) is said to be dominant and green (*y*) recessive. The genotype (genetic composition) of the F_1 peas is thus *Yy*, and their phenotype (physical appearance) is yellow. If one F_1 offspring is bred with another, giving rise to F_2 progeny, the genes for yellow and green seeds segregate in a characteristic manner such that the ratio between F_2 plants with yellow seeds and those with green seeds is 3:1.

Mendel's findings, apparently ahead of their time, were largely ignored until 1900, when Mendel's laws were rediscovered and their importance recognized. Shortly thereafter, the role of chromosomes as the carriers of genes was proposed. It was realized that most cells of higher plants and animals are diploid—containing two copies of each chromosome. Formation of the germ cells (the sperm and egg), however, involves a unique type of cell division (meiosis) in which only one member of each chromosome pair is transmitted to each progeny cell. Consequently, the sperm and egg are haploid, containing only one copy of each chromosome. The union of these two haploid cells at fertilization creates a new diploid organism, now containing one member of each chromosome pair derived from the male and one from the female parent. The behaviour of chromosome pairs thus parallels that of genes, leading to the conclusion that genes are carried on chromosomes.

The fundamentals of mutation, genetic linkage, and the relationships between genes and chromosomes were largely established by experiments performed with the fruit fly, *Drosophila melanogaster*. *Drosophila* can be easily maintained in the laboratory, and they reproduce about every two weeks, which is a considerable advantage for genetic experiments. Indeed, these features continue to make *Drosophila* an organism of choice for genetic studies of animals, particularly the genetic analysis of development and differentiation.

In the early 1900s, a number of genetic alterations (mutations) were identified in *Drosophila*, usually affecting readily observable characteristics such as eye colour or wing shape. Breeding experiments indicated that some of the genes governing these traits are inherited independently of each other, suggesting that these genes are located on different chromosomes that segregate independently during meiosis. Other genes, however, are frequently inherited together as paired characteristics. Such genes are said to be linked to each other by virtue of being located on the same chromosome. The number of groups of linked genes is the same as the number of chromosomes (four in *Drosophila*), supporting the idea that chromosomes are carriers of the genes.

Linkage between genes is not complete, however; chromosomes exchange material during meiosis, leading to recombination between linked genes. The frequency of recombination between two linked genes depends on the distance between them on the chromosome; genes that are close to each other recombine less frequently than do genes farther apart. Thus, the frequencies with which different genes recombine can be used to determine their relative positions on the chromosome, allowing the construction of genetic maps. By 1915, nearly a hundred genes had been defined and mapped onto the four chromosomes of *Drosophila*, leading to general acceptance of the chromosomal basis of heredity.

Genes and their relationship with Enzymes

Early genetic studies focused on the identification and chromosomal localization of genes that control readily observable characteristics, such as the eye colour of *Drosophila*. How these genes lead to the observed phenotypes, however, was unclear. The first insight into the relationship between genes and enzymes came in 1909, when it was realized that the inherited human disease phenylketonuria results from a genetic defect in metabolism of the amino acid phenylalanine. This defect was hypothesized to result from a deficiency in the enzyme needed to catalyse the relevant metabolic reaction, leading to the general suggestion that genes specify the synthesis of enzymes.

DNA-the Genetic Material

Understanding the chromosomal basis of heredity and the relationship between genes and enzymes did not in itself provide a molecular explanation of the gene. Chromosomes contain proteins as well as DNA, and it was initially thought that genes were proteins. The first evidence leading to the identification of DNA as the genetic material came from studies in bacteria. These experiments represent a prototype for current approaches to defining the function of genes by introducing new DNA sequences into cells.

The experiments that defined the role of DNA were derived from studies of the bacterium that causes pneumonia (*Pneumococcus*). Virulent strains of *Pneumococcus* are surrounded by a polysaccharide capsule that protects the bacteria from attack by the immune system of the host. Because the capsule gives bacterial colonies a smooth appearance in culture, encapsulated strains are denoted S. Mutant strains that have lost the ability to make a capsule (denoted R) form rough-edged colonies in culture and are no longer lethal when inoculated into mice.

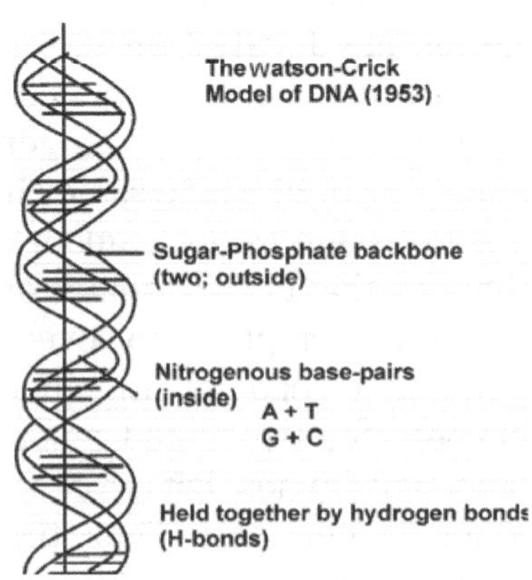

Fig. DNA-the Genetic Material

In 1928 it was observed that mice inoculated with Non-encapsulated (R) bacteria plus heat-killed encapsulated (S) bacteria developed pneumonia and died. Importantly, the bacteria that were then isolated from these mice were of the S type. Subsequent experiments showed that a cell-free extract of S bacteria was similarly capable of converting (or transforming) R bacteria to the S state. Thus, a substance in the S extract (called the transforming principle) was responsible for inducing the genetic transformation of R to S bacteria.

In 1944 Oswald Avery, Colin MacLeod, and Maclyn McCarty established that the transforming principle was DNA, both by purifying it from bacterial extracts and by demonstrating that the activity of the transforming principle is abolished by enzymatic digestion of DNA but not by digestion of proteins. Although these studies did not immediately lead to the acceptance of DNA as the genetic material, they were extended within a few years by experiments with bacterial viruses. In particular, it was shown that, when a bacterial virus infects a cell, the viral DNA rather than the viral protein must enter the cell in order for the virus to replicate. Moreover, the parental viral DNA (but not the protein) is transmitted to progeny virus particles. The concurrence of these results with continuing studies of the activity of DNA in bacterial transformation led to acceptance of the idea that DNA is the genetic material.

DNA Replication

The two strands of a DNA molecule could separate and serve as templates for synthesis of new complementary strands, the sequence of which would be dictated by the specificity of base pairing. The process is called semiconservative replication because one strand of parental DNA is conserved in each progeny DNA molecule.

Direct support for semiconservative DNA replication was obtained in 1958 as a result of elegant experiments, performed by Matthew Meselson and Frank Stahl, in which DNA was labelled with isotopes that altered its density. *E. coli* were first grown in media containing the heavy isotope of nitrogen (^{15}N) in place of the normal light isotope (^{14}N). The DNA of these bacteria consequently contained ^{15}N and was heavier than that of bacteria grown in ^{14}N. Such heavy DNA could be separated from DNA containing ^{14}N by equilibrium centrifugation in a density gradient of CsCl. This ability to separate heavy (^{15}N) DNA from light (^{14}N) DNA enabled the study of DNA synthesis. *E. coli* that had been grown in ^{15}N were transferred to media containing ^{14}N and allowed to replicate one more time. Their DNA was then extracted and analysed by CsCl density gradient centrifugation. The results of this analysis indicated that all of the heavy DNA had been replaced by newly synthesized DNA with a density intermediate between that of heavy (^{15}N) and that of light (^{14}N) DNA molecules. The implication was that during replication, the two parental strands of heavy DNA separated and served as templates for newly synthesized progeny strands of light DNA, yielding double-stranded molecules of intermediate density. This experiment thus provided direct evidence for semiconservative DNA replication, clearly underscoring the importance of complementary base pairing between strands of the double helix.

The ability of DNA to serve as a template for its own replication was further established with the demonstration that an enzyme purified from *E. coli* (DNA polymerase) could catalyse DNA replication *in vitro*. In the presence of DNA to act as a template, DNA polymerase was able to direct the incorporation of nucleotides into a complementary DNA molecule.

MOLECULAR STRUCTURE OF THE CELL MEMBRANE

On the molecular structure of the membrane before the isolation of plasma membranes, theories were generally based on indirect information. Since substances soluble in lipid solvents penetrate the plasma membrane easily, Overton postulated in. 1902 that the plasma membrane is composed of a thin layer of lipid. In 1926 Gorter and Grendell found that the lipid content of hemolyzed erythrocytes was sufficient to form a double layer of lipid molecules over the entire cell surface.

This theory was also supported by electrical measurements that indicated a high impedance at the plasma membrane. The high impedance is because of the faex fact that it is difficult for ions to penetrate a lipid layer. Tension at a water-oil interface is about 10 to 15 dynes per centimeter, whereas surface tension of cells is almost nil. It has been postulated that the low tension is due to the presence of protein layers on the lipid components. In fact, when a very small amount of protein is added to a model lipid-water system, the surface tension is lowered comparably.

In 1935 Danielli and Davson explained about all these properties. They proposed that the plasma membrane contained a *lipid bilayer,* with protein adhering to both lipid-aqueous interfaces. Other models containing globular lipid micelles, globular proteins, or a combination of both have been proposed a globular-bilayer transition has also been postulated.

These globular models do not account satisfactorily for the high electrical impedance. On the fine structure of the plasma membrane, electron microscopy has thrown some light and has revealed the numerous structural differentiations that this membrane and the underlying cell cytoplasm have in different cell types. To resolve the structure of the plasma membrane, extremely thin sections (~ 20 nm) must be used, otherwise one would observe different orientations of the plasma membrane with respect to the plane of the section.

The membrane appears most thin when it is exactly perpendicular to the plane of the section. Membranes of *6* to 10 mm have been observed at the surface of all cells. The membranes of two cells that are in close contact appear as dense lines separated by a space of 11 to 15 nm, which is strikingly uniform and contains a material of low electron density. This intercellular component can be considered as a kind of cementing substance.

MEMBRANE–RE-EVALUATION OF THE EM IMAGE

There are three layer in cell membrane outer dense layers of 2.0 nm and a middle clear one of about 3.5 nm led Robertson to postulate the so-called *unit membrane* model. The electron microscopic image was interpreted along the lines of the Danielli-Davson model, so that the clear layer corresponded to the hydrocarbon chains of the lipids, and the dense layers, to the proteins. This unit membrane model was also extended to all kinds of intracellular membranes. That this concept is an oversimplication which does not account for the many proteins traversing the membrane can be readily seen by observing some high resolution micrographs in which fine bridges cross the unit membrane. On the basis of some experiments the unit membrane has been re-evaluated recently methods designed to avoid the removal of proteins from membranes during the preparation for conventional electron microscopy or freeze-fracturing. With both methods a granular appearance of the membrane is obtained, which is more in keeping with the fact that a high proportion of the proteins traverse the membrane. According to these findings the unit membrane image is, to a great extent, artifactual.

Membrane Fluidity

For studying purpose, it can be classified as physical and biological.

The physical techniques are of two main classes:

1. Those that involve a minimal perturbation of the membrane, such as *X-ray diffraction* and *nuclear magnetic resonance (NMR) spectroscopy,*
2. Those that use certain added molecules to monitor specific sites of the membrane. Into this second class fall *fluorescence microscopy,* which uses fluorescent probes and *electron spin resonance (ESR)* spectroscopy.

In practice, these physical techniques can be subdivided on the basis of their ability to examine mainly either the lipid or the protein components, or both. ESR spectroscopy uses paramagnetic probes (*e.g.*, nitroxide-containing amphipathic molecules) which are introduced into the lipid bilayer. This can be done with natural and artificial membranes The data obtained provide information on lipid-protein packing, lateral diffusion of lipids, lipid-protein interactions, the fluidity of the membrane, and the rate of transmembrane rotation (so-called flip-flop) of molecules across the bilayer.

The *biological techniques* involve light and fluorescence microscopy and electron microscopy, including freeze-fracturing and radioisotope labeling methods. One of the simplest methods consists of binding gold or carbon particles to the cell surface and observing under the light microscope the movement of those particles on the surface. More information has come from studies in which different ligands, such as antibodies and plant lectins, interact with cell surface receptors. If these ligands are labeled with fluorescent dyes their movement can be followed by fluorescence microscopy. For example, in a classical experiment by Frye and Edidin mouse L cells and human transformed cells which have different surface antigens, were marked with the corresponding antibodies labeled with two distinct fluorescent dyes. These cells were then induced to fuse under the influence of Sendai viruses. While at the beginning both cell surfaces could be recognized by their differing labels, after 40 minutes considerable intermixing of the antigens had occurred, so that the two labels could no longer be recognized. This intermixing was retarded by temperatures below 20° C.

In surface antigens of cultured muscle cells, a diffusion rate of 1×10^{-9} cm per second was calculated. Also, the so-called clustering or capping effect observed in lymphocytes may result in the displacement of antigenic molecules within the surface membrane. In this case, upon treatment with a labeled antibody, the antigens are first randomly distributed, but after some time they become clustered, forming patches, and agglomerate at one pole of the cell (*i.e.*, capping) where pinocytosis of the antigen-antibody complexes takes place. This process is also inhibited by temperatures that produce solidification of the lipid bilayer.

A number of histochemical techniques using the electron microscope may give information about the fluidity of the membrane. For example, antigens labeled with ferritin or peroxidase can be used. Another extremely valuable tool is freeze-fracturing with which it is possible to demonstrate the movement of the particles in response to different experimental conditions that produce changes in the molecular structure of the membrane.

MEMBRANE FLUIDITY

The fluid mosaic model has been used to explain the fact that many receptors present in the cell membrane are coupled to enzymes such as adenylate and guanylate cyclase. We can say that they are macromolecules, in most cases of a protein nature, that act at the interface between the cell surface and the environment. A receptor has the dual function of recognizing a chemical signal (*i.e.*, one of the many regulatory agents, such as peptide hormones, neurotransmitters, prostaglandins, antigens plant lectins, and some bacterial toxins) and of initiating a biological response. The function of recognition is determined by the presence of a site that binds specifically to the ligand (*i.e.*, to the binding portion of a particular regulatory agent). This interaction produces a conformational change that may induce the translocation of ions or the activation of membrane enzymes, such as adenylate cyclase, thus producing cyclic AMP (cAMP).

In a cell there may be several kinds of receptors (at least eight in a fat cell) coupled to the same adenylate cyclase. The ligands do not compete for each specific receptor when binding, and the effect of multiple ligand-receptor bindings on adenylate cyclase is not additive. These findings have been interpreted as indicating that each ligand reacts with a unique receptor and that several receptors may be coupled to a certain number of enzyme molecules in the membrane. According to the *mobile hypothesis* both the receptors and the enzymes can diffuse independently within the plane of the membrane The greater degree of freedom that this model includes within the lipid realm of the membrane would allow different receptors to couple to a single enzyme.

Fusing cells in which the adenylate cyclase was inactivated but the receptor remained intact, with other cells which contain only the enzyme, resulted in receptor-enzyme coupling. In other words, the receptor of one cell (*e.g.*, the b-adrenergic receptor of an erythrocyte) was coupled to the adenylate cyclase of the other cell (*e.g.*, a tumor cell lacking the receptor). The coupling is manifested by a great increase in cAMP after receptor interaction with the specific ligand (in this case isoproterenol).

The diagram shown below interprets in a highly simplified manner the fusion of two different cells that results in coupling between a receptor from one cell and adenylate cyclase from the other.

The Fluid Mosaic Model

The main source of the knowledge of the molecular organization of biological membrane comes from an integration of the data on chemical analysis and from the application of several biophysical techniques.

The important concepts are:

- That the lipid and integral proteins are disposed in a kind of mosaic arrangement;
- That biological membranes are quasi-fluid structures in which both the lipids and the integral proteins are able to perform translational movements within the overall bilayer. The concept of fluidity implies that the main components of the membrane are held in place only by means of non-covalent interactions.

For better study of the molecular organization of the membrane, it is necessary to remember that, not only the lipids, but also many of the intrinsic proteins and glycoproteins of the membrane, are amphipathic molecules. The term *amphipathy,* coined by Hartley in 1936, refers to the presence, within the same molecule, of hydrophilic and hydrophobic groups these amphipathic molecules constitute liquid crystalline aggregates in which the polar groups are directed towards the water phase and the non-polar groups are situated inside the bilayer. The integral proteins of the membrane are intercalated to a greater or lesser extent into a rather continuous lipid bilayer. This arrangement is based on the fact that these integral proteins are also amphipathic, with polar regions protruding from the surface and non-polar regions embedded in the hydrophobic interior of the membrane. It is well recognized that a protein of appropriate size or a cluster of protein subunits may pass across the entire membrane (transmembrane proteins). Such traversing proteins could be in contact with the aqueous solvent on both sides of the membrane. A similar model has been postulated for the cholinergic receptor. This model also reflects the generally accepted view that the major proportion of the phospholipids in the membrane are arranged in a bilayer form.

This mosaic model of the membrane, with intercalating proteins, comes from the use of freeze-fracturing techniques in erythrocytes and other cell membranes. The red cell ghosts show a large number of particles, about 8 nm in diameter, that have been interpreted as representing proteins embedded within the plane of cleavage which passes through the middle of the lipid bilayer. There are between 500,000 and 600,000 such particles in a cell and more are attached to the inner half of the membrane (PF) than to the outer half (EF).

We mentioned earlier that these particles represent dimers of the polypeptide found in *band 3,* and that they may also correspond to anionic permeation channels. The degree of dispersion or aggregation of these particles within the membrane appears to be influenced by the *spectrin-actin system.* This suggests that some link between this system and the dimers of band 3 could exist in the living membrane. The mosaic model allows for the characteristic *asymmetry* of the membrane, in which specific components predominate in the outer half or the inner half of the membrane. This implies that there are constraints on the transmembrane rotation of those macromolecules that are well oriented for the carrying of information across the cell membrane. In this model the fluidity of the membrane also implies that both the lipids and the proteins have considerable freedom of movement within the bilayer.

The fluidity of the lipids depends:

- On the degree of saturation of the hydrocarbon chains
- On the ambient temperature. A considerable proportion of the lipids in the membrane are unsaturated, so that the melting point of the bilayer is below body temperature.

The Myelin Sheath and the Photoreceptors

Important information about the molecular structure of the cell membrane is derived from studies of natural multi-layered lipoprotein systems, such as the myelin sheath and the photoreceptors.

The Myelin Sheath: The myelin sheath is lipoprotein membrane that surrounds the axon of the nerve fibre. In peripheral nerves this sheath is formed by the Schwann cells. In central nerves the myelin sheath is produced by the activity of the oligodendroglial cells. It is a very special membrane structure in which the lipids are more abundant than the protein. In central myelin there are two main protein moieties: the proteolipid and the basic protein.

The Myelin Sheath has been known for over a century that the myelin sheath has a strong birefringence, which indicates a high degree of organization at a submicroscopic level. Studies with x-ray diffraction have revealed a spacing of 17 nm in amphibian and 18.0 to 18.5 nm in mammalian peripheral nerves. Within this period, the proportion corresponding to the lipid, protein, and water content has been estimated. The existence of lecithin-cholesterol and Sphingomyelin-cholesterol complexes is postulated.

The first type can be accommodated within the thickness of the lipid layer (L), but in the second, the longer sphingomyelin molecules must interdigitate in order to fill the same space. This model also accounts for the localization of protein (HP) and water (HL) and is in accord with the view that each x-ray diffraction period of 18 nm corresponds to two unit membranes, *i.e.*, to two lipid bilayers separated by a watery space (H), that originally corresponds to extracellular space.

It has confirmed by electron microscopic studies that myelin has a multi-layered membranous structure. In most cases, however, the x-ray diffraction period is reduced to 10 to 12 nm, thereby introducing a great many artifacts. Recently, with a highly polar embedding medium based on polymerized glutaraldehyde-urea, the periodicity obtained was of the same order as by x-ray diffraction. With this technique the lipids are not extracted; this explains the lacks of collapse of the structure.

In nerve conduction to function as an insulator, preventing the dissipation of energy into the surrounding medium. It might act not only as a dielectrics (insulating) material but also as a kind of resonant conductor in which the energy waves resonating in the lipid layers between the protein membranes could pass with maximum speed and minimum loss of energy.

Photoreceptors

The retinal rods and cones are highly differentiated cells that have at their outermost segment a lipoprotein structure that is specialized for photo-reception. Studies with the polarization microscope suggest a submicroscopic organization consisting of transversely oriented protein layers alternating with lipid molecules arranged longitudinally along the axis of the photoreceptor. This type of layered organization has been demonstrated by electron microscopy in fragmented rod outer segments and in thin sections of the retina. These observations indicate that the rod consists of a stack of superimposed disks (several hundred) along the axis. These disks are really flattened sacs made of two membranes, which surround a thin space of 3 nm and become continuous at the edges. The space between the rod sacs is 5 to 12 nm. The cone outer segments, with minor differences, have a similar structure.

To osmotic change rod sacs are highly sensitive. In hypotonic solutions they swell considerably, and the inner space between the membranes becomes very large. The rod outer segments can easily be isolated in the frog and be submitted to chemical analysis. The protein composition is simpler than in other membranes. The photopigment protein, *rhodopsin,* represents 40 per cent of the total mass of the rod sacs; another 40 per cent is lipid. Rhodopsin is an integral type of protein requiring detergents for extraction. In the pline of the membrane the rhodopsin molecules have a liquid-like distribution. X-ray diffraction studies have led to the interesting conclusion that this molecule may change its position within the lipid layer. In the dark (non-stimulated condition) the molecule is about one third embedded in the lipid and two thirds exposed to the water surface. Upon activation by light (bleached rhodopsin) the molecule becomes more deeply embedded within the lipid bilayer. This change is presumably caused by changes in surface charges of the molecule. The process of photoreceptors transform light energy into another type of energy that can be conducted as nerve impulse. It is based on a cycle of chemical reactions which involve the visual pigments present in the protein membranes of the rod and cone sacs. This multilayered structure is a very effective system that facilitates the maximum absorption and utilization of light by the chromophoric groups present in the visual pigments (retinenes). The acute sensitivity of the photoreceptors, which can react to a single photon, can be explained by the fact that the possibility of striking a sensitive molecule is increased by a factor of hundreds or thousands by the molecular organization of the photoreceptor. It is known that the interaction with photons of light produces liberation of Ca^{2+} ions which, in turn, inhibit the sodium current that is characteristic of the photoreceptor in the dark condition. In this way photoreceptors act as energy transducers, transforming light energy into electrical signals.

Cell Permeability

For functioning of the living cell and maintaining of satisfactory intercular physiologic conditions permeability is fundamental This function determines which substances can enter the cell; many of which may be necessary to maintain its vital processes and the synthesis of living substances. It also regulates the outflow of excretory material and water from the cell. The presence of a membrane establishes a net difference between the *intracellular* fluid and the *extracellular* fluid in which the cell is bathed. This may be fresh or salt water in unicellular organisms grown in ponds or the sea, but in multicellular organisms the internal fluid, *i.e.*, the blood, the lymph, and especially the *interstitial* fluid, is in contact with the outer surface of the cell membrane one of the functions of the cell membrane is to maintain a balance between the osmotic pressure of the intracellular fluid and that of the interstitial fluid.

In higher organisms, the osmotic pressure of the body as a, whole is regulated principally by the kidneys, and the osmotic pressure of the interstitial fluid is about the same as that of the intracellular fluid. In plants, the intracellular fluid has a higher osmotic pressure than the extracellular fluid. The cell is protected from bursting by a rigid cellulose wall. Animal cells generally lack the turgidity that characterizes plant cells. The unfertilized eggs of some marine animals, such as the sea urchin, behave like genuine osmometers. Since they are spheroid, one can, by measuring the diameter, determine the volume and the changes that the egg undergoes with changes in the osmotic pressure of the medium. The osmotic equilibrium is maintained by means of a contractile vacuole, in many unicellular. This "organelle" extracts water from the protoplasm, releasing its contents into the external medium.

Electrical Potentials Acroes

In all cells there is a difference in ionic concentration with the extracellular medium and an electrical potential across the membrane. These two properties are intimately related, since the electrical potential depends on an unequal distribution of the ions on both sides of the membrane. The interstitial fluid has a high concentration of Na^+ and Cl^-; and the intracellular fluid, a high concentration of K^+ and of larger organic anions (A^-). Using fine microelectrodes with a tip of 1 mm or less, investigators are able to penetrate through the membrane into a cell and also into the cell nucleus and to detect an *electrical potential* (also called the *resting,* or *steady,* potential), which is always negative inside. The values of the membrane potential vary in different tissues between -20 and -100 millivolts or mv.

Passive Permeability

The passage of molecules and ions across membranes may be the result of two principal mechanisms. Permeability may be *passive,* if it occurs only because of physical laws such as *diffusion* or *active* if it requires energy for transport. There are several kinds of permeability which will be considered in the following sections. Without an intervening membrane, when two solutions of different concentration are mixed a process of intermixing called *diffusion* occurs. For example, if a concentrated solution of sugar is placed in contact with water, there will be a net movement (*i.e.*, flux) of the solute from the region of higher concentration to that of a lower concentration. In this case the higher the difference in concentration between the two solutions (*i.e.*, the *concentration gradient),* the more rapid the rate of diffusion. The presence of a lipoprotein membrane, such is the plasma membrane, greatly modifies this diffusion or passive permeability. Overton demonstrated that substances that dissolve in lipids pass more easily into the cell, and Collander and Barlund, in their classic experiments with the cells of the plant *Chara,* demonstrated that the rate at which substances penetrate depends on their solubility in lipids and the size of the molecule. The more soluble they are, the more rapidly they penetrate, and with equal solubility in lipids the smaller molecules penetrate at a faster rate.

The permeability (P) of molecules across the membrane is:

$$P @ \frac{KD}{t},$$

with K, the partition coefficient; D, the diffusion coefficient, which depends on the molecular weight; and t, the thickness of the membrane The partition coefficient in most cell membranes is similar to that of olive oil and water. The *partition coefficient* can be measured by mixing the solute with an oil-water mixture and then waiting until the phases are separated. The coefficient is the concentration of the solute in the oil, divided by the concentration in the aqueous phase. Today, the *diffusion coefficient* can be easily determined by using radioactive solutes and measuring their rate of entry into the cytoplasm at various external concentrations. From the experiments. It is evident that many molecules may enter the cell by simple diffusion if the external concentration is higher than the internal. On the other hand, waste products accumulated in the cytoplasm may be eliminated by diffusion.

Concentration and Electrical Gradients

The diffusion of ions across membranes is even more difficult because the same is depends not only on the concentration gradient but also on the electrical gradient present in the system. Within the cell there is a large concentration of non-diffusible anions (A^-) and that an electrical gradient across the membrane is established.

Donnan predicted in 1911 that if a theoretical cell, having a non-diffusible negative charge inside, is put in a solution of KC1, K^+ will be driven into the cell by both the concentration and the electrical gradients; Cl^-, on the other hand, will be driven inside by the concentration gradient, but will be repelled by the electrical gradient. As shown by Donnan, the equilibrium concentrations will be exactly reciprocal:

$$\frac{[K^{\cdot}{}_{in}]}{[K^{+}{}_{out}]} @ \frac{[Cl^{-}{}_{out}]}{[Cl^{-}{}_{in}]}.$$

According to Donnan equilibrium involving only physical forces (*i.e.*, without expenditure of energy by the membrane) was confirmed by the demonstration that the membrane potential was negative on the inside and was accompanied by a high K^+ and a low Cl^- concentration.

The relationship between the concentration gradient and the resting membrane potential is given by the Nernst equation:

$$E @ RT \log \frac{C_1}{C_2},$$

where E is given in millivolts, R is the universal gas constant, and T is the absolute temperature.

Donnan equilibrium for KG can now be expressed as follows:

$$E @ RT \log \frac{[K^{\cdot}{}_{in}]}{[K^{\cdot}{}_{out}]} @ RT \log \frac{[Cl^{-}{}_{out}]}{[Cl^{-}{}_{in}]}$$

Any increase in the membrane potential will cause an increase in the ion asymmetry across the membrane, and vice versa. While the first measurements of membrane potentials and ion concentration seemed to confirm this type of *passive* or *diffusion equilibrium,* more precise determinations in different cell types demonstrated that this was not the case. As mentioned in the next section, this discrepancy may be explained by the involvement of the active transport of ions.

Active Transport of Ions

Extra than the diffusion movement of neutral molecules and ions across membranes, cell permeability includes a series of mechanisms that require energy. These medianisms are generally described as *active transport.* Adenosine triphosphate (ATP), which is produced mainly by oxidative phosphoryl-ation in mitochondria, is generally used as the source of energy. For this reason, active transport is generally related to, or coupled with cell respiration.

There is explained active transport against a concentration gradient in which water has to be moved upstream (*i.e.*, against gravity). The osmotic work to be done is expressed by the Nernst equation, ft charged molecule crossing through an electrochemical gradient also may imply expenditure of energy. For example, to maintain a low intracellular concentration (Na^+, the cell must extrude sodium against a gradient (*i.e.*, higher Na^+ concentration outside). In addition, it must do this against an electrochemical barrier since the membrane is negative inside and positive outside.

A Sodium Pump

During transportation of ion against an electrochemical gradient an extra consumption of oxygen is required. It is calculated that 10 per cent of the resting metabolism of a frog muscle is used for transport of sodium ions. This consumption may increase to 50 per cent in some experimental conditions in which the muscle is stimulated. The resting membrane potential is due to active transport may be demonstrated in plant and animal cells that have been metabolically blocked by anoxia or specific poisons. In this case leakage of K^+ occurs and the potential may decrease to zero. This is clearly observed when anoxia is combined with the poisoning of glycolysis. As suggested by Krogh in 1946. The membrane potential is not really at equilibrium but in a "steady state" involving the constant expenditure of energy.

There is a example of active transport has been provided by experiments with isolated frog skin. The epithelium is specialized to transport Na^+ from the pond water to the interstitial fluids, and by this mechanism the frog can trap this essential ion for use in different tissues. The isolated skin can be kept alive for many hours and used as a wall between two chambers in which the ionic concentration and other factors are changed experimentally.

By means of this preparation, a difference in potential across the skin has been demonstrated: the inside surface is positive with respect to the outside.

The sodium ions are transported from the outer towards the inner surface, and the current produced is due to the flux of sodium. At was also observed that the antidiuretic hormone of the neurohypo-physis stimulates the transport of sodium and water. Similar findings have been observed in the isolated toad bladder and the kidney tubules of *Necturus*.

Ionic transport is intense in the salivary and sweat-producing glands, and even more so in the glands of the stomach that produce H^+ and Cl^- that must be replaced by the blood. Transport of ions is fundamental to maintenance of the osmotic equilibrium of the cell, the required concentration of anions and cations and the special ions needed for the functioning of the cell. Together with the extrusion of Na^+, which is continuously pumped out by the cell, there is an exit of water molecules. In this way the cell keeps its osmotic pressure constant.

Potassium ions, which are concentrated inside the cell, must pass against a concentration gradient. This can be achieved by a "pumping" mechanism at the expense of energy. As explained earlier, Na^+ also is transported by an active process, which is sometimes called the "sodium pump." The passive (downhill) fluxes are distinguished from the active (uphill) fluxes. Notice that the active pumping out of Na^+ is the main mechanism for maintaining a negative potential inside the membrane of -50 mv. This diagram demonstrates that the distribution of ions across the membrane depends on the summation of two distinct processes:

1. Simple electrochemical diffusion forces which tend to establish a Donnan equilibrium (*i.e.*, passive transport)
2. Energy-dependent ion transport processes (*i.e.*, active transport)

It may be observed that the Na^+— K^+ pump drives ions extruding Na^+ and taking in K^+ (active fluxes). At the same time the passive Na^+ influx depends on a large driving force (D.F.), resulting from the concentration and voltage gradients, and only a slight permeability of the membrane to Na^+ (P_{Na}). Similarly, the net passive K^+ efflux results not from a small driving force, but from a greater permeability to K^+ (P_k).

Ionic Transport through Charge Pores In the Membrane

The molecular machinery involved in ionic transport is located within the cell membrane. This has been demonstrated in two key materials. For example, if red blood cells are hemolyzed so that only the cell membrane remains, they can be filled again with appropriate solutions containing ions and ATP, and Na^+ is transported and K^+ is taken up as in a normal cell.

The giant axon of the squid, which has a diameter of about 0.5 mm, can be emptied of the axoplasm and then refilled with solutions of different electrolytes. The transport of ions against a concentration gradient, steady potentials, and even action potentials with the conduction of impulses can be obtained in this preparation in which most of the axoplasm is lacking and the excitable membrane left alone. The utilization of radioscopes demonstrated that ions can enter into the cell rapidly without obvious osmotic effects.

It was then suggested that an ionic interchange across the membrane could take place through electrically charged pores. Knowledge about the diameter of the different ions in the hydrated state is particularly pertinent. In this respect it is interesting to remember that the sodium ion, although smaller than K^+ and Cl^- in weight, is large in the hydrated condition and enters with more difficulty into the cell.

The pore could be envisioned as the interstice between four adjacent protein subunits which could form a hydrophilic channel across the membrane. The ion channels for Na^+ and K^+ are represented in the polarized or resting condition, or in the state of depolarization that occurs during the action potential. It is postulated that the Na^+ and K^+ channels are closed in the polarized resting membrane.

The rest of the diagram shows the opening of the Na^+ pores at the peak of depolarization of the action potential and the opening of the K^+ pores during repolarization. In red blood cell the total area of the pores has been estimated to be on the order of 0.06 per cent of the surface area. This means that a 0.7 nm pore would be surrounded by a Non-porous square 20×20 nm. These findings indicate that the cell uses only a minute fraction of its surface area for ionic interchange.

Anion Transport in Erythrocytes

From tissues to the lung the red cell membrane is endowed with an extremely active anion permeation mechanism to transport CO_2 This system can exchange chloride and bicarbonate ions across the membrane. Using several chemical probes that inhibit anion permeability, knowledge has been gained about the molecular localization of this mechanism.

Some of the probes can be transported by the anion system and, under certain conditions, can be fixed covalently to it. In this way it is possible to determine not only the number of permeation sites but also the polypeptide in which they are localized. This approach has permitted the identification of the *polypeptide of band 3* with the site involved in the transport. This protein spans the membrane and may exist as a dimer (or tetramer). We have seen earlier that it also corresponds to the 8 nm particles observed in freezefractured membranes.

The proposed model for anion transport is that of a continuous proteinaceous aqueous channel across the membrane having, near the outer surface, one anion binding site with three positive charges, and a hydrophobic barrier to limit the free diffusion of anions.

Passage through the barrier could occur only as a consequence of the binding of the anion. In the above model it is assumed that the segment containing the binding site could exist in two conformations, one facing outward, the other inward. In this way the segment could act as a gate in the proteinaceous channel, swinging between the two positions; this would permit the binding site to interact with anions coming from the outside or from the interior of the cell. In this model it is observed that the anion binding site is located in a 65,000 daltons segment of the protein which remains in the membrane after a 35,000 daltons segment has been separated by a proteolytic enzyme.

Bibliography

A.K. Mandal and G.L. Gibson : *Forest Genetics and Tree Breeding*, CBS Publicatoin, Delhi, 2002.

Abha Bhardwaj and Samresh Choudhuri: *Molecular Cell Biology*, Wisdom Press, Delhi, 2012.

Ajay Sharma : *A Text Book of Forest Trees Entomology*, International Book Dist, 2007.

Amar Tyagi: *Cell Biology*, Anmol Publications, Delhi, 2006.

Arunima Mukherjee: *Cell Biology*, Oxford Book Company, Jaipur, 2009.

B.D. Sharma and J.C. Rana.: *Plant Genetic Resources of Western Himalayas: Status and Prospects*, Bishen Singh Mahendra Pal Singh, Dehradun, 2005.

B.P. Singh and Umesh Srivastava.: *Plant Genetic Resources in Indian Perspective Theory and Practices*, Indian Council of Agricultural Research, 2004.

C.P. Trivedi: *Vedic Cell Biology with Life Energy and Rebirth*, Parimal Publications, Delhi, 2007.

Channarayappa: *Cell Biology*, Universities Press, 2010.

David S Latchman: *Basic Molecular and Cell Biology*, I K International Publications, Delhi, 2006.

Falguni Chakrabarty : *Adaptation of the Santals to the Hill-Forest Environment*, APH Publication, Delhi, 2012.

G.P. Gupta: *Plant Cell Biology*, Discovery Publications, Delhi, 2004.

Giridhar A Kinhal and R Jagannatha Rao : *Adaptive Management of Medicinal Plants and Non Timber Forest Products*, BSMPS Publication, Delhi, 2008.

H.S. Bhamrah and Kavita Juneja: *Molecular Cell Biology*, Anmol Publications, Delhi, 2002.

Harry G. Champion and S.K. Seth : *A Revised Survey of the Forest Types of India*, Natraj Publication, Delhi, 2005.

Herbert Kendall Hayes, Forrest Rhinehart Immer and David Clyde Smith.: *Methods of Plant Breeding*, Biotech Books, Delhi, 2009.

J.P. Pascal & B.R. Ramesh : *A Field Key to the Trees and Lianas of the Evergreen Forests of the Western Ghats*, Inst. Francais De Pondichery, 1998.

K.V. Mohanan.: *Essentials of Plant Breeding*, PHI Learning, Delhi, 2007.

Kumkum Thakur: *Methods of Cell Biology and Molecular Biology*, International Scientific Publishing Academy, Delhi, 2011.

Nicholl : *Introduction to Genetic Engineering*, Cambridge Univ Press, Delhi, 1998.

Nolta : *Genetic Engineering of Mesenchymal Stem Cells*, Springer Publication, Delhi, 1998.

P Gomathinayagam, K Vanangamudi, M Subramanian, K Natarajan and K R V Sathya Sheela.: *Lexicon of Plant Breeding and Genetics*, International Book Distributer, Delhi, 2008.

P K Shivaraj Kumar: *Genetic Analysis in Cell Biology*, RBSA Publications, Delhi, 2007.

Pratik Satya : *Genomics and Genetic Engineering*, New India Publication, Delhi, 2007.

Praveen Taank : *Advances in Forestry Research in India*, Cyber Tech Publication, Delhi, 2008.

Preeti Joshi : *Genetic Engineering and Its Applications*, Agrobios Publication, Delhi, 2003.

S.K. Singh: *Manual of Cell Biology*, Campus Books, Delhi, 2012.

Sachin Padgaonkar: *Basics of Cell Biology*, Anmol Publications, Delhi, 2008.

T. Devasena: *Cell Biology*, Oxford University Press, Jaipur, 2012.

T. Saravanan and K. Vanangamudi.: *Principles and Methods of Plant Breeding*, International Book, 2005.

Yasin Jeshima Khan, Srivignesh Sundaresan. P. T. Prathima, J. Nanjundan and M. Arumugam Pillai.: *Merging Plant Breeding With Crop Biotechnology*, New India Publishing Agency, Delhi, 2012.

Index